Cellular Basis and Aetiology of Late Somatic Effects of Ionizing Radiation

Cellular Basis and Aetiology of Late Somatic Effects of Ionizing Radiation

Edited by

R. J. C. HARRIS

*Head of the Division of Experimental Biology and Virology
Imperial Cancer Research Fund, London*

A Symposium held in London 27–30 March 1962
under the Auspices of UNESCO and the IAEA

1963

ACADEMIC PRESS
London and New York

ACADEMIC PRESS INC. (LONDON) LTD.
Berkeley Square House
Berkeley Square
London, W.1

U.S. Edition published by
ACADEMIC PRESS INC.
111 Fifth Avenue
New York 3, New York

COPYRIGHT © 1963 BY UNESCO

Library of Congress Catalog Card Number: 63–12815

Printed in Great Britain at
The Aberdeen University Press

LIST OF PARTICIPANTS

PETER ALEXANDER, *Chester Beatty Research Institute, Institute of Cancer Research, London, England* (pp. 259, 277)

Z. M. BACQ, *Université de Liège, Liège, Belgium*

I. BERENBLUM, *The Weizmann Institute of Science, Rehovoth, Israel* (p. 41)

R. BRINKMAN, *Radiopathologisch Laboratorium der Rijksuniversiteit, Groningen, Holland* (p. 179)

GEORGE W. CASARETT, *School of Medicine and Dentistry, The University of Rochester, New York, U.S.A.* (p. 189)

HERMAN B. CHASE, *Brown University, Rhode Island, U.S.A.* (p. 309)

H. COTTIER, *Medical Research Center, Brookhaven National Laboratory, Upton, New York, U.S.A.* (pp. 27, 113)

H. J. CURTIS, *Biology Department, Brookhaven National Laboratory, Upton, New York, U.S.A.* (pp. 251, 267)

J. F. DANIELLI, *Department of Zoology, King's College, University of London, England*

F. DEVIK, *Statens Radiologisk-fysiske Laboratorium, Montebello, Oslo, Norway* (p. 135)

V. DRÁŠIL, *Institute of Biophysics, Czechoslovak Academy of Sciences, Brno, Czechoslovakia* (p. 145)

A. R. GOPAL-AYENGAR, *Biology and Medical Divisions, Atomic Energy Establishment, Trombay, Bombay, India*

H. L. GRAY, *Research Unit in Radiobiology, Mount Vernon Hospital and the Radium Institute, Northwood, Middlesex, England*

A. GLÜCKSMANN, *Strangeways Research Laboratory, Cambridge, England* (p. 121)

A. HADDOW, *Chester Beatty Research Institute, Institute of Cancer Research, London, England*

R. J. C. HARRIS, *Imperial Cancer Research Fund, Mill Hill, London, England*

P. L. T. ILBERY, *Department of Preventive Medicine, University of Sydney, Sydney, Australia* (p. 83)

A. KEPES, *Department of Natural Sciences, UNESCO, Paris, France*

LIST OF PARTICIPANTS

P. C. KOLLER, *Chester Beatty Research Institute, Institute of Cancer Research, London, England* (pp. 35, 59, 93, 335)

E. I. KOMAROV, *Division of Isotopes, International Atomic Energy Agency, Vienna, Austria*

H. B. LAMBERTS, *Radiopathologisch Laboratorium der Rijksuniversiteit, Groningen, Holland* (p. 207)

L. F. LAMERTON, *Physics Department, Institute of Cancer Research, Royal Cancer Hospital, London, England* (pp. 149, 213)

J. LEJEUNE, *Faculté de Médecine, Institut de Progénèse, Paris, France* (pp. 103, 247)

H. MAISIN, *Cliniques Universitaires St. Raphael, Institut du Cancer, Louvain, Belgium* (pp. 17, 319)

PETAR N. MARTINOVITCH, *Institute of Nuclear Sciences, " Boris Kidrič ", Beograd, Yugoslavia* (pp. 221, 327)

W. V. MAYNEORD, *Department of Physics, The Royal Marsden Hospital, London, England*

P. B. MEDAWAR, *National Institute for Medical Research, Mill Hill, London, England*

R. H. MOLE, *M.R.C. Radiological Research Unit, Harwell, England* (pp. 3, 107, 161, 273)

H. J. MULLER, *Zoology Department, Indiana University, Bloomington, Indiana, U.S.A.* (p. 235)

L. NÉMETH, *State Oncological Institute, Budapest, Hungary* (p. 57)

E. E. POCHIN, *Medical Research Council, Department of Clinical Research, University College Hospital Medical School, London, England*

J. ROTBLAT, *St. Bartholomew's Hospital Medical College, Charterhouse Square, London, England* (p. 313)

W. L. RUSSELL, *Biology Division, Oak Ridge National Laboratory, Oak Ridge, Tennessee, U.S.A.* (p. 229)

O. C. A. SCOTT, *Radiobiology Department, Mount Vernon Hospital and the Radium Institute, Northwood, Middlesex, England* (p. 301)

ARTHUR C. UPTON, *Biology Division, Oak Ridge National Laboratory, Oak Ridge, Tennessee, U.S.A.* (pp. 67, 171, 285)

V. ZELENY, *Division of Isotopes, International Atomic Energy Agency, Vienna, Austria*

PREFACE

In this volume are published the papers and the discussion of the third joint UNESCO and IAEA sponsored Symposia on topics of radiobiology related to fundamental cell biology.

The first of these symposia was held in Venice in June 1959, and the second in Moscow in October 1960. Both concentrated on the nature of the initial processes at the sub-cellular level which are responsible for initiating the chain of events that leads to cell damage. In this, the third of these symposia, it was decided to enter a less extensively studied area of radiobiology and one to which cell biology has not so far made a very large contribution.

This symposium deals with the mechanisms which are responsible for the late somatic effects of ionizing radiation; emphasis was placed on the discussion of the nature of early biological changes (i.e. occurring within days of irradiation) which are responsible for the eventual appearance of cancer, leukaemia and non-neoplastic late effects, particularly those which bring about the observed shortening of life-span. At the present time very little is known about the aetiology of these disturbances. Somatic mutations, activation of latent viruses and anatomical and morphological changes (e.g. in blood vessels) due to cell killing at the time of irradiation, as well as modification of the structure of extra cellular connective tissue components may all contribute. In radiation induced cancer, one of the many unresolved problems is whether the cell (or cells) which starts the tumour must actually be irradiated or whether radiation damage in the surrounding tissue can initiate malignancy.

A knowledge of the cellular basis of late effects of radiation is of more than academic interest since only by understanding mechanisms will it be possible to predict with confidence the magnitude of the hazards to human beings from low doses of radiation. Late somatic effects are, of course together with genetic changes, the principal hazards of the peaceful uses of atomic radiations. In deciding the level of exposure that is acceptable, considerations of the late somatic effects are predominant. It has frequently been pointed out that prediction of the risks from very small doses of radiation and, in particular, the problem of whether there is a threshold cannot be decided from animal experimentation even if attempted on a vast scale. Only an understanding of the basic mechanisms will make it possible to decide the shape of the dose response curves at low doses. These considerations seemed to justify devoting a whole symposium to this narrow and relatively unexplored field. We hoped to gather a group of some thirty scientists actively engaged in this field and I believe that with the exception of workers in the U.S.S.R. who found themselves unable to participate, the group was fairly representative both on the basis of discipline and nationality.

The symposium was divided into six three-hour sessions each of which dealt with specific topics. The final and seventh session was devoted to general discussion and was led by Professor Z. M. Bacq. For this general discussion the technique was adopted of posing a number of controversial statements and asking only those to speak who disagreed with the formulation of this statement. This method of running the general discussion was first adopted by Professor Bacq in Moscow where it proved eminently successful. In each of the first six sessions there was only one formal presentation, which was in the form of an introduction to the subject, and which took not more than 30 minutes. The remaining time was devoted to free discussion. The participants were asked beforehand to give an indication of specific scientific communications which they wished to make in the various sessions, and the titles of these contributions were known to the Chairman of each session. These contributions usually occupied between five and ten minutes and consisted of presentation of new data directly relevant to the subject under discussion. The Chairman's task was to plan each session around the list of contributions. In this book the specific contributions have been written up in the form of papers and are not printed in the form in which they were given. The general discussion which surrounded these contributions was tape recorded and after suitable editing is reproduced here.

Although no formal committee was set up to help in the planning of this conference, my task of organizing it was greatly assisted by the advice of colleagues and friends, of whom I particularly wish to mention Dr. Vladimir Zeleny, Professor P. C. Koller and Professor L. F. Lamerton. As already stated, the organization of the individual sessions was left to the Chairmen, all of whom took great trouble and discharged their duties splendidly. The success of the meeting was in a large measure due to the efforts of the Chairmen. In the general organization, I received a great deal of help from the staff of the IAEA and I am particularly indebted to Dr. Zeleny, who was indefatigable in straightening out many difficulties. The IAEA also carried out the arduous task of transcribing the tape recording. It is a particular pleasure to thank the many members of the staff of the Chester Beatty Research Institute and, in particular, Mr. N. P. Hadow, Mr. K. Moreman, and Miss M. Samuel, all of whom put in a great deal of work in the general running and administration of the meeting. We must also acknowledge with gratitude the help of the National Science Foundation of the U.S.A. who generously made available travel funds to some of the U.S. participants.

December 1962   P. ALEXANDER

CONTENTS

LIST OF PARTICIPANTS..................................... v

PREFACE.. vii

SESSION I
LEUKAEMOGENESIS: QUANTITATIVE ASPECTS AND CO-FACTORS
Chairman: E. E. POCHIN

Introduction by R. H. MOLE.................................... 3
Effects of Low X-Ray Doses (0·1–1 r) on Haematopoiesis Shown by Cytological and Haematological Study and ^{59}Fe Incorporation—A Four Months' Survey. By H. MAISIN......................... 17
Leukaemogenic Effect of Whole-body ^{60}CO-γ-Irradiation Compared with ^{3}H-Thymidine and ^{3}H-Cytidine: Preliminary Report on the Development of Thymic Lymphomas in C57BL/6J Mice. By H. COTTIER, E. P. CRONKITE, E. A. TONNA, AND N. O. NIELSEN.... 27
The Effect of Single and Fractionated Doses of Irradiation on the Haematopoietic Systems of C57BL Mice. By P. C. KOLLER AND VALERIE WALLIS... 35
New Evidence on the Mechanism of Radiation Leukaemogenesis. By I. BERENBLUM AND N. TRAININ............................ 41
A Comparative Study of the Late Effects of Certain Radiomimetic Drugs and X-Rays. By L. NÉMETH........................ 57
Serial Transplantation of Haematopoietic Tissue in Irradiated Hosts. By P. C. KOLLER AND S. M. A. DOAK...................... 59

SESSION II
LEUKAEMOGENESIS: ROLE OF VIRUSES AND CYTOLOGICAL ASPECTS
Chairman: H. CURTIS

Introduction by ARTHUR C. UPTON............................ 67
Chromosomes in Murine Pre-Leukaemia. By P. L. T. ILBERY, P. A. MOORE, S. M. WINN, AND C. E. FORD........................ 83

The Chromosomes in Virus-Induced Murine Leukaemias. By P. C.
  KOLLER, E. LEUCHARS, C. TALUKDAR, AND V. WALLIS.......... 93
Chromosome Abnormalities and the Leukaemic Process. By J. LEJEUNE 103
Quantitative Aspects of Experimental Leukaemogenesis by Radiation.
  By R. H. MOLE... 107
Effects of Ionizing Radiation on Cellular Components: Electron Microscopic Observations. By H. COTTIER, B. ROOS, AND S. BARANDUM 113

SESSION III

CARCINOGENESIS

Chairman: A. HADDOW

Introduction by A. GLÜCKSMANN................................ 121
The Effect of X-Irradiation Compared to an Apparently Specific Early
  Effect of Skin Carcinogens. By F. DEVIK..................... 135
Long-term Consequences of ^{90}Sr in Rats and the Problem of Carcinogenesis. By K. SUNDARAM....................................... 139
The Influence of Radiation on the Survival of Mice Injected with
  Ascites Tumour Cells. By V. DRÁŠIL.......................... 145
Radiation-induced Bone Tumours—Fractionation Studies. By J. P. M.
  BENSTED, N. M. BLACKETT, V. D. COURTENAY, AND L. F.
  LAMERTON... 149
Carcinogenesis as the Result of Two Independent Rare Events. By R.
  H. MOLE.. 161
Preliminary Studies on Late Somatic Effects of Radiomimetic Chemicals
  By A. C. UPTON, J. W. CONKLIN, T. P. MCDONALD, AND K. W.
  CHRISTENBERRY.. 171

SESSION IV

NON-NEOPLASTIC LATE EFFECTS

Chairman: I. BERENBLUM

Introduction by R. BRINKMAN................................... 179
Concept and Criteria of Radiologic Ageing. By G. W. CASARETT...... 189
Initial X-Ray Effects on the Aortic Wall and Their Late Consequences.
  By H. B. LAMBERTS.. 207
The Response of Tissues to Continuous Irradiation. By L. F. LAMERTON 213

The Cholesterol Concentration in Adrenal Glands of Hypophysectomized-grafted Rats (with and without Destroyed Median Eminence) after Whole-body Exposure to a Lethal Dose of X-Rays. By P. N. Martinovitch, D. Pavić, D. Sladić-Simić, and N. Živković 221
The Cellular Basis and Aetiology of the Late Effects of Irradiation on Fertility in Female Mice. By W. L. Russell and E. F. Oakberg 229

SESSION V

MECHANISMS OF LIFE-SPAN SHORTENING

Chairman: W. L. Russell

Introduction by H. J. Muller.................................. 235
Detection of Segmentary Heterochromia in Foetuses Irradiated *in utero*. By J. Lejeune and M.-O. Rethore.................. 247
Chromosome Aberrations in Liver Cells in Relation to Ageing. By H. J. Curtis, and Cathryn Crowley........................ 251
The Failure of the Potent Mutagenic Chemical, Ethyl Methane Sulphonate, to Shorten the Life-span of Mice. By Peter Alexander and Miss D. I. Connell.................................... 259
Life-span Shortening from Various Tissue Insults. By H. J. Curtis and Cathryn Crowley.................................... 267
Does Radiation Age or Produce Non-specific Life-shortening? By R. H. Mole.. 273
Differences between Radiation-induced Life-span Shortening in Mice and Normal Ageing as Revealed by Serial Killing. By Peter Alexander and Miss D. I. Connell........................ 277
Age-specific Death Rates of Mice Exposed to Ionizing Radiation and Radiomimetic Agents. By A. C. Upton, M. A. Kastenbaum, and J. W. Conklin.. 285

SESSION VI

THE MODIFICATION OF THE LATE EFFECTS OF IONIZING RADIATION

Chairman: C. H. Gray

Introduction by O. C. A. Scott.................................. 301
Apparent Radiation Protection Against Local Ageing Effects in the Skin of Mice. By Herman B. Chase.......................... 309

Dependence of Radiation-induced Life-shortening on Dose-rate and Anaesthetic. By P. J. LINDOP AND J. ROTBLAT 313

Effects of Post-irradiation Injection of Yeast Sodium Ribonucleate and its Nucleotides on the Differential Count of Bone-Marrow. By H. MAISIN 319

The Effects of Total-body X-Irradiation on the Reproductive Glands of Infant Female Rats. By D. SLADIĆ-SIMIĆ, N. ŽIVKOVIĆ, D. PAVIĆ, AND P. N. MARTINOVITCH 327

Peripheral Blood Studies upon Some Isogenic Chimaeras. By A. J. S. DAVIES, ANNE M. CROSS, AND P. C. KOLLER 335

SESSION VII

GENERAL DISCUSSION

Chairman: Z. M. BACQ

General Discussion, with Introduction by THE CHAIRMAN 341

AUTHOR INDEX 353

SESSION I

Chairman: E. E. POCHIN

LEUKAEMOGENESIS: QUANTITATIVE ASPECTS AND CO-FACTORS

By R. H. MOLE

Effects of Low X-Ray Doses on Haematopoiesis shown by Cytological and Haematological Study and ^{59}Fe-Incorporation— A Four-months' Survey

By H. MAISIN

Leukaemogenic Effect of Whole-body ^{60}Co-γ-Irradiation Compared with ^{3}H-Thymidine and ^{3}H-Cytidine; Preliminary Report on the Development of Thymic Lymphomas in C57BL/6J Mice

By H. COTTIER, E. P. CRONKITE, E. A. TONNA, AND N. O. NIELSEN

The Effect of Single and Fractionated Doses of Radiation on the Haematopoietic Systems of C57BL Mice

By P. C. KOLLER AND VALERIE WALLIS

New Evidence on the Mechanism of Radiation Leukaemogenesis

By I. BERENBLUM AND N. TRAININ

A Comparative Study of the Late Effects of Certain Radiomimetic Drugs and X-Rays

By L. NEMETH

Serial Transplantation of Haematopoietic Tissue in Irradiated Hosts

By P. C. KOLLER AND S. M. A. DOAK

LEUKAEMOGENESIS: QUANTITATIVE ASPECTS AND CO-FACTORS

R. H. MOLE

M.R.C. Radiobiological Research Unit, Harwell, England

INTRODUCTION

Leukaemogenesis has been singled out from carcinogenesis and given twice the time in the programme for the reasons, I suppose, that even small doses of radiation, as received in diagnostic radiology, appear to be leukaemogenic in man, that enough radiation-induced human leukaemia has occurred to suggest quantitiative dose-response relationships, and that the vast amount of experimental work which has been done might have been expected by now to have turned up something definite and important about mechanisms.

THE PHENOMENOLOGICAL DEFINITION OF LEUKAEMIA

Leukaemia, like many other indispensable terms in current use in medicine, and indeed in biology generally, is merely a descriptive word for a group of phenomena whose nature is not properly understood but which can readily be recognized by a trained observer. The gross symptoms of leukaemia may be highly variable and the underlying common feature is a progressive and ultimately lethal proliferation of cells which are morphologically not too unlike one or other (or more than one) of the different kinds of cells which make up the normal haematopoietic, lymphopoietic and reticular tissues of the body. Individual cases of leukaemia are usually named according to the predominant cell-type and it is a basic assumption that the particular cell (or cells) from which the leukaemia started is (or are) related to the predominant cell in the same way as stem-cells are related to their differentiated progeny, although in leukaemia, both generally and in specific cases, it is always uncertain just how far back in the cell lineage one has to go to find the cell (or cells) in which the leukaemia originated. The varieties of leukaemia are given specific names but, as in taxonomy in general, the boundaries between named species and even the grouping of particular sets of species into genera or families is often somewhat arbitrary.† If this is forgotten it becomes all too easy to talk

---

† The demonstration of hybridization between somatic cells in tissue culture (Sorieul and Ephrussi, 1961) may make it possible to apply to the problems of classification of leukaemia a similar criterion to that used in the classification of species, the possibility or otherwise of interbreeding.

about leukaemia as if it were one "thing" and thus to asume that there must necessarily be a common basic mechanism for every case of leukaemia. In this and many other ways the problems of leukaemia are just the same as the problems of cancer and it is generally, though not quite universally, accepted that leukaemia is in fact cancer of the haematopoietic tissues.

It is indeed possible to imagine that all leukaemias originate in totipotent cells which normally are capable of giving rise to fully differentiated daughter-cells as different as small lymphocytes, granulocytic leucocytes, erythrocytes and macrophages. The predominant cell in different cases of leukaemia would then depend on the particular route of biological development of the daughter-cells of the "leukaemic" stem-cell and how far along that route their mock-differentiation has proceeded. There might then be a single basis for all leukaemias, the particular morphological features of the predominant cells of a particular case being an "accidental" consequence of some factor determining cell differentiation. On the other hand, most workers nowadays would believe that there are several really different kinds of leukaemia. The epidemiological evidence certainly suggests that the causes of acute leukaemia, of chronic myeloid leukaemia and of chronic lymphatic leukaemia in man are distinct (Court Brown and Doll, 1959) and it is important to note that radiation has been shown to increase the first two kinds of leukaemia but not the third.

BIOLOGICAL CHARACTERS OF DEVELOPED LEUKAEMIA

Grafting leukaemic cells into genetically acceptable donors is followed by their multiplication. The cell in developed leukaemia has permanent and inheritable properties and it must be genetically different from normal cells.

Much more interesting is the progressive transformation of these properties as the leukaemic cells continue to multiply. This occurs not only on transplantation but also in the primary host of origin where leukaemic cells are not by any means always identical (cf. Hauschka, 1961). In this respect leukaemic cells are qualitatively different from normal cells for not only do biochemical and metabolic properties change but the chromosomal constitution can also change. There is a characteristic plasticity of chromosomal number and morphology which just does not, and indeed could not, occur in normal cells if ordinary ideas of genetic determination are true.

Individual cases of leukaemia often have a quality of uniqueness. The experimenter with inbred strains of mice can easily forget this but a competent clinical haematologist in a hospital with, say, a dozen cases of leukaemia can usually tell by looking at the cells in a blood smear from which case it came. No cytologist, however competent, can do this with a set of blood smears from normal people.

Thus the cells of developed leukaemia have a genetic character different from the normal and the unsolved question is how this difference is acquired. The cause of a genetic change is not necessarily primarily genetic.

It is as well to note that the evidence suggests that leukaemic cells of chronic leukaemias divide more slowly (not faster) than the corresponding normal cells, and that the accumulation of abnormal cells in these varieties of leukaemia is due to their abnormally long survival time. As is true of all forms of neoplasia, the body's control of the rate of cell division of leukaemic cells is an imperfect control. It is a false definition of cancer to say that it is uncontrolled growth.

LEUKAEMIA AS A RARE EVENT

Leukaemogenesis, like carcinogenesis, must be a rare event in cellular terms. If leukaemia originates in one cell—and, since a developed leukaemia can sometimes be transmitted by grafting a single cell, this hypothesis cannot be dismissed out of hand—then only one cell in the 1·5 kg of active bone-marrow in a human adult need be changed in order that myeloid leukaemia should develop. If all 10^{11} myeloid cells capable of proliferation are susceptible of leukaemia induction then the probability of action of an inducing agent causing a 100% incidence of leukaemia is 10^{-11} per cell (Brues, 1959). If only one in 10,000 of these cells is susceptible of a leukaemic change, the probability would be 10^{-7}: how far back one goes in a cell lineage before finding "susceptible" cells is quantitatively very important. If leukaemia is thought to originate not in a single cell but in a field of cells, the required probability for a whole field of 1 mm^3 would still be about 10^{-6}.

There is also an element of rareness about the kind of leukaemogenic (or carcinogenic) events which follow irradiation. Single doses of 5 to 50 r may produce gross degrees of physiological damage (Table I) but doses of several hundred to several thousand r or more are needed to produce large ($\sim$50%) incidences of leukaemia or cancer. Radiation must be much more efficient in killing cells or in interfering with their ability to multiply (their so-called reproductive integrity) than in causing malignant transformation of individual cells or of small foci of cells.

This observational fact should make one very cautious about applying to leukaemogenesis and carcinogenesis the results and ideas derived from the radiobiological experiments on cell populations. In cell-survival experiments information on the effect of dose-rate and LET and oxygen tension is derived from the whole population of cells which retain the ability to divide. On the other hand, if irradiation of cells can induce leukaemic (or cancerous) changes directly, what is relevant to leukaemogenesis (or carcinogenesis) is the information coming from the small fraction of the whole surviving population in which these particular changes have occurred. Until there is some experimental means for distinguishing this fraction of cells separately, there will be no way of knowing if conclusions based on reproductive integrity also apply to the postulated malignant transformation.

TABLE I. *Some somatic effects in mammals of small doses of X- or γ-radiation* (from Mole, 1962)

Organ	Species	Age at exposure	Dose (r) approx.	Structure or function examined	Magnitude of effect
Ovary	Mouse	10 days	10	Primitive oocytes	50% Depletion
		7–14 days	85 (0·5 r/hr)	Fertility when adult	Reduced to ~10%
		Adult	50	Reproductive capacity	More than halved
Testis	Mouse	Adult	20	Spermatogonia (late A—early B)	50% Depletion
Thymus	Rat	Weanling	20	Cortical lymphocytes	50% Depletion
Bone marrow	Rat	Adult	40	Mature normoblasts	50% Depletion
		Adult	50	24-hr uptake of ^{59}Fe in blood	Halved
Eye	Mouse	Adult	15–30	Lens opacities	Detectable
Stomach	Rat	Adult	40	Emptying time	Doubled

The vast amount of information on chromosomal changes after irradiation is also of uncertain relevance. To be observed, cells have to be dead yet it is only viable cells which are relevant; if the carcinogenic change is a rare event the only relevant chromosomal change will be a rarely occurring one and again there is no way of knowing which it is.

HUMAN RADIATION-INDUCED LEUKAEMIA

No species has ever been, or ever can be, examined so intensively and in such large numbers as the human. Not only does this provide a sound foundation of fact but the human evidence also covers a range of effect which is inaccessible to experiment and therefore uniquely significant to any academic study of leukaemogenesis. Acute leukaemia and chronic myeloid (or granulocytic) leukaemias can be caused by radiation. There is no evidence that chronic lymphoid leukaemia and other kinds of haematosarcoma can be so caused but this could be merely because these conditions have a longer latent period.

The quantitative aspects of human leukaemogenesis by radiation have been thoroughly considered by many authorities (Brues, 1958; Court Brown, 1958; Cronkite *et al.*, 1960; Heyssel and Brill, 1961; Murray and Hempelmann, 1961), especially the distinction between the linear dependence of leukaemia incidence on dose (Lewis, 1957) and the curvilinear which relates incidence to the square of the dose. If we are thinking of the smallest possible effects of radiation, the distinction may be illusory since a small part of any curve is a straight line, but from the point of view of mechanism the distinction is important. It may be worthwhile to recall those aspects of the human observations that do not fit easily with any simple hypothesis.

The best documented and most detailed study is still that of Court Brown and Doll (1957) on ankylosing spondylitics given part body X-irradiation in fractionated doses spread over several weeks and at dose-rates of $\sim$10–50 r/min. The study is continuing and more data and more accurate deductions should soon be forthcoming. The published data give a linear regression of leukaemia incidence on dose which of itself suggests that a dose of 54 r would cause no leukaemia (Court Brown and Doll, 1958) but are of course compatible with the non-threshold linear hypothesis. They appear also just as compatible with a squared-dose hypothesis (Brues, 1959; Burch, 1960). The leukaemogenic effect of a given exposure seemed to be greater the older the individual at the time of irradiation (Doll, 1962). There were unexplained differences between different radiotherapy centres. Cases began to appear when a latent period of 2 years or so had elapsed, but the way in which the risk of leukaemia changes with time thereafter is not yet properly known (cf. Wise, 1961).

The data from Japan on leukaemia in those exposed to atomic bomb explosions are of first importance because exposure to radiation was not correlated with other factors which might conceivably affect leukaemia incidence or radiosensitivity. When radiation is given to people with ankylosing spondylitis, with thymic enlargement, with carcinoma of the cervix, or because they are pregnant and there is a possible obstetric difficulty, this is not necessarily so.

Generally, experience in Hiroshima and Nagasaki has been similar (Heyssel et al., 1960; Tomonaga, 1962). In each the dose-response relationship seems broadly linear with a similar and considerable uncertainty at doses of less than 100 r. However the two regression lines have very different slopes (Fig. 1). The over-all increase in nine years of observation has exceeded the whole life-time expectancy of leukaemia by three-fold and six-fold respectively so that the radiation exposure has really caused leukaemia, not merely accelerated the appearance of leukaemias which were going to occur anyway. In each city the proportional increase in incidence rate for a given exposure seems greater in younger people than in older, the opposite of what occurred in ankylosing spoldylitis (cf. Doll. 1962). There may be differences between the cities in the rate of change with time since exposure of the risk of contracting leukaemia; there are also quite complex changes with time in the relative incidence of chronic granulocytic leukaemia and the different forms of acute leukaemia.

A linear relation between radiation dose and incidence of leukaemia (or cancer) is often considered crucial evidence for a somatic mutation theory and it is often argued, for instance in connection with the data on leukaemia from Japan, that since at the lowest dose only a very few cases of leukaemia are recorded and leukaemic incidence is not significantly raised above the unexposed level, therefore it is hazardous to take the linear dose-response as the true one even though it fits the higher dose data. This argument seems

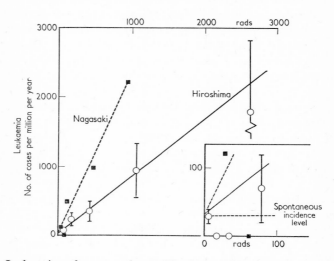

FIG. 1. Leukaemia and exposure dose at Hiroshima and Nagasaki. The Hiroshima data come from Heyssel *et al.* (1960); the Nagasaki data from Tomonaga (1962). The bottom left-hand corner of the figure is shown in the inset at the right at a ten times greater scale.

illogical: when a whole population with a continuously varying radiation exposure is divided into a number of dosage groups in order to find out what the dose-response relationship is like, it will always be the case in practice that the lowest-dosage group will have an incidence not significantly greater than the unexposed level for this will be the consequence of the way in which any investigator will approach his analysis of the population. If, on a first analysis, the lowest-dosage group gives a significantly raised incidence, then clearly it will be worth subdividing the group in order to get more evidence about the shape of the dose-response curve at still lower doses. The major reasons for reserve in accepting that a linear dose-response relation has been established in the Japanese data (Fig. 1) are the problems in establishing the dose experienced by individuals, the breadth of the dose-range in each dosage group and the difficulty about the control group with which the irradiated groups are to be compared (cf. Fig. of Tomonaga, 1962).

In other human populations which have been studied, exposure dosage is either not known, as in radiological specialists in U.S.A. and in Great Britain (cf. Court Brown, 1958), or is clearly not the only determining factor, as in infants and children given radiotherapy (Murray and Hempelmann, 1961). Thymic enlargement in infants is not a necessary precursor of radiation-induced leukaemia but whether it predisposes to radiation-induced leukaemia is not yet known. The size of the radiation port or field, *i.e.* the amount of tissue irradiated (which in infants can vary from a small fraction to a large fraction of the whole body) is quite probably important but under given conditions of irradiation there did not seem to be much dependence of leukaemia incidence on dose. Firm conclusions are difficult to draw since in one thorough follow-up of 1,600 infants and children given radiotherapy sufficient to cause four cases of thyroid cancer (approximately a 100-fold

TABLE II. *Leukaemia and cancer in children irradiated in utero*

		Rate per 10,000 live births			Rate per 10,000 live births X-rayed group only	
		Not X-rayed (a)	X-Rayed (b)	Relative risk (b/a)	1 or 2 films	Pelvimetry
Leukaemia	First-born	4·7	6·9	1·5		
	Others	3·6	4·6	1·3		
	All	4·0	6·1	1·5	3·8	6·9
Cancer of C.N.S.		1·6	2·5	1·6	2·7	2·3
Other cancers		1·8	2·5	1·4	2·7	2·1

Selected data from Tables 4, 8 and 9 of MacMahon (1962).

increase over the normal) there was unaccountably no appreciable increase in leukaemia.

The vexed question of the leukaemogenic effect of whole-body irradiation of the human foetus *in utero* seems to have been largely clarified by MacMahon (1962) who followed up three-quarters of a million children for at least 6 years each. Diagnostic radiology of the mother was followed by a statistically real risk of leukaemia (Table II) but at 40% above the unirradiated level the risk was smaller by half than that suggested by the original observations of Stewart and co-workers. The satisfactory conclusion was reached that the results of all previous studies, both positive and negative, are statistically compatible with those of MacMahon. The actual dose received by the foetus is not known and will certainly have varied considerably from time to time and from hospital to hospital for technical reasons. The leukaemia incidence was twice as high in children exposed to pelvimetry (three or more films) than in children given a smaller exposure (one or two films). However, the difference was not statistically significant and in fact was absent for other forms of cancer which were also generally increased in incidence by 40% above the unirradiated level. Thus there is no satisfactory evidence relating leukaemia incidence to degree of exposure, and no way of knowing whether the greater proportion of the leukaemia occurred in a small proportion of the children who happened to have received doses at the upper end of the dose range. It is possible to deduce that the probability of induction of leukaemia per r was very roughly of the order of 10^{-5} per year and that, since the increase in leukaemia in irradiated children appeared to stop after about 7 years of age, the risk is not indefinitely prolonged.

In some countries health legislation may require fluoroscopy of the chest of pregnant women and exposure of the developing foetus will not always be avoided. Comparison of the mothers of 102 children who died from leukaemia and the mothers of 309 control children showed that 4·9% of the first group, but none of the second group had been fluoroscoped (p = 0·008) (Anne-Marie Méry, 1961).

The other new information on the possible leukaemogenic effect of diagnostic radiology comes from Denmark (Faber, 1962, personal communication). In people over 40 years of age with acute leukaemia there was a higher frequency of diagnostic irradiation in the past than in those with chronic myeloid or chronic lymphatic leukaemia, although the interpretation of the observations must take into account the additional finding that acute leukaemia and chronic myeloid leukaemia usually showed themselves at different times after irradiation. More interestingly still there seemed to be a marked correlation between the kind of disease the individual was suffering from when he was X-rayed and the time interval before the appearance of acute leukaemia, though not of chronic myeloid leukaemia. Those with rheumatic diseases (cf. ankylosing spondylitis) or osteoarthritis showed a shorter time-interval (latency) than those with benign gynaecological disorders or with cancer (unspecified). An association between rheumatic disease in general and leukaemia (all types combined) has already been reported (Abbatt and Lea, 1958). As Faber says, the Danish numbers are small but maybe we cannot expect to demonstrate a straightforward relation between leukaemia incidence and radiation without a deeper understanding of the real nature of the diseases being studied.

This brief survey of human information suggests several broad conclusions. First, different kinds of human leukaemia show sufficiently different aspects of behaviour towards induction by irradiation for it to be necessary to consider them separately even though there may be quite serious difficulties in classifying individual cases. Unfortunately, the current international nomenclature of leukaemia is sadly deficient (Mathé, 1962) and this is a serious handicap in epidemiological studies. Second, there is good evidence that the leukaemogenic effect of radiation does not depend simply on dose but that the circumstances of the irradiation as well as the reactivity of the irradiated individual are also quantitively important. Third, "small" doses of radiation of the magnitude encountered in diagnostic radiology have a real, though low, probability of causing leukaemia, at least for whole-body irradiation *in utero*. Nevertheless this does not necessarily disprove the contention that serious tissue damage is a necessary factor in radiation-induced leukaemia. The doses involved may damage some tissues quite markedly (Table I) and no investigation seems to have been made of the haematological damage in foetuses exposed to diagnostic irradiation. Lastly it still remains true that the exposure dose-rate was high in all the situations in which radiation is known to have caused leukaemia in man.

KINDS OF MECHANISMS FOR LEUKAEMOGENESIS BY RADIATION

One basic kind of hypothesis is that the late somatic effects of radiation are due to some direct cellular effect of radiation, such as somatic mutation. Another basic kind of hypothesis is that cancers, including leukaemias, follow gross radiation-induced disturbances in physiology or tissue structure. It seems to me a serious mistake to put these hypotheses in opposition to each other. They are not logical opposites and both could easily be true: tissue damage could well facilitate the expression of direct cellular effects by changing the cellular environment or encouraging cell division.

It is easy to waste a great deal of time in wholly unnecessary argument about *the* mechanism of cancer or leukaemia. Surely we should recognize that multiple correlation is a characteristically biological phenomenon (Huxley, 1958), as we would all accept if we discussed the mammary tumour agent of the mouse and the dependence of tumour manifestation on genetic constitution and hormonal influence. The nature of scientific proof entails that experimental conditions can be arranged in which the tumour yield depends critically on one factor only, but since such an experiment can be done for each of the three factors separately it is not possible to single out one of them as *the* aetiological factor except in some private, arbitrary sense.† I think we should avoid arguing about whether one hypothesis or the other is true but instead try to define their relative importance and fields of action.

It is probably a fair summary to say that the somatic mutation hypothesis of carcinogenesis appeals especially to those who have no first-hand acquaintance with cancer or leukaemia, a category which includes its originator Boveri. Its advantages are that many kinds of mutation can indeed be

---

† The word " co-factors " in my title is similarly to be deplored.

caused by radiation and that there are no problems in accepting a leukaemogenic effect of doses of whole-body irradiation as small as 5 r or less. It is sometimes thought that in its favour is the association of mongolism and leukaemia and against it the lack of any firm correlation between leukaemia in identical twins, but these may well be arguments of dubious force. If somatic mutation is indeed the basic mechanism of leukaemia it is a qualitatively, different kind of mutation, as pointed out earlier, since the cellular change is not to a stable genotype but to a progressively altering one. There seems no necessary correlation, as there would be for ordinary genetic mutation, with the stable genotype characteristic of the mongol or of identical twins. Brues's (1958) arguments against the somatic mutation theory are, as he emphasized, the arguments against one particular kind of mutation, point-mutation, as a sole cause; he was concerned to demolish the idea of a linear relation between radiation dose and carcinogenesis.

Those who do have a first hand acquaintance with human cancer and leukaemia may develop a very strong emotional bias against the somatic mutation hypothesis from the feeling that, if it is true, nothing can be done against cancer (Rous, 1959). Although there is no direct evidence in favour of the hypothesis this kind of prejudice is no evidence against. There has been much discussion as to whether there is a "threshold" for leukaemia induction: the reasons why this question is a theoretical one and can never be answered by observation are given elsewhere (Mole, 1958). The data themselves can never prove any particular hypothesis; they can only be compatible or incompatible with it.

The other general hypothesis, which relates leukaemogenesis to radiation-induced tissue damage, is obviously true in the sense that a high incidence of leukaemia in experimental animals as well as in man occurs only when the radiation exposure has been enough to cause severe tissue damage, though just what is the nature of the critically important tissue damage is not at all clear. Is it possible to show that leukaemia does not occur unless there is a sufficient degree of tissue damage? This is very like trying to prove a general negative and if the idea is accepted that leukaemia, whatever the nature of its initiation, starts in a small focus somewhere, then even if leukaemia can be caused to occur without obvious tissue damage the problem of sampling is going to prevent the sceptical biologist from denying that there might well have been sufficient tissue damage somewhere where he did not happen to look.

The experimenter with irradiated animals is always going to be biased towards theories of tissue damage because an experiment can involve only a relatively small number of animals and to get enough leukaemia to be "significant" we know already that tissue damage must be caused. The human epidemiologist is always going to be biased towards theories of direct cellular action just because he deals with a few cases of leukaemia in thousands or hundreds of thousands of individuals. What seems to be needed to disprove the *necessity* of tissue damage is an experiment in which leukaemogenesis (carcinogenesis) was initiated by irradiation of cells *in vitro*. And the difficulties of demonstrating direct cellular action in this way are probably clear to everyone.

The difficulties may be even greater if the general ideas of Armitage and Doll (1957) and of Fisher (1958) are accepted. In order to account for the change with age in the age-specific mortality rate of leukaemias in man it is postulated (Court Brown and Doll, 1959) that some forms of leukaemia, e.g. chronic myeloid leukaemia, may be due to two successive cellular events while others, e.g. chronic lymphoid leukaemia, may be due to three successive cellular events. The events are mathematical abstractions though it is easy to think of them as in some way equivalent to other, more biological formulations of successive processes, such as initiation and promotion. If two or, even worse, three successive events or processes are to be made to occur *in vitro* the experimental problem is clearly even more complex than if only one event is needed.

Although the theories are hardly more than speculative, they are compatible with the facts of bone-cancer production by bone-seeking isotopes (Mole, 1962). If they are broadly true, the dose-response curve for radiation-induced leukaemogenesis at low incidence rates, 10^{-6}, and the duration of risk following a single exposure will depend critically on whether the first or second type of event is more easily caused to occur by radiation. If radiation causes the first type of event with a higher probability than the second, then the effect of irradiation will be to "age" the individuals in the irradiated population in the sense that, though not visibly different from an unirradiated population of the same age, the age-specific leukaemia incidence rates will be those characteristic of an older population. The increased risk of development of leukaemia would not decrease with time (unless additions are made to the hypothesis, cf. Burch, 1960). Some experimental evidence has been interpreted in this sense (Upton *et al.*, 1960; Lindop and Rotblat, 1961) but on the whole the human data suggest that the risk of development of leukaemia after exposure to radiation reaches a peak a number of years later and then decreases. This is what would be expected if radiation caused the second type of event with a higher probability than the first kind of event, but a second corollary, that the proportional risk would be the same at all ages and the absolute risk progressively greater the older the individual at the time of exposure, while possibly true of the ankylosing spondylitics, may or may not be true in the Japanese (Doll, 1962). If radiation can cause both kinds of event then the dose response relation ought to be expected to change as the range of radiation-dose changes, a most important practical deduction.

CONCLUSION

Possible mechanisms for leukaemogenesis by radiation are listed in Table III. At present there is scientific proof only for an indirect mechanism, defined as a mechanism causing leukaemia to originate in unirradiated cells (Kaplan et al., 1956; Law and Potter, 1956; Barnes et al., 1959).

TABLE III. *Mechanisms of radiation-induced leukaemia* (cf. Mole, 1962)

1 Radiation acts directly on cells. Leukaemia stems from the affected cells.
   1.1 Direct action on cellular genetic mechanism.
   1.2 Direct genetic or non-genetic effect increasing susceptibility to infection or other exogenous agent.

2 Radiation acts indirectly to increase the probability of an event which has some chance of occurring all the time. Leukaemia does not originate specifically in irradiated cells.
   2.1 Leukaemia results from a natural instability in the process of cell multiplication. Radiation, by killing cells, increases the rate of cell multiplication or by altering cell environment favours the instability.
   2.2 Radiation increases susceptibility to infection or exogenous agents by interfering with immune reactions, detoxication mechanisms or in other ways.

REFERENCES

ABBATT, T. D., and LEA, A. T. (1958). *Lancet ii*, 880.
ARMITAGE, P., and DOLL, R. (1957). *Brit. J. Cancer* 11, 161.
BARNES, D. W. H., FORD, C. E., ILBERY, P. L. T., JONES, K. W., and LOUTIT, J. F. (1959). *Acta Un. int. Cancr.* 15, 544.
BRUES, A. M. (1958). *Science* 128, 693.
BRUES, A. M. (1959). In "Low-level Irradiation" (A. M. Brues, ed.), p. 73, Publication No. 59, American Association for the Advancement of Sciences, Washington, D.C.
BURCH, P. R. J. (1960). *Nature, Lond.* 185, 135.
COURT BROWN, W. M. (1958). *Brit. med. Bull.* 14, 168.
COURT BROWN, W. M., and DOLL, R. (1957). Special Report Series No. 295 Medical Research Council, H.M.S.O., London.
COURT BROWN, W. M., and DOLL, R. (1958). *Lancet i*, 162.
COURT BROWN, W. M., and DOLL, R. (1959). *Brit. med. J. i*, 1063.
CRONKITE, E. P., MALONEY, W., and BOND, V. P. (1960). *Amer. J. Med.* 28, 673.
DOLL, R. (1962). *Brit. J. Radiol.* 35, 31.
FABER, M. (1962). *Strahlentherapie*, in the press.
FISHER, J. C. (1958). *Nature, Lond.* 181, 651.
HAUSCHKA, T. S. (1961). *Cancer Res.* 21, 1020.
HEYSSEL, R., and BRILL, A. B. (1961). In "Radioactivity in Man" (G. H. Meneely, ed.). Charles C. Thomas, Springfield, Illinois.
HEYSSEL, R., BRILL, A. B., WOODBURY, L. A., NISHIMURA, E. T., GHOSE, T., HOSHINO, T., and YAMASAKI, M. (1960). *Blood* 15, 313.
HUXLEY, J. (1958). "Biological Aspects of Cancer", Allen and Unwin, London.
KAPLAN, H. S., HIRSCH, B. B., and BROWN, M. B. (1956). *Cancer Res.* 16, 434.
LAW, L. W., and POTTER, M. (1956). *Proc. nat. Acad. Sci., Wash.* 42, 160.
LEWIS, E. B. (1957). *Science* 125, 965.

Lindop, P. J., and Rotblat, J. (1961). *Proc. roy. Soc.* **154B**, 332.
MacMahon, B. (1962). *J. nat. Cancer Inst.* **28**, 1173.
Mathé, G. (1962). *Bull. World Hlth Org.* **26**, 585.
Méry, Anne-Marie (1961). Contribution à l'étude des facteurs étiologiques des leucémies de l'enfant. Thèse, Paris.
Mole, R. H. (1958). *Brit. med. Bull.* **14**, 184.
Mole, R. H. (1959a). United Nations Peaceful uses of Atomic energy, Proceedings of the Second International Conference, Geneva, September 1958, **22**, 145.
Mole, R. H. (1959b). *Brit. J. Radiol.* **32**, 497.
Mole, R. H. (1962). In "Some Aspects of Internal Irradiation" (T. F. Dougherty, ed.). Pergamon Press, Oxford.
Murray, R. W., and Hempelmann, L. H. (1961). In "Radioactivity in Man" (G. R. Meneely, ed.). Charles C. Thomas, Springfield, Illinois.
Rous, P. (1959). *Nature, Lond.* **183**, 1357.
Sorieul, S., and Ephrussi, B. (1961). *Nature, Lond.* **190**, 653.
Tomonaga, M. (1962). *Bull. World Hlth Org.* (in press).
Upton, A. C., Kimball, A. W., Furth, J., Christenberry, K. W., and Benedict, W. H. (1960). *Cancer Res.* **20**, No. 8 (part 2) 1.
Wise, M. E. (1961). *Hlth Phys.* **4**, 250.

DISCUSSION

UPTON: What would be the expectation if radiation could produce both kinds of events?
MOLE: It depends entirely on the relative probabilities. If you accept the basic idea of Armitage and Doll which they propounded theoretically just to account for the distribution of age-specific mortality rates in man for every different kind of cancer they could lay their hands on, then cells which have undergone the first kind of change gradually accumulate during life and further have some selective advantage so that they multiply progressively. However, no ostensible tumour appears until a second change occurs and then there is an inevitable progression towards what we recognize as a real cancer. If the probability of producing the first kind of change with radiation is much higher than producing the second, then all you are doing with radiation is just ageing the individual by producing more of these cells which he would have got anyway if he had lived another 10 years or 5 years, whatever it might be. If, on the other hand, the probability is much higher of producing the second type of change, then you are not increasing the risk of getting leukaemia right through life, you are just increasing the risk temporarily. If cells don't undergo the second kind of change, then nothing is going to happen; if they do suffer the change presumably leukaemia shows itself fairly soon. It is interesting that in Japan the actual incidence of leukaemia in the survivors at both Hiroshima and Nagasaki is now in one city three times, in the other six times, what you would expect from the natural life-span incidence of leukaemia in unirradiated population, so that radiation has produced a real excess of leukaemia, it hasn't just altered the distribution of leukaemia in time or with age. That seems to suggest that radiation, if you accept this very mathematical idea, must produce the second kind of change more easily than the first, but that doesn't mean to say it doesn't also produce the first change. If you assume that this first change leads to clones of cells which proliferate and that when radiation does something to cells in such clones, from that moment on the thing inevitably becomes an ostensible leukaemia or cancer, then you can only assume that there is going to be a persistent increased risk for the rest of life if these doubly altered cells can persist for the rest of life without undergoing progressive multiplication. Now I think the basic idea of both Fisher, and Armitage and Doll is that this doesn't happen, but that when the second

change happens then the thing really gets under way. This doubly-altered cell doesn't stay latent because if so it would make nonsense of all their age-specific mortality rate distributions. Once the second change happens the cell has some reason for undergoing cell division and turning into a real tumour. This means, then, that when an individual is exposed to radiation some cells which have previously suffered the first change may suffer the second change. Then you are going to get the risk of cancer or leukaemia occurring in the next few years, but once those cells are exhausted there won't be any further increased risk above the normal for the age of the individual concerned.

ROTBLAT: Your data, I think from Japan, show an increase in ageing.

MOLE: In Japan the data are remarkably consistent with the idea that radiation affects a special kind of cell and that there are more of these to affect the older you are. However, in Japan, the incidence of leukaemia in people under the age of 20, the proportional increase, has been greater than in people who are older, but I don't know myself just how to interpret that.

MULLER: As I understand it, what you call the second kind of change cannot occur unless the first kind has already occurred, not merely that the second kind must occur also. In other words, this couldn't be the type of genetic change which would remain latent until the first kind of change had occurred.

MOLE: I agree, the two incidents have to occur in the right order.

MULLER: But have you also considered the other possibility of each event being necessary regardless of the other?

MOLE: If you do that, then it doesn't allow you to fit the observed change of age-specific mortality with age for different kinds of cancer. You've got to have some kind of directional effect to allow for the increase with age.

BERENBLUM: I would just like to say that there seems to me an inconsistency in this kind of reasoning because it presupposes that all the progeny of any one cell continue to be present later on to undergo the second change. This situation, in fact does not exist; if it did then neoplasia would be the normality. What in fact occurs, statistically speaking, is as follows. The number of stem-cells—that is, cells capable of progressive division — is statistically more or less constant throughout life. This is what is meant by normality, and is true whether for skin or for haemaopoietic tissue or for anything else. But, for the number to remain constant statistically speaking 50% of the progeny must die and only one remains as the stem-cell for the next generation. If this goes on like this under normal conditions then the final progeny of stem-cells capable of acting as stem-cells is still the same as originally. Therefore the chances of this cell undergoing a second change is theoretically no greater than before. I think this ought to be kept in mind in this kind of reasoning.

MOLE: Well, I've got two things to say about that: first of all I think that neoplasia *is* normal. One-fifth of us is going to die of it. Secondly, I think this argument of constancy of stem-cell number is not necessarily applicable. All the hypothesis proposes is that once altered stem-cells have a particular selective advantage; they are still capable of acting as stem-cells and producing cells which have a normal physiological function. As the altered stem-cells increase in number, the number of unaltered stem-cells correspondingly diminishes.

BERENBLUM: I said that under normal conditions the number of stem-cells that exists in the body for any particular system of cells is more or less constant. In a tumour it is not constant, and that is the difference.

MOLE: Yes, but the hypothesis deals with events before there is a tumour. This hypothetical first change is not supposed to produce a recognizable tumour in any sense of the word at all, it merely provides a reservoir of cells which can undergo a second change.

EFFECTS OF LOW X-RAY DOSES (0·1–1 r) ON HAEMATOPOIESIS SHOWN BY CYTOLOGICAL AND HAEMATOLOGICAL STUDY AND ^{59}Fe INCORPORATION—A FOUR MONTHS' SURVEY

H. MAISIN

Cliniques Universitaires St. Raphael, Institut du Cancer, Louvain, Belgium

SUMMARY

In our strain of rats, total-body irradiation with 0·1 r and 1 r, administered between 7 and 84 days before, does not modify the mortality following subsequent sublethal X-ray exposure. Although such a low dose does not affect the chance of survival of the rats from a later lethal dose, the number of cells per mg of marrow and the erythropoiesis as measured by the ^{59}Fe incorporation in the red cells are, however significantly increased between the 4th and 8th weeks after 1 r. ^{59}Fe Incorporation in the red cells alone, 4 and 6 weeks after 0·1 r is also increased. The number of reticulocytes in the peripheral blood significantly increases during the 2 to 4 weeks after 1 r and 0·1 r; the red and white cells do not change. The significance of these results is discussed.

Various authors have shown that mice irradiated with sublethal X-ray doses, become radioresistant to a subsequent sublethal, or even lethal, X-ray dose. The phenomenon occurs according to the authors, between 10 and 20 days after the administration of the sublethal dose. Thus Betz (1950) has shown that mice irradiated with 500 r ($LD_{10(30)}$ in his strain) would become more resistant to 700 r (a LD_{100} by the 10th day) administered 15 days later. Paterson et al. (1952) gave the mice one-half of their $LD_{50(30)}$; after 20 days, these mice required 546 r to die in the proportion of 50%. Dacquisto (1959) was able to show in Swiss mice that the $LD_{50(30)}$ which is 487 r, is increased to 560 r if he gave them 50 r WBR 10 days before, and to 617 r if the 50 r are given 17 days before.

We have tried to confirm these data in our L-strain rats. Thus, we planned (Maisin, 1961) a series of experiments, administering doses of 50 or 400 r—$LD_{10(30)}$—between 7 and 42 days before a LD_{15}, LD_{50}, or $LD_{90(30)}$, but we did not get enhancement of the resistance of our rats. On the contrary, rats receiving a conditioning, or prior, dose of 400 r died more rapidly and in a much higher proportion than the controls. So, after 450 r (LD_{15}), the mortality at 30 days varied between 70 and 95% and after 550 r, LD_{90}, 100% of the rats were dead on the 10th day, half of them by intestinal syndrome when otherwise they die only by medullary syndrome. A conditioning dose of 50 r had no influence on a subsequent dose of 450 r but was more or less prejudicial against 500 r, LD_{50} and 550 r.

We also gave 10 r and 5 r but we were unable to induce radioresistance. However, 50 r whole-body irradiation increases significantly the total number of nucleated marrow-cells per mg (Maisin, 1961). This increase occurs between 7 and 21 days after the administration of the dose.

Knowlton and Hempelmann (1949) have shown that X-ray doses ranging from 5 r to 325 r enhance the mitotic activity in the adrenal gland, jejeunum, lymph node and epidermis of the mouse. They call this phenomenon, overcompensation. It appears between a few hours and 10 days after irradiation depending on the organ, and is more apparent in less radiosensitive tissues.

Pape and Jellinke (1950) and Pape (1951) stated that X-ray doses (1–20 r) to the spleen led to cellular change and to enhanced resistance of mice to whole-body X-radiation delivered later. He considered this protective action as a cellular resistance of various tissues and due to proliferation of lymphoid and reticulo-endothelial tissue originating from the irradiated spleen.

In the present paper, we describe the results obtained after whole-body irradiation of 1 r or 0·1 r. We show that one X-irradiation of such low dose is unable to induce radioresistance against the medullary syndrome produced by a subsequent $LD_{50(30)}$, although the number of nucleated marrow-cells per mg and the erythropoiesis (as measured by ^{59}Fe incorporation in the red cells) and number of reticulocytes in the peripheral blood are statistically increased during a certain period.

EXPERIMENTAL CONDITIONS

The homozygous L-strain rats were male and at the beginning of the experiment were 4 months old and weighed 130 to 145 g. They were housed in pairs. The animals were distributed in series of 40; 20 of them receiving a first irradiation of 1 r or 0·1 r which can be called the conditioning dose, 20 serving as controls. Later, all 20 animals of each series were irradiated with 500 r X-rays which is $\pm LD_{50(30)}$.

The different experimental conditions: conditioning dose, sublethal dose, time-interval in days between the conditioning and sublethal doses, the number of rats in each group on the day of the administration of the conditioning dose and of the sublethal dose are reported in Tables I and II. The sublethal dose was always administered on the same day to conditioned rats and to the controls.

The rats were fasted for 24 hours before and after exposure to the sublethal dose. They were not fasted for the administration of 1 or 0·1 r. For all the irradiation, we used a General Electric Maxitron–250 X-ray apparatus operating at 250 kV, 25 mA; 0·25 mm Cu + 1 mm Al filter. For the administration of the $LD_{50(30)}$, the anticathode to rat distance was 80 cm and the output, measured in air with a Victoreen Radocon, was 47 r/min. The rats were irradiated in fours on a turntable. When we administered 1 r or 0·1 r, the distance was 1·20 m and 3·2 m and the irradiation time 2″9 and 1″9 respectively.

TABLE I. *Experimental conditions and* $LD_{50(30)}$

Conditioning dose		Interval between conditioning dose and sublethal dose (days)	Sublethal dose		Percentage mortality 30 days after sublethal dose
Number of rats	Dose		Number of rats	Dose	
20	1 r	7	20	500 r	60
20	none		20		70
20	1 r	14	20	500 r	75
20	none		20		40
20	1 r	21	20	500 r	45
20	none		20		65
20	1 r	28	20	500 r	45
20	none		20		60
20	1 r	42	20	500 r	50
20	none		20		40
20	1 r	56	19	500 r	43
20	none		20		40
20	1 r	84	20	500 r	35
20	none		20		45

TABLE II. *Experimental conditions and* $LD_{50(30)}$

Conditioning dose		Interval between conditioning dose and sublethal dose (days)	Sublethal dose		Percentage mortality 30 days after sublethal dose
Number of rats	Dose		Number of rats	Dose	
20	0·1 r	14	20	500 r	35
20	none		20		45
20	0·1 r	21	20	500 r	45
20	none		20		40
20	0·1 r	42	20	500 r	40
20	none		20		30
20	0·1 r	56	19	500 r	32
20	none		18		28

The condition of all the rats was recorded daily until 30 days after the sublethal exposure.

In a second series of experiments we just gave the conditioning dose and on the day normally devoted to the administration of the sublethal dose we sacrificed the rats to count the number of nucleated cells per mg of bone-marrow and to evaluate their erythropoiesis by ^{59}Fe incorporation into the red cells. We used male rats of the same age and weight, and counted the number of marrow cells per mg in the femur using a method already published (Maisin, 1959). A tibia, marrow-smear from each rat was also made but we have not yet examined all of them and thus, it is too early to report the results. To evaluate the erythropoiesis, we injected subcutaneously, 3 days before, $\frac{1}{3}$ µc of ^{59}Fe with a specific activity of 4–5 c/g Fe, as ferric citrate in isotonic, neutral and sterile solutions. The blood was withdrawn from the inferior vena cava and the haematocrit recorded by a micro-method. The activity in the red cells was counted in a vial scintillation counter per ml, per minute and per 100,000 counts of ^{59}Fe injected.

We choose to kill the animals 72 hours after injection of the iron because at this time, in our control non-irradiated rats, the activity in the red cells has reached a plateau (Maisin, 1959).

Before killing the rats we also counted the number of red and white cells per mm^3 and the number of reticulocytes per 1,000 red cells in the peripheral blood of the tail. The reticulocyted were stained by the method of Heilmeyer.

We killed 6 rats in succession, 14, 28, 42, 56, 84, and 98 days after they had received 1 r; after 0·1 r we killed them after an interval of 14, 28, 42 and 56 days. Each time, we sacrificed 6 non-irradiated controls of the same age, maintained in the same conditions from the beginning of the experiment.

RESULTS AND DISCUSSION

First of all the administration of the conditioning or prior dose of 1 r or 0·1 r did not have any influence on the weight of the animals up to the day of administration of the $LD_{50(30)}$.

Mortality

The conditioning dose did not have any influence on the mortality of the rats before the administration of the second dose; indeed, there were very few deaths: one in a group receiving 1 r (Table I), one in two groups receiving 0·1 r and two in a control group (Table II).

The conditioning dose did not have more effect on the behaviour of our rats against 500 r, however long the length of time between the two doses (Table I and II, Fig. 1); indeed on one occasion the conditioned animals died sooner and on another the controls, but the difference between the two groups was never statistically significant. More rats survived in the series of 0·1 r and in the controls (Table II) and similarly in the last series of Table I for the rats receiving 1 r and the controls. We think that this decreased mortality may be explained by a more progressive feeding of the rats by purina chow during their growth.

Not only the percentage of mortality at the 30th day after 500 r but also the incidence of mortality after 500 r is the same in the conditioned animals and in the controls. Indeed the mortality curves (which are not presented here) are similar.

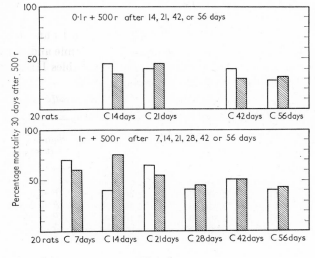

Fig. 1.

These results are not suggestive of any kind of radioresistance in our rats previously irradiated with low doses. We already know (Maisin, 1961) that higher doses of 5, 10, 50 or 400 r administered between 7 and 42 days before do not induce radioresistance in our rats however large the X-ray dose: $LD_{10(30)}$, $LD_{50(30)}$ $LD_{90(30)}$. Can it be reasoned from these results that rats react differently from mice against conditioning or prior irradiation? It is perhaps very dangerous to think so. We are more inclined to believe that the differences obtained by certain authors are due to the fact that the unconditioned animals were not always subjected to the experiments on the same day as the conditioned ones (Betz, 1950; Paterson et al., 1952), or that some authors came to too rapid a conclusion from their data (cf. Dacquisto (1959), in whose best experiment the χ_c^2 is < 0.2).

In view of these results, it was interesting to investigate the cellular behaviour of the marrow. We chose the marrow because all our animals irradiated with 500 r died of medullary syndrome. It was also interesting because different authors had bound radioresistance with higher mitotic activity not only in radiosensitive organs but even more in radioresistant ones (Knowlton and Hempelmann, 1949). As bone-marrow is highly radiosensitive it was of the utmost interest to evaluate its reaction to low doses particularly in our rats because such doses do not involve the appearance of radioresistance. We thought it best to estimate the marrow activity by using three tests: the variation of the total number of marrow cells; the incorporation of ^{59}Fe in the red cells (Maisin, 1959) and the enumeration of reticulocytes in the peripheral blood.

Number of marrow cells

The results obtained in the rats pre-irradiated with 1 r and 0·1 r and in the control non-irradiated rats after different periods of time and their statistical significance (evaluated by t-test) are presented in Tables III and IV.

TABLE III. *Total number of nucleated bone-marrow cells per mg marrow*

Time after irradiation (days)	1 r	Controls	Significance (t-test)
14	2,004,000	2.104,000	n.s.
28	1,855,000	1,417,000	$P < 0.05$
42	2,501,000	2,103,000	$P < 0.3$
56	2,304,000	2,013,000	$P < 0.05$
84	2,184,000	1,763,000	$P < 0.1$
98	1,971,000	1,695,000	$P < 0.02$

TABLE IV. *Total number of nucleated bone-marrow cells per mg marrow*

Time after irradiation (days)	0·1 r	Controls	Significance (t-test)
14	2,066,500	2,074,500	n.s.
28	1,883,000	1,832,000	n.s.
42	1,849,500	1,639,000	$P < 0.2$
56	1,685,000	1,642,500	n.s.

After 1 r, the total number of cells per mg of marrow is statistically increased at the 28th, 56th and 98th days; after 14 days, there is no difference. The difference in number of cells in favour of the rats irradiated with 1 r is also interesting by the 84th day and slightly significant by the 42nd day.

Generally speaking, the number of cells is increased between 28th and 98th days in favour of the rats irradiated with 1 r.

After 0·1 r, the numbers of cells per mg of marrow are higher between 28 and 56 days but the differences are not significant except slightly after 42 days.

Erythropoietic activity

The red-cell incorporation of ^{59}Fe into the peripheral blood of the rats pre-irradiated with 1 r or 0·1 r and of the controls with their statistical significance are recorded in Tables V and VI.

TABLE V. ^{59}Fe *activity in red cells* (counts/min per ml for 10^5 cells)

Time after irradiation (days)	1 r	Controls	Significance (t-test)
14	8,632	7,968	n.s.
28	10,841	8,005	$P < 0·01$
42	9,725	7,225	$P < 0·05$
56	9,989	6,427	$P < 0·01$
84	8,401	7,771	$P < 0·3$
98	7,398	6,896	n.s.

TABLE VI. ^{59}Fe *activity in red cells* (counts/min per ml for 10^5 cells)

Time after irradiation (days)	0·1 r	Controls	Significance (t-test)
14	7,600	6,333	$P < 0·3$
28	9,011	6,720	$P < 0·05$
42	8,921	6,722	$P < 0·1$
56	9,678	8,341	$P < 0·2$

The incorporation of ^{59}Fe in the red cells of rats pre-irradiated with 1 r is higher than in the controls; the results are highly significant after 28, 42 and 56 days, slightly after 84 days. For the 0·1 r pre-irradiated rats, the results are higher but the enhancement is not so large as after 1 r. Anyway, the difference of ^{59}Fe incorporation between the pre-irradiated animals and the controls is significant in favour of the pre-irradiated rats after 28 and 42 days and slightly significant after 14 and 56 days.

Reticulocyte counts

The results are collected in Tables VII and VIII. After 1 r, the reticulocytes are significantly increased by the 14th day, increased by the 28th day and slightly increased from the 42nd day to the 98th day. After 0·1 r, the reticulocytes are also significantly increased by the 14th and the 28th day. The increase is slightly significant on the 42nd day and present, but not significant on the 56th day.

TABLE VII. *Differential cell count in peripheral blood*

Time after irradiation (days)	Red cells per mm³ blood		White cells per mm³ blood		Reticulocytes per 1000 red cells		
	1 r	Controls	1 r	Controls	1 r	Controls	t-test
14	6,613,000	6,250,000	15,366	13,683	28·5	18	$P < 0.01$
28	5,720,000	5,793,000	19,383	12,366	27	20	$P < 0.1$
42	6,450,000	6,456,000	15,220	15,720	36·8	31	$P < 0.3$
56	6,340,000	6,075,000	15,600	11,500	37	29	$P < 0.2$
84	5,956,000	5,740,000	12,483	12,000	19	15·8	$P < 0.2$
98	6,157,000	6,940,000	9,970	10,116	43	35	$P < 0.2$

TABLE VIII. *Differential cell count in peripheral blood*

Time after irradiation (days)	Red cells per mm³ blood		White cells per mm³ blood		Reticulocytes per 1000 red cells		
	0·1 r	Controls	0·1 r	Controls	0·1 r	Controls	t-test
14	7,120,000	7,156,000	14,216	16,020	41	25	$P < 0.01$
28	6,546,500	6,646,500	13,133	15,683	23·6	16	$P < 0.05$
42	5,976,500	5,612,000	14,766	13,866	16·6	13·2	$P < 0.2$
56	5,368,000	5,928,000	19,200	13,383			n.s.

Finally the results of the red and white cell counts per mm³ in the peripheral blood are also recorded in Tables VII and VIII. The results for pre-irradiated rats are not very different to those for the controls, whatever the X-ray dose.

What kind of explanation can be found for these last experimental results? First of all, these low doses of X-rays activate the marrow for a certain time, 1 r more than 0·1 r, but even so low a dose administered for so short a time as 1″9 has an influence on the marrow activity. The erythropoietic activity (as measured by ^{59}Fe incorporation) shows more precise enhancement than the number of cells of the marrow, much more after 1 r, less so after 0·1 r. It is thus possible that, when the differential count of the white and red precursors was made, the difference in the number of marrow cells was essentially due to the red-cell precursors. After 0·1 r, when the ^{59}Fe incorporation showed less increase, the total number of marrow cells was only slightly enhanced. It is, however possible that the reticulo-endothelial system could participate in this increase in the number of marrow cells as Pape (1951) has pointed out. It is curious that notwithstanding the higher erythropoietic activity after 1 r, the reticulocyte count is enhanced by about the same amount in the 1 r and 0·1 r groups. This last test could be a very sensitive one.

Actually we know different causes for an enhancement of erythropoietic activity (excluding radiation), pre- (Valentine and Pearce, 1952; Stohlman et al., 1955) and post-irradiation bleeding, or post-irradiation administration of p-aminopropiophenone (Stohlman et al., 1955) but in all these experiments, the increase of erythropoietic activity is a result of hypoxia induced by bleeding or PAPP. After 1 r or 0·1 r, we cannot speak of hypoxia.

It is also very curious to note that the increase in the number of marrow cells follows a very great gradient of X-ray dose. We have already pointed out that 50 r also significantly increases the number of marrow cells, at least between 7 and 21 days after (Maisin, 1961). One hundred roentgen is too much (Maisin, 1959). The fact that the increase in the number of cells is already so substantial after no more than one roentgen could explain even pathological modifications occurring much later in marrow and lymph nodes; for example the absence of threshold or a very low one in post-irradiation leukaemogenesis occurring after repeated low doses (Lorenz et al., 1955). A similar mechanism occurring after small X-ray doses in other organs not even radiosensitive (Knowlton and Hempelmann, 1949) could also explain the appearance of cancer in those organs (Lorenz et al., 1955).

Finally we can be assured that the increase in cell numbers which must be a consequence of mitotic activity is not synonymous with radioresistance and that in the rat, radioresistance is not a question of the size of the pre-irradiation dose for we gave a very large gradient of dose without ever obtaining radioresistance.

REFERENCES

BETZ, H. (1950). *C.R. Soc. Biol., Paris* **144**, 1439.
DACQUISTO, M. P. (1959). *Radiation Res.* **10**, 118.
KNOWLTON, N. P., and HEMPELMANN, L. H. (1949). *J. cell. comp. Physiol.* **33**, 73.
LORENZ, E., HOLLCROFT, J. W., MILLER, E., CONGDON, C. C., and SCHWEISTHAL, R. (1955). *J. nat. Cancer Inst.* **15**, 1049.
MAISIN, H. (1959). Syndrome médullaire après irradiation, Arscia, Bruxelles.
MAISIN, H. (1961). Congrès Association Radiobiologistes de l'Euratom, Rapallo (in press).
PAPE, R. (1951). *Radiol. Austriaca* **4**, 35.
PAPE, R., and JELLINKE, N. (1950). *Radiol. Austriaca* **3**, 43.
PATERSON, E., GILBERT, C. W., and MATTHEWS, J. (1952). *Brit. J. Radiol.* **25**, 427.
STOHLMAN, F., Jr., CRONKITE, E. P., and BRECHER, G. (1955). *Proc. Soc. exp. Biol., N.Y.* **88**, 402.
VALENTINE, W. N., and PEARCE, M. L. (1952). *Blood* **7**, 1.

DISCUSSION

ALEXANDER: Did you do a sham irradiation, Dr. Maisin? Is it possible that you could get such a result if you just put the animals under the X-ray machine but did not switch it on? Could it be a stress phenomenon?

MAISIN: We didn't always do a sham irradiation, but we sometimes did, and we never got any increase in the total number of cells, or in erythropoiesis.

LAMERTON: Did you say that you got about the same extent of stimulation if you used 50 r? If so, then you have something which is apparently invariant over a wide range of dose.

MAISIN: I would not say that with different doses we got a stimulation which was greater after 1 r than after 50 r. 100 r is too much, but 50 r can cause an increase. On the other hand the stimulation lasts longer after 1 r.

LAMERTON: But we also know that after 50 r there is a decrease in the radioactive-iron uptake very soon after irradiation.

MAISIN: Yes, very soon after. We didn't repeat your experiment this time, but formerly we did and we got the same results. I know, too, that you did an experiment showing that even with 25 r or 5 r there is a decrease. I have the feeling that later the animals react in another way.

BRINKMAN: In some irradiation rooms you can smell ozone and we have often been troubled by getting results which were due to the presence of ozone and not to the irradiation itself. Could there be something like this in your experiments?

MAISIN: I don't think so because with modern machines the ozonization of the room is not large. Moreover, the sham-irradiated animals were kept in the same room.

BACQ: There is one thing which troubles me. If you had more red-cell precursors in the marrow; if you had more reticulocytes in the circulating blood for a long time, then, over a long period, you must get polycythaemia unless the mean life-span of the cells is decreased after this type of irradiation. Have you measured the mean life-span of these red cells?

MAISIN: Not yet.

ROTBLAT: I notice that there was a gradual decrease in mortality after the additional 500 r.

MAISIN: If you look at the Figure you can see that certain rats receiving 1 r or 0·1 r have a decreased mortality, but you see the same phenomenon in the controls. In fact it happens in all the groups which we fed on purina chow. Thus you cannot speak of a gradual decrease in mortality. The rats which did not receive purina chow were more sensitive. We think that it is a question of nutrition.

MOLE: I think these results are quite startling. One of my colleagues ran into trouble over the estimation of peroxide in irradiated animals because he found that just picking the animals up made a difference. It is very important if the effect of a very small dose of irradiation like this can be revealed, so I do hope you will do the experiment in what is called a "double blind trial", where the people who examine the animals don't know which have been irradiated. I think that, to be really convincing, this is what you will have to do.

MAISIN: We have got systematic results. Most of the experiments were repeated, not all of them but some of them, but we agree that the double blind method can be applied.

UPTON: How many replicate experiments do the data represent? Was it the same number each time?

MAISIN: Each time we counted 6 animals for each dose, always with controls and always for the irradiated animals. We repeated this at least two or three times for all the results which were statistically different. At first I was sceptical myself.

LEUKAEMOGENIC EFFECT OF WHOLE-BODY ^{60}CO-γ-IRRADIATION COMPARED WITH ^{3}H-THYMIDINE AND ^{3}H-CYTIDINE: PRELIMINARY REPORT ON THE DEVELOPMENT OF THYMIC LYMPHOMAS IN C57BL/6J MICE†

H. COTTIER, E. P. CRONKITE, E. A. TONNA, AND N. O. NIELSEN

Medical Research Center, Brookhaven National Laboratory, Upton, New York, U.S.A.

SUMMARY

Groups of 32 female mice (C57BL/6J) 6 weeks of age were given:
(1) a total of 480 r whole-body ^{60}Co-γ-irradiation in increments of 160 r at 7 day intervals and at a dose-rate of 22 r/min, or
(2) three subcutaneous injections of ^{3}H-thymidine (3×10 μc/g body weight; specific activity 1·9 c/m mole) at weekly intervals, or
(3) three subcutaneous injections of ^{3}H-cytidine (3×10 μc/g body weight; specific activity 1 c/m mole) at weekly intervals, or
(4) three subcutaneous injections of both ^{3}H-thymidine and ^{3}H-cytidine (dosage as mentioned under (2) and (3)) at weekly intervals, or
(5) received no treatment.

The mice have been followed to date for 250 days. Only whole-body ^{60}Co-γ-irradiated mice have developed thymic lymphomas. The difference is statistically significant ($P<0.01$).

These experiments will be helpful in determining the possible hazards involved in the use of these radioactive compounds and also may lead to a better understanding of basic phenomena underlying radiation leukaemogenesis and carcinogenesis, since ^{3}HTDR gives predominantly intranuclear radiation and ^{3}HCR predominantly cytoplasmic irradiation.

The *in vivo* application of tritiated thymidine (^{3}HTDR) has been shown to produce morphologically detectable acute radiation damage when sufficient amounts are given (Johnson and Cronkite, 1959; Grisham, 1960; Cronkite *et al.*, 1961; Sauer and Walker, 1961; Bateman and Chandley, 1962; Smith *et al.*, 1962). The lowest dose reported to cause demonstrable injury is between 1 and 5 μc ^{3}HTDR (specific activity 1·9 c/m mole) per g body weight in: mouse spermatogonia (Johnson and Cronkite, 1959); regenerating rat liver cells (Grisham, 1960); and rat liver cells (Post and Hofman, 1961). No data are available on possible radiation effects of ^{3}H-cytidine (^{3}HCR).

Radiation dosimetry in an animal injected with ^{3}HTDR is very difficult if not impossible. The β dose-rate expressed in r per tritium disintegration as

† Research supported by the U.S. Atomic Energy Commission.

a function of the distance from a point-source has been computed (Robertson and Hughes, 1959; Robertson et al., 1959), but application of these data to *in vivo* dosimetry (e.g. autoradiography) is not readily feasible since it involves such largely unsolved problems as distribution and availability of the precursor in the tissues, magnitude and rate of its uptake by various cell types synthesizing DNA, proliferative scheme of all cell lines, geometry and coincidence error in autoradiography, unequal intranuclear distribution of the label and its possible re-utilization when the labelled cells disintegrate. Tentative estimations of the radiation dose delivered, based on grain counts in autoradiographs therefore depend on number of debatable assumptions (Lajtha and Oliver, 1959; Cronkite et al., 1961).

In contrast to the substantial amount of information available on *acute* radiation damage caused by application of ^{3}HTDR very little is known about *late* effects of this radioactive compound. One of the important questions is, whether and at what dose level ^{3}HTDR may exert a tumorigenic effect. Since theoretical prediction in this respect is not practicable on the basis of dosimetry and comparisons with the tumour-promoting effect of whole-body X- or γ-irradiation, long-term experiments with animals injected with tritiated nucleosides are essential. Lisco and co-workers (1961) reported that in CAF-mice as little as 1 μc ^{3}HTDR (specific activity 360 mc/m mole) per g body weight, given in a single injection, may stimulate tumorigenesis. However, the data presented by these authors does not at present conclusively prove the carcinogenic effect of ^{3}HTDR.

This preliminary report deals with the time-course of development of thymic lymphomas in female C57BL/6J mice given fractionated whole-body ^{60}Co γ-irradiation as compared to animals treated with large doses of ^{3}HTDR and/or ^{3}HCR and untreated controls.

MATERIALS AND METHODS

Two hundred and twenty-four female, 6-week-old C56BL/6J mice (Roscoe B. Jackson Memorial Laboratory, Bar Harbor, Maine) weighing 16 to 18·5 g (average 17·3 g) were assigned by random selection to seven groups of 32. They were kept in cages containing 8 animals each and had *ad libitum* access to water and food (Purina Laboratory Chow for mice, Ralston Purina Company, St. Louis, Missouri).

Groups I-III (32 animals in each) were given fractionated whole-body ^{60}Co-γ-irradiation (total dose 480 r, given in three fractions of 160 r at 7-day intervals; dose-rate 22 r/min; mice kept in air during irradiation).

Group IV (32 animals) was given three single subcutaneous injections of ^{3}H-thymidine at weekly intervals (3 × 10 μc/g body weight, based on average weight of each cage. Specific activity 1·9 c/m mole (Schwarz Bioresearch, Mt. Vernon, N.Y.); solution for injection: 100 μc/ml in saline for first injection, 200 μc/ml for second and third injections).

Group V (32 animals) was given three subcutaneous injections of ³H-cytidine at weekly intervals (3 × 10 μc/g body weight, based on average weight of each cage. Specific activity 1 c/m mole (Schwarz Bioresearch, Mt. Vernon, N.Y.); solution for injection: 100 μc/ml in saline for first injection, 200 μc/ml for second and third injections).

Group VI (32 animals) received three subcutaneous injections of both ³HTDR and ³HCR at weekly intervals (dosage as in Group IV and V).

Group VII had no treatment.

The animals were allowed to die spontaneously or were killed in a definite moribund state.

RESULTS

During the first months after initiation of the experiment, some animals of each group (19 out of 224 in total) developed ulcerative skin lesions due to mites around the base of the tail. These mice were then kept in separate cages. All animals were given two successive dipping treatments at an interval of 2 weeks in a solution of 2% Aramite and 0·1% Nacconal in water. By this procedure, the skin lesions were prevented from spreading to the colony in general.

The animals that died spontaneously or were killed in a moribund state are listed in Table I. Except for Groups I-III all mice dying showed generalized

TABLE I. *Number of mice killed in moribund state or dead with or without thymic lymphoma*

Group	Number of animals	Treatment ‡	Number of mice killed in moribund state or dead (Days after start of experiment)									
			1–25	26–50	51–75	76–100	101–125	126–150	151–175	176–200	201–225	226–256
I	32	3 × 160 r	—	—	—	—	1** 1*	1**	—	1**	1**	1†
II	32	3 × 160 r	—	—	—	—	—	—	—	3**	2**	2**
III	32	3 × 160 r	—	—	—	—	—	—	2**	3**	2**	—
IV	32	3 × ³HTDR	—	—	—	—	1*	1*	—	—	—	—
V	32	3 × ³HCR	—	—	—	—	1*	1*	—	—	—	—
VI	32	3 × ³HTDR+³HCR	—	—	—	—	1*	1*	—	—	—	—
VII	32	None	—	—	—	—	1*	—	—	—	—	—

* Dying with skin lesions and generalized infection.
** Dying with thymic lymphoma.
† Dying with mammary gland carcinoma.   ‡ See text.

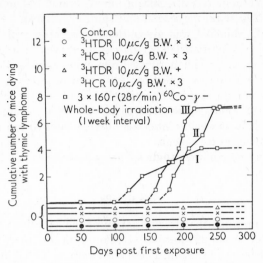

Fig. 1. Cumulative number of mice killed in a moribund state or dead with thymic lymphoma plotted against the days post-exposure. Each line represents a group of 32 animals. B.W. = body weight.

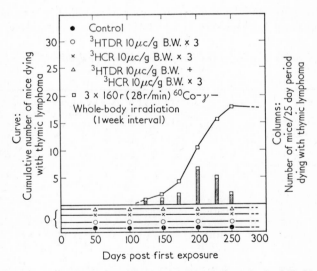

Fig. 2. Cumulative and periodical (25-day periods) number of mice killed in a moribund state or dead with thymic lymphoma plotted against time post-exposure. The ^{60}Co-γ-irradiated mice are pooled in one group of 96 animals, the other lines represent groups of 32 animals each. B.W. = body weight.

infection originating in the skin ulcers mentioned above. No predilection of one of the experimental groups for this skin disease was observed, nor was there an apparent relation between this condition and the development of thymic lymphomas.

The number of animals dying, or killed when moribund, with thymic lymphomas is shown in Figs. 1 and 2 as a function of time after initiation of the experiment. The difference between the number of 18 thymomas in the ^{60}Co-γ-irradiated groups I–III observed within 250 days versus none in all the other groups including controls is statistically significant ($P < 0.01$, Chi-square test). The incidence of thymic lymphomas per cage (groups of 8 animals) ranged from zero to 4. One ^{60}Co-γ-irradiated mouse developed a mammary gland carcinoma within 8 months after exposure.

DISCUSSION AND CONCLUSIONS

From these data it can be concluded that under the conditions of the experiment the incidence of thymic lymphomas in mice within 250 days after fractionated whole-body ^{60}Co-γ-irradiation (3×160 r at 7 days intervals) was significantly higher than after three subcutaneous injections at weekly intervals of ^{3}HTDR ($3 \times 10\ \mu c/g$ body weight) or ^{3}HCR ($3 \times 10\ \mu c/g$ body weight) or both. Up to 250 days after initiation of the experiment no difference in the development of thymic lymphomas could be observed between the untreated controls and the animals given tritiated pyrimidine nucleosides as mentioned above. Whether ^{3}HTDR, ^{3}HCR or both, given to female 57CBL/6J mice in three weekly doses of 10 $\mu c/g$ body weight at the age of 6 to 8 weeks promotes the incidence of this particular neoplastic disease, can only be evaluated when the animals have been observed over their entire life-span. The most important period in this respect will be around 400 days after first exposure, the average latent period reported before thymic lymphomas appear in low-dose irradiated female C57BL mice (Kaplan and Brown, 1952). No comment of course is as yet possible as to an eventual influence of these tritiated compounds on life-span.

If it is true that in CAF mice a single injection of ^{3}HTDR (specific activity 360 mc/m mole in a dose of 1 $\mu c/g$ body weight does have a tumorigenic effect (Lisco et al., 1961), then our results give rise to some questions. The lowest dose of ionizing radiation given to mice as a *short* whole-body irradiation reported to have given a tumorigenic effect, other than the production of ovarian tumours, is of the order of 200 r (Kaplan and Brown, 1952; Furth et al., 1959). To produce increased numbers of tumours by *continuous* whole-body γ-irradiation, except for the induction of ovarian tumours, daily doses in the order of 0·11 r were necessary (Lorenz et al., 1955). If a single dose of 1 μc

^{3}HTDR/g body weight should be sufficient to rise the incidence of neoplastic disease (Lisco et al., 1961), it is interesting to note that three doses given at weekly intervals and each ten times higher, as used in our experiment has not as yet, up to 250 days after injection, produced any detectable thymic lymphomas. It must be mentioned however, that we are not dealing with the same strains of mice, and that Lisco and co-workers (1961) injected the ^{3}HTDR into animals of different ages. In addition to this, these authors followed their mice over a longer period of time than we have to date. It is evident that more data are necessary to clarify the problem. Experiments like these, aimed at detecting possible tumorigenic and/or leukaemogenic effects of tritiated nucleosides, have not only practical implications in determining the possible hazards involved in the use of these radioactive compounds but they are also of interest for the understanding of basic phenomena related to radiation leukaemogenesis and carcinogenesis. With ^{3}HTDR, radiation is delivered predominantly within the nuclear material sparing most of the cytoplasm. To some extent the contrary is true for ^{3}HCR, since autoradiographic findings at later stages following isotope injection often reveal a much higher grain count over the cytoplasm. Thus it may be expected that experiments such as those of the present investigation, may provide some information as to the relative importance of nuclear versus cytoplasmic damage in the pathogenesis of radiation-induced neoplastic growth.

REFERENCES

BATEMAN, J. A., and CHANDLEY, A. C. (1962). *Nature, Lond.* **193**, 705.
CRONKITE, E. P., BOND, V. P., FLIEDNER, T. M., KILLMANN, S. A., and RUBINI, J. R. (1961). I.A.E.A. Symposium on the detection and use of tritium in the physical and biological sciences, Vienna, 3–10 March 1961 (in press).
FURTH, J., UPTON, A. C., and KIMBALL, A. W. (1959). *Radiation Res. Suppl.* **1**, 243.
GRISHAM, J. W. (1960). *Proc. Soc. exp. Biol., N.Y.* **105**, 555.
JOHNSON, H. A., and CRONKITE, E. P. (1959). *Radiation Res.* **11**, 825.
KAPLAN, H. S., and BROWN, M. B. (1952). *J. nat. Cancer Inst.* **13**, 185.
LAJTHA, L. G., and OLIVER, R. (1959). *Lab. Invest.* **8**, 214.
LISCO, H., BASERGA, R., and KISIELESKI, W. E. (1961). *Nature, Lond.* **192**, 571.
LORENZ, E., HOLLCROFT, J. W., MILLER, E., CONGDON, C. C., and SCHWEISTHAL, R. (1955). *J. nat. Cancer Inst.* **15**, 1049.
POST, J., and HOFMAN, J. (1961). *Radiation Res.* **14**, 713.
ROBERTSON, J. S., and HUGHES, W. L. (1959). *Proc. nat. biophys. Conf.* 278.
ROBERTSON, J. S., BOND, V. P., and CRONKITE, E. P. (1959). *Int. J. appl. Radiation Isotopes* **7**, 33.
SAUER, M. E., and WALKER, B. E. (1961). *Radiation Res.* **14**, 633.
SMITH, W. W., BRECHER, G., STOHLMAN, F., Jr., and CORNFIELD, J. (1962). *Radiation Res.* **17**, 201.

DISCUSSION

GRAY: Have you any idea what the probability of survival of any of your cells was which took up the thymidine? Because if there is a specific initiating event it is obviously being expressed only when the cells divide.

COTTIER: We have but very little information on this as far as long-term effects are concerned. Tonna, Shellabarger and Cronkite (unpublished data) found that 25 µc of tritiated thymidine given to newborn rats did not influence their growth-rate up to the age of 9 months. At this time there were still some labelled cells found in various tissues. Our mice treated with tritiated nucleosides as mentioned above seemed to behave quite normally. They did not undergo a weight loss comparable to that observed in the ^{60}Co-γ-irradiated group. One mouse died accidentally 3 weeks after the last injection of tritiated thymidine; the thymus still contained a number of labelled cells. The other material is not yet processed.

UPTON: May I ask Dr. Cottier what proportion of cells in the bone-marrow and thymus were labelled? A number of workers have shown now that, to induce thymic lymphoma, whole-body irradiation was necessary. If one gives a pulse of thymidine one doesn't succeed in labelling all dividing cells although if you give several pulses you can get that later.

COTTIER: In order to establish a meaningful labelling index one should be able to recognize specific dividing cell classes as such. With respect to the lymphatic tissue this in itself is a difficult problem. After a single injection of tritiated thymidine the "large lymphoid precursors" in the thymus or in germinal centres of lymph nodes and spleen show a labelling index ranging from 50 to 70%. Thus it seems very probable that a number of potential stem-cells were left unlabelled by this method. On the other hand there is evidence that labelled cells continue to proliferate even after doses of tritiated nucleosides such as given here.

LAMERTON: Would it not be true, Dr. Cottier, that almost all the cells would take up labelled cytidine and only a proportion labelled thymidine? Knowing the range of the beta particle from cytidine quite a proportion of the nuclei of these cells would not have been irradiated.

COTTIER: The situation after administration of tritiated cytidine is indeed difficult to evaluate. Disregarding the fact that a fraction of this labelled precursor contributes to DNA synthesis, there are still other possibilities of nuclear irradiation since

(1) at least a good portion of macromolecular RNA is formed in the nucleus and subsequently transferred to the cytoplasm, and

(2) the outer shell of the nuclear chromatin may be irradiated by tritium located in the adjoining cytoplasm.

BACQ: Is it logical, or not, to suppose that precursor cells, which do not take up tritiated thymidine, are those which normally synthesize much larger amounts of thymidine? So that when you measure the percentage of cells which take up thymidine what you have actually shown is the inhomogeneity of the biochemical behaviour of these cells.

COTTIER: The so-called pool of DNA-precursors naturally available within the cells is considered by most authors to be rather small. But there is little knowledge about the actual size of this pool in specific cell types. By continuous infusion of tritiated thymidine one can label practically 100% of dividing cell classes.

MULLER: Isn't it true that some cells cannot incorpotate thymidine into their DNA?

COTTIER: It has been reported that in some mouse leukaemias the neoplastic cells may lack the enzyme that converts thymidine into its monophosphate (Shooter, Bianchi and

Crathorn, 8th Congr. Europ. Soc. Haematol., Vienna 1961), but there is little additional information. Intestinal epithelium obtained by biopsy from sprue-patients and incubated *in vitro* with tritiated thymidine in some instances did not label (Cronkite and Killmann, unpublished data). This may be another example of cells being unable to incorporate tritiated thymidine into their DNA.

DEVIK: The differences quoted by Dr. Cottier between his results and Dr. Lisco's results with respect to the dose may not be as large as they seem. We injected a series of mice with different amounts (1 μc/g tritiated thymidine or 10 μc/g) and measured the total amount of activity. This was not ten times higher in the second group; it was quite variable in different organs, within the gut it was almost ten times as high but in most tissues it was much less. It may be possible that the dose difference is smaller, and one more point which Dr. Gray mentioned is that with a higher dose you may kill all the cells. We think this is apparent from other results.

COTTIER: We do not know as yet what degree of cell death was caused by the administration of tritiated nucleosides in doses such as those given in our experiment. It is not excluded that the lack of increased leukaemogenesis after this treatment may, at least in part, be due to a disintegration of labelled cells. On the other hand the statistical significance of the tumorigenic action Lisco's group obtained with the injection of only 1 μc tritiated thymidine per g body weight (specific activity 360 mc/m mole) is open to criticism.

THE EFFECT OF SINGLE AND FRACTIONATED DOSES OF IRRADIATION ON THE HAEMATOPOIETIC SYSTEMS OF C57BL MICE

P. C. KOLLER AND VALERIE WALLIS

Chester Beatty Research Institute, Institute of Cancer Research, London, England

INTRODUCTION

Kaplan and his co-workers (1953a) have shown that in 85 to 95% of mice of C57BL strain, lymphoid leukaemia develops in the thymus 50 to 100 days after periodic total-body irradiation. Evidence has been obtained by these investigators and others (cf. Miller, 1961) showing that the target-organ is the thymus, and that the haematopoietic tissue is also involved in the leukaemogenic process. When isogenic bone-marrow cell suspension is injected into irradiated C75BL mice, the development of leukaemia can be prevented (Kaplan et al., 1953b; Miller, 1961).

One of our aims is to study the nature of this haematopoietic "principle" and its mode of operation. The present report deals with the single and cumulative effects of X-irradiation on the haematopoietic system of C57BL mice, during the pre-leukaemic period.

METHODS

28 to 30-day-old C57BL mice (males and females) were irradiated with four, weekly doses of 180 r (total-body) at a dose-rate of 60 r/min. The mice were divided into four groups, one received only one dose of 180 r, another two, a third group three, and the fourth, four doses of 180 r. Mice were sacrificed 1 and 6 days after each irradiation. Peripheral blood samples were taken and total white-cell counts were made on all mice sacrificed. For the study of radiation-induced cell injuries, bone-marrow material was prepared. Mice were injected with colcemid (dose: 2% of the body weight of an 0·02% solution) intraperitoneally and sacrificed 60 minutes after injection. Bone-marrow cells in metaphase were divided into two classes: normal and abnormal the latter included cells which either showed chromosome fragments or contained more than the normal diploid number of chromosomes (40). By scoring cells in metaphase, not all radiation injuries can be detected, and the figures obtained must be considered as an underestimate.

RESULTS

Cell composition of peripheral blood

Table I shows the effect of various doses of total-body irradiation at 1, 6, 30 and 60 days after exposure to X-rays. It can be seen that about 30 days are required for the restoration of the normal blood values. The data also show that the number of white cells in the peripheral blood is reduced to nearly the same level after each dose and that the period of recovery is about the same. The number of fractionated doses, i.e. the total amount of radiation, it seems, does not affect the recovery time.

TABLE I. *White cell counts in irradiated* C57BL *mice*

Time after X-rays (days)	1 × 180 r	2 × 180 r	3 × 180 r	4 × 180 r	Control
1	2,050	6,600	2,600	2,700	7,050 (5 weeks)
6	3,200	3,800	4,600	2,500	
30	14,900	4,400	7,450	7,400	10,500 (7 weeks)
60	13,600	10,800	13,300	10,000	

Chromosome analysis in the bone-marrow

Bone-marrow and spleen of irradiated mice were taken at varying intervals and preparations were made for cytological analysis. Dividing cells were, however, extremely rare in the spleen samples and did not provide sufficient data for comparative study. Though many metaphases were seen in the bone-marrow samples, in only a fraction of these cells were the chromosomes spread well enough to count. Table II shows the frequency of abnormal cells observed in the cell population of the marrow of irradiated mice. It was found that 1 day after receiving 1 × 180 r and 2 × 180 r, nearly half the cells at metaphase were abnormal, with fragmented chromosomes. In the samples taken 6 days after the mice were exposed to these doses, the frequency of injured cells was low; in the females, no abnormal cells were seen out of 40 cells analysed.

When the marrow cells of mice receiving three or four doses of 180 r are considered, it can be seen that, at least in the females, the frequency of abnormal cells is about five times less than in the marrow of females which received only one or two doses of 180 r. This finding is of special interest, because it may indicate that the cell population after two doses of 180 r had undergone a change, due perhaps to selection, as a result of which it became more radiation resistant.

The mitotic injuries shown by the marrow, lead to cell death and account for the depletion of the white cells in the peripheral blood.

TABLE II. *Percentage of cells with chromosome abnormalities*

Dose	Time after X-rays (days)	Males		Females	
1 × 180 r	1	41·2	(17†)	33	(3†)
	5	0	(15)	0	(20)
2 × 180 r	1	44·4	(45)	43·3	(30)
	6	5·7	(35)	0	(20)
3 × 180 r	1	0	(9)	8·3	(24)
	6	—		0	(22)
4 × 180 r	1	12·5	(8)	0	(13)
	6	0	(13)	6·6	(30)

† Figures in brackets are number of cells analysed.

Mitotic index

The rate of mitosis has been determined in the bone-marrow samples by analysing 2,000 cells, and counting the number of cells in prophase and

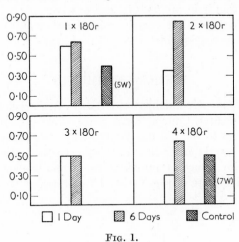

Fig. 1.

metaphase. The data obtained are given in Fig. 1. The mitotic index, $\frac{\text{(number of dividing cells} \times 100)}{\text{(total number of cells counted)}}$, in 5- and 7-week-old C57BL mice is given as control value: it is 0·40 and 0·50 respectively. It can be seen that the number of cells in division, 1 and 6 days after receiving 1 × 180r , is higher than in the control. This increase might indicate (i) recovery from the mitosis-suppressing effect of irradiation and (ii) repair process. The figures obtained from marrow of of mice which received 2 × 180 r and 4 × 180 r are below the control value; they seem to indicate that 1 day after this radiation dose, the duration of mitosis suppression was longer than after 1 × 180 r. The interesting finding concerns the higher mitotic index in the 6-day marrow samples, i.e. 1 day before another fraction of 180 r was delivered, as compared with the mitotic index found in the marrow of unirradiated control mice. Further investigation is required to determine the statistical significance of this difference and to correlate the phenomenon, if it is true, with the reduction in the number of cells showing mitotic injury after three of four doses of 180 r.

CONCLUSION

The extensive studies of Kaplan and associates (Kaplan and Brown, 1952; Kaplan *et al.*, 1953a, b) established the fact that in mice of C57BL strain the important factor in the leukaemogenic process is not the total radiation dose, but the method of fractionation of the dose. The cumulative effects of fractionated doses, which have been the subject of this preliminary report are, very likely, significant contributory factors in the pre-leukaemic process. Further investigation is in progress to find out how the injuries inflicted on the haematopoietic system in the early stage of leukaemic transformation, are contributing to the process which results in lymphoid leukaemia.

ACKNOWLEDGMENTS

The authors are greatly indebted to Dr. J. F. A. P. Miller for his help and to the Medical Research Council for a grant. This investigation has been supported by grants to the Chester Beatty Research Institute (Institute of Cancer Research: Royal Cancer Hospital) from the Medical Research Council, the British Empire Cancer Campaign, the Anna Fuller Fund, and the National Cancer Institute of the National Institutes of Health, U.S. Public Health Service.

REFERENCES

KAPLAN, H. S., and BROWN, M. B. (1952). *J. nat. Cancer Inst.* **13**, 185.
KAPLAN, H. S., BROWN, M. B., and PAULL, J. (1953a). *Cancer Res.* **13**, 677.
KAPLAN, H. S., BROWN, M. B., and PAULL, J. (1953b). *J. nat. Cancer Inst.* **14**, 303.
MILLER, J. F. A. P. (1961). *Advanc. Cancer Res.* **6**, 291.

DISCUSSION

DRASIL: I should like to draw attention to some results by Dr. Karpfel. He found many chromosome aberrations in bone-marrow cells which were injected into irradiated mice. Furthermore he observed that the percentage of cells with chromosome aberrations is much higher in allogenic (homologous) chimaeras than in syngeneic (isologous) chimaeras. These results suggest an interaction between the irradiated host and the injected bone-marrow cells, derived from a non-irradiated donor.

KOLLER: How do you know that the cells showing chromosome-breaks are the non-irradiated donor cells?

DRASIL: The mice were given 1,000 or 1,200 r total-body irradiation, without bone-marrow therapy very few dividing cells were seen in the marrow of such animals 24 and 44 hours after irradiation. In those mice which were given bone-marrow from non-irradiated donors, the number of cells in division was high. On this basis we may assume that the dividing cells showing the chromosome injuries were derived from the non-irradiated donor.

KOLLER: Dr. Ford of Harwell, in a very thorough work, demonstrated that bone-marrow cells of irradiated mice show chromosome injuries and such cells may appear for several days after irradiation. Ford's observation suggests that we should be very careful in our interpretation regarding the identity of cells with injured chromosomes in the marrow of irradiated animals, which received non-irradiated donor cells.

NEW EVIDENCE ON THE MECHANISM OF RADIATION LEUKAEMOGENESIS†

I. BERENBLUM AND N. TRAININ
The Weizmann Institute of Science, Rehovoth, Israel

The work to be presented here arose from an attempt to apply some of the principles of experimental carcinogenesis to radiation leukaemogenesis. Such an approach has the advantage of making use of the accumulated evidence, the special techniques, and some of the theorizing, from the broad field of carcinogenesis. There is, however, a possible danger to be guarded against, which hardly needs stressing before this audience—of looking upon radiation merely as another carcinogenic tool among the hundreds already known, and of underestimating the importance of the unique properties of ionizing radiation in relation to the over-all biological response.

A few introductory remarks about the mechanism of carcinogenesis in general, before turning to the specific problem of radiation leukaemogenesis, should make the relationship clearer.

One of the crucial problems of carcinogenesis is the existence of a very long latent period, during which no morphological evidence of neoplasia is detectable. Even when one extrapolates the growth-curve of a tumour to its morphological zero point, when, presumably, the neoplastic change is still confined to a single cell, there is still a very long latent period preceding it, which must be accounted for in the carcinogenic process. The state of "pre-neoplasia" is, thus, more a functional than a morphological problem.

We now know that, in the case of skin at least, the changes during the long latent period do not constitute a single continuous process, but are made up of separate components—a sort of "biological chain reaction", by analogy with the familiar "biochemical chain reactions" in cellular metabolism.

In skin carcinogenesis, two consecutive processes—"initiation" and "promotion"—are now clearly defined, each of which can be separately induced by artificial means. One of the most effective initiating agents for mouse skin is urethane, or ethyl carbamate (Salaman and Roe, 1953); the most potent promoting agent for mouse skin is croton oil (Berenblum, 1941).

Urethane alone is not carcinogenic for mouse skin; croton oil alone is to some extent carcinogenic if the treatment is very prolonged. That their

† A review based on work supported in part by a grant from the Joseph and Helen Yeamans Levy Foundation and by research grant C-5455 of the National Cancer Institute, U.S. Public Health Service.

combined action is not additive, is shown by the fact that tumours develop rapidly and in large numbers when the promoting action *follows* the initiating action, but not in reverse (Berenblum and Haran, 1955). Urethane also acts as initiator for mouse skin when it is administered systemically (Haran and Berenblum, 1956); croton oil acts as promoting agent only when applied topically. A complete carcinogen, such as 3.4-benzpyrene of 9, 10-dimethyl-1.2-benzanthracene, applied *repeatedly* to the skin, naturally possesses both initiating and promoting properties; however, when applied *a single time*, it usually acts as initiator only.

Initiating action is very rapid; promoting action, very slow. The initiating process appears to be irreversible, the same number of tumours arising from subsequent promoting action, whether the latter is given immediately following the initiating stimulus or only after an interval of several months (Berenblum and Shubik, 1947a). The number of tumours resulting from the two-stage process is a function of the concentration of the *initiating* agent; the speed of their appearance is dependent more on the efficiency of the *promoting* action (Berenblum and Shubik, 1947a, b).

The nature of initiating action is not known, though a mutation-like change could explain the quantitative relationships observed, and would be consistent with the rapid action. There is no proof, however, that the change is "mutational" in the genetic sense. Promotion, on the other hand, operates very differently: it is not only slow in action but, in its early stages, apparently reversible.

The most plausible explanation to account for the results obtained for skin carcinogenesis is that initiation causes an irreversible change in a normal epidermal stem-cell, converting it into a "dormant tumour cell", which then needs a further, prolonged stimulus—promoting action—to enable it to develop into a progressively growing tumour (Berenblum, 1954). Without the added promoting action, the growth equilibrium, which exists in normal skin epithelium and also under conditions of reparative hyperplasia, would apply equally to the dormant tumour cell and its progeny. Promoting action presumably causes a delay in maturation of the dormant tumour cell and its progeny, leading to the formation of a critical sized colony, after which, the process becomes self-perpetuating (Berenblum, 1960).

Whether ths scheme is applicable to carcinogenesis in general, and if so, whether it offers a complete explanation of the mode of action of carcinogens or merely an over-simplified model, could only be answered if such a two-stage mechanism were also demonstrated in a number of systems other than of the skin. For technical reasons, this proved to be difficult, especially for carcinogenesis of internal organs.

When, therefore, Kawamoto *et al.* (1958) reported that urethane had a "co-leukaemogenic", or synergistic, action *vis-à-vis* X-ray leukaemogenesis, we decided to explore the possibility of radiation and urethane having an initiation-promotion relationship. This was tested by the "reversal" type of experiment, mentioned above.

Using 450 r, in divided doses of 90 r at 5-day intervals, for total-body radiation, and 100 mg, in divided doses of 20 mg at 5-day intervals, for intraperitoneal injections of urethane, the following results were obtained when tested in C57BL/6 mice (see Table I): (i) confirmation of the results of Kawamoto *et al.*, that radiation leukaemogenesis, in a low-leukaemia strain of mice, is augmented by simultaneous administration of urethane, and (ii) a similar augmentation observed when the urethane treatment is begun 2 weeks *after* completion of the radiation treatment, *but not when the sequence is reversed* (Berenblum and Trainin, 1960, 1961a). (The "leukaemias" produced were essentially thymic lymphomas, which tended, in some cases only, to be accompanied later by a blood picture of lymphatic leukaemia.)

TABLE I. *Promoting action of urethane in radiation leukaemogenesis in* C57BL/6 *mice*

Treatment†	Incidence of leukaemia
Radiation alone	23/75 (31%)
Radiation and urethane concurrently	29/50 (58%)
Radiation followed by urethane	42/70 (60%)
Urethane followed by radiation	21/74 (28%)
Urethane alone	2/61 (3%)

† Radiation: 90 r × 5, at 5-day intervals = 450 r.
Urethane treatment: 20 mg × 5, at 5-day intervals = 100 mg, injected intraperitoneally.
Interval between completion of one treatment and commencement of the other: 2 weeks.

Though the results supported a two-stage mechanism for leukaemogenesis, the experimental conditions were obviously not ideal, since the radiation, which was intended to serve as initiating factor only, was by itself highly leukaemogenic. Experiments were then performed using single irradiations, at different dose levels, instead of divided doses, in the hope that a suitable range might be found where leukaemia failed to develop by initiation alone or by promotion alone, but did develop by initiation followed by promotion. The results (see Table II) suggest that such an experimental set-up is possible, though further refinements are needed to reach the ideal conditions.

Yet another possibility had to be considered, before accepting the results as evidence of a two-stage mechanism. It seemed rather surprising that urethane, an *initiator* for skin carcinogenesis, should act as a *promoter* for leukaemogenesis. Could it be that urethane did not act as promoter in the strict sense of the term, but indirectly, by further depressing the activity of the bone-marrow (cf. Kaplan and Brown, 1951).

To test this alternative possibility, the effect of intravenous injections of isologous normal bone-marrow suspension was investigated in relation to the "two-stage" experimental system. It was found (Berenblum et al., 1961) that while such treatment interfered with the over-all leukaemogenic action of

TABLE II. *Leukaemia induction in C57BL/6 mice resulting from single doses of radiation followed or preceded by urethane*

Treatment	Leukaemia incidence	Expected incidence on assumption of simple additive action
150 r alone	1/84 (1%)	
150 r followed by urethane	11/60 (18%)	6%
300 r alone	5/64 (8%)	
300 r followed by urethane	11/47 (23%)	13%
400 r alone	4/70 (6%)	
400 r followed by urethane	27/80 (34%)	11%
Urethane alone	5/110 (5%)	
Urethane followed by 150 r	4/65 (6%)	6%
Urethane followed by 300 r	7/61 (11%)	13%

total-body radiation, as had been previously demonstrated (Kaplan et al., 1953), it did not interfere with the "promoting" phase brought about by urethane (see Table III).

We were thus led to conclude, despite our own earlier scepticism, that the phenomenon observed did, in fact, represent a two-stage mechanism of leukaemogenesis, in which low doses of radiation acted as initiating factor and urethane as promoting factor. The significance of this conclusion was two-fold. (i) It provided a broader basis for the fundamental study of the mechanism of carcinogenesis in general, and (ii) it offered a more refined tool for the study of leukaemogenesis. The latter is, naturally, the one that concerns us most here, in this Symposium.

Experiments were then set up, using the radiation-urethane two-stage system as an analytical tool, to explore some of the complicating factors known to be implicated in radiation leukaemogenesis, e.g. (a) the inhibition resulting from shielding a part of the body (Kaplan and Brown, 1951), and the role of bone-marrow in radiation leukaemogenesis (Kaplan et al., 1953); (b) the prevention of leukaemogenesis by thymectomy, and its reversal by subsequent implantation of unirradiated normal thymus (Kaplan et al., 1956); and (c) the involvement of a specific leukaemia virus in radiation leukaemogenesis (Gross, 1959; Lieberman and Kaplan, 1959).

Regarding the "bone-marrow effect" in radiation leukaemogenesis, one experiment, involving injections of isologous bone-marrow following urethane treatment, after an initial total-body radiation, has already been described. A further experiment is in progress, in which the action of urethane is being tested in relation to partial-body radiation. From as yet incomplete results, it would appear that urethane can produce at least a partial "reversal" of the inhibition of shielding.

TABLE III. *Influence of bone-marrow injections on X-ray leukaemogenesis in C57BL/6 mice, with and without urethane treatment*

Group	Primary treatment†	Secondary treatment†	No. of mice used	Leukaemia incidence	Leukaemia incidence, excluding generalized nonthymic leukaemia
1	Irradiation	Urethane	85	20/72 (27%)	19/72 (26%)
2	Irradiation	Urethane plus bone-marrow	85	18/72 (25%)	18/72 (25%)
3	Irradiation	None	97	7/97 (7%)	6/97 (6%)
4	Irradiation plus bone-marrow	None	80	1/76 (1%)	0/76 (0%)
5	Irradiation plus bone-marrow	Urethane	85	14/73 (19%)	13/73 (18%)

† Irradiation: single total-body exposure to 400 r. Urethane: 0·2 ml of 10% solution (20 mg) injected intraperitoneally and repeated ten times at weekly intervals (totalling 200 mg). A suspension of isologous bone-marrow, injected intravenously once only in groups 4 and 5 (20–24 hr after irradiation), and repeated ten times in group 2 (24 hours after each urethane injection). Interval between irradiation and first urethane injection was 2 weeks.

Another experiment, designed to investigate the role of the thymus in relation to the two-stage system, showed that the inhibition resulting from thymectomy could not be reversed by subsequent urethane treatment. Further work is in progress, to determine the phase in the two-stage process at which re-implantation of normal thymus becomes effective in reversing the inhibition caused by the initial thymectomy. It is still too early to report on the findings of this experiment.

The most interesting series of experiments to be reported, however, are those designed to investigate the possible role of a virus in radiation leukaemogenesis.

The relevant data from previous work are (i) the demonstration by Gross (1951, 1958) of a viral agent in leukaemic tissue of high-leukaemia strains of mice, that is capable of transmitting the disease to newborn mice of a low-leukaemia strain; (ii) the subsequent observation by Gross (1959) that such an agent can also be demonstrated in leukaemic tissue of a low-leukaemic strain (C3H) in which the disease was induced by radiation; and (iii) the observation by Lieberman and Kaplan (1959) that such an agent can be detected in irradiated C57BL mice shortly before the induced leukaemia becomes clinically demonstrable.

There are at least four possible ways of interpreting these results: (i) that the virus is already present in a low-leukaemia strain of mice, but that the target-organ (?thymus) is in a non-responsive state, and that radiation causes the target-organ to become responsive; (ii) that a leukaemia virus is present in insufficient amounts, and that radiation causes it to reach an effective concentration, possibly by neutralizing the action of inhibitors; (iii) that mice of a low-leukaemia strain carry a precursor-virus, and that radiation converts it into an active virus; and (iv) that the radiation actually produces a complete leukaemia virus *de novo*. In the light of the two-stage mechanism, a further possibility needs to be considered, namely (v) that initiating action (e.g. by low doses of radiation) causes the liberation of a precursor-virus *de novo*, and that promoting action (e.g. by urethane) causes it to be converted into an active virus.

In an attempt to analyse these possible interpretations, use was made of the radiation-urethane two-stage system in a modified form, whereby the action of radiation and that of urethane might be tested in separate animals. Any transmissible factor—virus or precursor-virus, liberated or already present in the irradiated animals—could thus be transferred to normal animals, which could then be submitted to urethane treatment. Several large-scale experiments of this kind, with variations in technique, are in progress, some of which have already yielded definitive results.

In the first experiments, several hundred adult C57BL/6 mice were given a single total-body exposure of 400 r; 24 hours later, the animals were killed, and 7 tissues (bone-marrow, lymph nodes, spleen, thymus, lung, brain and skin) were excised and minced, and each injected into separate batches of adult C57BL/6 mice (about 50 mice per batch). One week later, urethane treatment was given to the recipients—a series of 15 weekly intraperitoneal injections, totalling 300 mg per animal. Controls included: (*a*) similar minced tissue transfers without subsequent urethane treatment to the recipients; (*b*) transfer of minced tissues from non-irradiated animals, followed by urethane treatment to the recipients; (*c*) urethane treatment without previous tissue injection; and (*d*) untreated controls.

A second experiment, identical with the first, except that the interval between the irradiation and the excision of tissues for transfer was 30 days instead of 24 hours, was set up at the same time.

The results of these two experiments, almost complete, of which a preliminary report was previously published (Berenblum and Trainin, 1961b), are summarized in Table IV.

It will be noted that the untreated controls remained free from leukaemia, while those treated with urethane alone, or with non-irradiated tissues plus urethane, developed leukaemia in about 5% of animals. The mice receiving *irradiated* tissues plus urethane, on the other hand, yielded about 20% leukaemia, the type of tissue used being apparently not a decisive factor, nor the interval between the radiation of the donor mice and the time of transfer of tissues to the recipients. In the groups receiving tissues from irradiated mice *without subsequent urethane treatment*, the incidence of leukaemia was very low indeed.

TABLE IV. *Incidence of leukaemia in C57BL/6 mice injected with tissues from irradiated and non-irradiated donors, and subsequently with urethane (plus controls)*

Tissues injected	Injected 1 day after tissue irradiation		Injected 30 days after tissue irradiation		Non-irradiated tissue
	Subsequent urethane injections	No urethane injections	Subsequent urethane injections	No urethane injections	Subsequent urethane injections
Bone-marrow	9/40 (22·5%)	1/47 (2%)	8/45 (18%)	0/51 (0%)	1/44 (2%)
Lymph nodes	7/41 (17%)	1/48 (2%)	8/39 (21%)	2/46 (4%)	2/41 (5%)
Spleen	5/39 (13%)	0/48 (0%)	6/39 (15%)	2/48 (4%)	2/49 (4%)
Thymus	11/35 (31%)	0/49 (0%)	2/24 (8%)	2/48 (4%)	4/40 (10%)
Lung	6/40 (15%)	0/45 (0%)	8/36 (22%)	2/50 (4%)	2/47 (4%)
Brain	7/37 (19%)	0/48 (0%)	7/40 (17·5%)	0/52 (0%)	3/41 (7%)
Skin	6/30 (20%)	1/49 (2%)	7/46 (15%)	0/53 (0%)	2/43 (5%)
Total	51/262 (19%)	3/334 (1%)	46/269 (17%)	8/348 (2%)	16/305 (5%)

Controls: Urethane alone (without previous tissue injections)   5/110 (5%)
Untreated controls   1/141 (0·7%)

The results suggest that total-body irradiation, in mice of a low-leukaemia strain, causes the rapid appearance of a "transmissible" factor which, though itself not capable of transmitting the disease to other mice, does render such recipients subject to leukaemia formation by urethane treatment *over and above that which can be accounted for by the action of urethane alone.*

The possibility of the "transmissible" factor being a simple chemical substance liberated by the radiation, can be ruled out, since the agent appears to be present in as effective a concentration after 30 days as after 1 day following the radiation. The possibility that the "transmissible" factor is a dormant tumour cell, is also rendered unlikely by the fact that it is distributed throughout all the tissues of the body, which have been tested, within 24 hours of radiation. This raised the question of whether the factor might actually be a virus or precursor-virus type of entity.

The fact that all the tissues tested appeared to be equally effective, could mean one of two things: (i) that the "target-organ" for the liberation of the transmissible factor is widespread in the body, or (ii) that the site of formation may be restricted to one organ, but that the liberated agent is rapidly dispersed throughout the body.

TABLE V. *Incidence of leukaemia in C57BL/6 mice injected with tissues irradiated in vitro and subsequently injected with urethane (Preliminary data: experiment in progress only 8 months)*

Irradiated tissues	Urethane to recipients	No urethane to recipients
Brain	3/43	0/50
Liver	2/40	0/50
Kidney	1/45	0/50
Spleen	0/47	0/48
Thymus	2/41	0/48
Lymph nodes	3/45	0/50
Bone-marrow	3/42	0/47
Lung	1/43	0/49
Total	15/346 (4%)	0/392
(Urethane control)	1/215 (0.5%)	

Donor tissues irradiated *in vitro* with 1,000 r. Urethane to recipients: ten weekly i.p. injections of 0·2 ml of 10% solution, totalling 200 mg per animal.

To find out which of these two explanations is the correct one, a further experiment was undertaken, essentially the same as the previous one, except that the various tissues, taken from normal animals, were irradiated *in vitro*. Other modifications in technique included the following: (*a*) dose of *in vitro* radiation: 1,000 r; (*b*) tissues used: brain, liver, kidney, spleen, thymus, lymph nodes, bone-marrow and lung; and (*c*) urethane treatment to the recipients: 10 weekly doses, totalling 200 mg per animal.

The experiment has only been in progress about 8 months, and the figures, to date, provide no more than an indication of the trend of the experiment. It will be noted, however, that all the leukaemias so far recorded (see Table V) belong to the series in which injections of irradiated tissues were followed by urethane treatment. In our urethane controls, i.e. without prior tissue injections, the leukaemia incidence for the equivalent, short, time of action was 0·5%.

If these results are substantiated, one would have to conclude that many different tissues of the body are capable of yielding the transmissible factor in response to radiation treatment.

The next experiment was the crucial one—a repetition of the original *in vivo* system, but using cell-free extracts instead of tissue mince. Only two tissues—brain and blood plasma—were chosen, and the preparation of cell-free material consisted of four consecutive centrifugations at $7,000 \times g$. The tissues were derived from animals that had been irradiated with 500 r 24 hours previously; the cell-free extracts were injected intraperitoneally into normal recipients which, a week later, were submitted to ten weekly intra-peritoneal injections of 20 mg of urethane. Two control groups received the extracts without subsequent injections of urethane, and another control group received urethane alone.

This experiment is even less advanced than the previous one described above, and only a few leukaemias have so far appeared. The results seem encouraging, as will be noted from Table VI, but the final answer must naturally await the completion of the experiment.

TABLE VI. *Incidence of leukaemia in C57BL/6 mice injected with cell-free extracts of tissues from irradiated mice, and recipients injected with urethane* (*Preliminary data: experiment in progress only 6 months*)

Cell-free extract of	Urethane to recipients	No urethane to recipients
Brain from irradiated mice	5/96	0/99
Brain from non-irradiated mice	1/83	0/89
Plasma from irradiated mice	0/83	0/97
Plasma from non-irradiated mice	1/85	0/87
—	0/105	—

Irradiated donors received a single total-body exposure of 500 r.
Cell-free extracts prepared by 4 centrifugations at $7,000 \times g$: 1 ml c.f.e. of brain and 0·6 ml c.f.e. of plasma, resp., injected i.p.
Urethane treatment: ten weekly i.p. injections of 0·2 ml of 10% solution, totalling 200 mg per animal.
Interval between c.f.e. injection and 1st urethane injection: 1 week.

It might nevertheless be worth considering at this stage, how far one may venture to speculate, in the light of the results so far available, about the relative merits of the five possible explanations, mentioned earlier, concerning the presence of a virus in irradiated mice of a low-spontaneous-leukaemia strain.

The possibility that the promoting factor (e.g. urethane) operates on the secondary target-organ (?the thymus), or indirectly on it via the bone-marrow, rather than by some direct influence on the transmissible factor itself, is much weakened by the evidence presented here. Any indirect action of urethane—i.e. by depressing the bone-marrow activity, seems to have been excluded. On the other hand, no evidence of a specific, direct action of urethane on the thymus has so far been reported, though quite a marked depression of the thymus by urethane has recently been observed in our laboratory, by L. Fiore-Donati and A. M. Kaye (unpublished). But whether this plays a role in promoting action is doubtful. Further experiments are in progress to answer this question; and in the meantime, any conclusions on the subject would be premature.

The possibility that the virus is already present in normal C57BL mice, but in insufficient amounts, and that radiation causes its concentration to reach an effective level for leukaemia to express itself, would leave out of account the role of promoting action, unless one were to assume that the action of urethane and that of radiation are identical—which has been excluded.

That mice of a low-leukaemia strain might possess a precursor-virus, and that radiation causes its conversion into a complete virus—a possibility previously considered on hypothetical grounds by Kaplan (1961), not only lacks any supporting evidence, but also leaves out of account the role of a promoting phase in leukaemogenesis, so effectively performed by urethane.

The same criticism may be levelled against the possibility that radiation casues the formation of a complete leukaemia virus *de novo*. There is, in addition, the fact that the virus, obtained from tissues of a low-leukaemia strain following multiple radiations, is only demonstrable when injected into newborn mice, whereas the transmissible factor, in the present experiments involving single radiations, becomes effective for leukaemia-induction when injected into *adult* mice to which urethane is subsequently given.

On the balance, it would seem that the evidence so far available supports the possibility that initiating action (e.g. by a single dose of radiation) leads to the formation of a precursor-virus *de novo*, and that promoting action (e.g. by urethane) causes it to be converted into an active virus.

If this proved to be true, it would have far-reaching implications, not only in relation to the mode of action of ionizing radiation, but also in relation to carcinogenesis in general.

In conclusion, we should like to refer briefly to two of the most intriguing problems in radiation leukaemogenesis: (i) the fact that divided small doses of radiation are more effective than a single large dose for leukaemia induction, and (ii) the question of whether there is a threshold dose for radiation leukaemogenesis.

When all the data from our own experiments, involving single doses of radiation, at different dose levels, with and without urethane treatment, respectively, are recorded on a graph, plotting leukaemia incidence against dose of radiation, it appears (see Fig. 1) that there is a threshold dose for radiation alone, but apparently not for radiation followed by urethane treatment.

Assuming that for *initiating action*, radiation is equally effective whether it is given as a single dose or in divided doses (for which, therefore, there would be no threshold dose) whereas for *promoting action* (which is known to be a very slow process), radiation is relatively ineffective when given once only, but

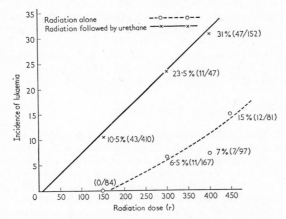

FIG. 1. Incidence of leukaemia in C57BL mice treated with single doses of X-rays, at different dose levels, with and without subsequent injections of urethane.

very effective when given in divided doses over a period of weeks, then an explanation would be found both for the conflicting evidence about the existence or non-existence of a threshold dose for leukaemogenesis as a whole, and also for the difference in effectiveness between single and divided doses of radiation for leukaemogenesis.

We realize, of course, that further data, based on very much larger numbers of animals, would be needed before this highly speculative explanation could be accepted. But as a working hypothesis, it may serve the purpose of stimulating further experimentation on the subject.

REFERENCES

BERENBLUM, I. (1941). *Cancer Res.* **1**, 807.
BERENBLUM, I. (1954). *Cancer Res.* **14**, 471.
BERENBLUM, I. (1960). *Med. J. Australia ii*, 721.
BERENBLUM, I., and HARAN, N. (1955). *Cancer Res.* **9**, 268
BERENBLUM, I., and TRAININ, N. (1960). *Science* **132**, 40.
BERENBLUM, I., and TRAININ, N. (1961a). Proc. 3rd Australasian Conf. Radiobiology, p. 98, Butterworths, London.

BERENBLUM, I., and TRAININ, N. (1961b). *Science* **134**, 2045.
BERENBLUM, I., and SHUBIK, P. (1947a) *Brit. J. Cancer* **3**, 109.
BERENBLUM, I., and SHUBIK, P. (1947b). *Brit. J. Cancer* **1**, 383.
BERENBLUM, I., REWALD, F. E., and TRAININ, N. (1961). *J. nat. Cancer Inst.* **27**, 1361.
GROSS, L. (1951). *Proc. Soc. exp. Biol., N.Y.* **76**, 27.
GROSS, L. (1958). *Brit. Med. J. ii*, 1.
GROSS, L. (1959). *Proc. Soc. exp. Biol., N.Y.* **100**, 102.
HARAN, N., and BERENBLUM, I., (1956). *Brit. J. Cancer* **10**, 57.
KAPLAN, H. S. (1961). *Acta Union Internat. Cancer.* **17**, 143; see also in discussion, p. 207.
KAPLAN, H. S., and BROWN, M. B. (1951). *J. nat. Cancer Inst.* **12**, 427.
KAPLAN, H. S., BROWN, M. B., and PAULL, J. (1953). *J. nat. Cancer Inst.* **14**, 303.
KAPLAN, H. S., CARNES, W. H., BROWN, M. B., and HIRSCH, B. B. (1956). *Cancer Res.* **16**, 422.
KAWAMOTO, S., IDA, N., KIRSCHBAUM, A., and TAYLOR, G. (1958). *Cancer Res.* **18**, 725.
LIEBERMAN, M., and KAPLAN, H. S. (1959). *Science* **130**, 387.
SALAMAN, M. H., and ROE, F. J. C. (1953). *Brit. J. Cancer* **7**, 472.

DISCUSSION

CURTIS: You showed that radiation can have an initiating action. Have you reversed this process using croton oil as an initiator and then following with radiation?

BERENBLUM: The only data available are those of Shubik and co-workers, using beta radiation for initiation and croton oil for promotion in skin carcinogenesis.

KOLLER: Prof. Berenblum has certainly tackled a very complicated problem in using C57BL mice to elucidate the two factors involved in leukaemogenesis. The explanation which Kaplan and co-workers gave is, that irradiation should not be considered as a factor which *produces* the leukaemia, but as a means of creating conditions in the cortical cells of the thymus such that a latent virus might proliferate and turn the cells leukaemic. Kaplan also showed that if he put bone-marrow into the irradiated animals he could prevent the leukaemia developing. Prof. Berenblum said that when he used a single dose of radiation and urthane and followed these with bone-marrow treatment leukaemia was not prevented in these animals. Now I would like to ask him two questions. Does he know the effect of this dose of urthane on the bone-marrow? How soon after urethane was the bone-marrow given to the animals because urethane can destroy the donated bone-marrow and thus frustrate its effects?

BERENBLUM: The bone-marrow suspension was given 24 hours after the urethane. A shorter interval would have allowed residual free urethane in the body to act on the injected bone-marrow. After 24 hours, most of the urethane is gone. According to the results of Kaplan, 24 hours should be quite sufficient to interfere with radiation leukaemogenesis.

KOLLER: What was the histological picture?

BERENBLUM: According to Rosin and co-workers, urethane depresses the metabolic activity of bone-marrow.

UPTON: May I compliment Prof. Berenblum on what I consider to be a very exciting and decisive set of experiments. Some of us have toiled for a number of years on this subject and I am really quite thrilled to hear this presentation. I must say, however, that I favour, perhaps on inadequate grounds, the notion that urethane was acting as a promoting agent on the host rather than on the completion of an inactive virus. I wonder whether administration of urethane to the filtrate (or cell-free extract) from the donor animal before passage into either a non-urethane treated recipient or, in parallel, a urethane-treated recipient, might help to elucidate this question?

BERENBLUM: This is an experiment which has to be done. Unfortunately, each experiment involves more than a thousand mice, and we cannot do all we should like to do at once. The possibility you favour has certainly to be considered. I hope I did not give the impression that the evidence presented proves the other one.

COTTIER: Prof. Berenblum did you consider a lowering of immune response as a consequence of urethane administration as a possible factor in these experiments?

BERENBLUM: My colleague, Dr. Trainin, is about to undertake this work, while on leave abroad.

MOLE: May I ask a question just to clarify my own mind? In Prof. Berenblum's classical experiments on skin carcinogenesis initiation is a change that, once produced, persists more or less indefinitely. Now here, if I understood him, initiation is the possible release of a possible agent. Is this the same thing?

BERENBLUM: This is a very searching question. I wish I knew the answer. All I can say is that initiation persists here as well, but it looks as if it is a different mechanism. It does not persist within a living cell. The fact that we get the same result 30 days later, or one day later, after initiating action is consistent with persistence. We appear to have moved from the cellular to the sub-cellular level. It is something new to us too, and we are very intrigued by it.

MOLE: May I pursue this? You refer to the fact that divided doses are more leukaemogenic than single doses and I take it you imply that they have some kind of promoting action. If you accept this idea then continuing irradiation presumably will always lead to a continued increase in leukaemia.

BERENBLUM: Yes, I agree. This is the implication. Further experiments along these lines are in progress.

UPTON: I would like to take up the point Dr. Mole raised. I think that one must balance against initiation and promotion the killing of cells versus the effects on the host environment that interfere with the expression of the carcinogenic transformation. As one increases the dose beyond a certain point one may well run into this kind of inhibitory effect and I dare say that fractionation would influence the survival of the cells which are initiated.

CURTIS: As to Dr. Upton's remark, we have some experiments which definitely show that the incidence of leukaemia decreases with repeated doses. It falls much below what you can get at earlier stages.

BERENBLUM: The point I want now to raise is a highly speculative one, which I haven't had time to discuss earlier: Whereas in our C57BL mice the urethane produced about 5% of leukaemia, there are reports in the literature of higher incidences of leukaemia produced by urethane in Swiss mice. We did try to speculate as to how this could be (see Fig. 1 p. 51). We know that if we start with C57BL mice we need both initiating action (X-rays) and promoting action (urethane) to induce leukaemia. At the other extreme, if we start with a strain of animals that already carries the leukaemia virus, like AKR, we do not need anything at all for leukaemia induction. Far from increasing the incidence of leukaemia, X-rays actually reduce the incidence in AKR mice. We also know that urethane only speeds up leukaemia development without increasing the incidence in AKR mice. Why not postulate that Swiss mice are at an inbetween level? They may already have the precursor-virus, and therefore will not require initiating action. But they will need urethane to complete the leukaemogenic process.

KOLLER: Prof. Berenblum's speculations tempt me to speculate on his speculations. He says that in C57BL mice the target-cells would be in the thymus. We know that by using X-rays alone we cannot produce leukaemia. If he adds urethane, he increased the incidence of the leukaemia; now I believe that with X-rays we damage the bone-marrow too, because from the bone-marrow I speculate that cells might aid the regeneration of the thymocytes in the cortical layer of the thymus. But if, on top of the irradiation, we add the urethane, then this bone-mattow in the host animal is more depleted and more damaged and cannot repopulate the thymus. This damages the thymocytes much more excessively and the regenerated cells then undergo malignant transformation. I believe this is the role of urethane.

BERENBLUM: There are several points which argue against this. One is that if you give urethane plus excess bone-marrow, you still get your leukaemogenesis.

KOLLER: I asked you when you gave your bone-marrow and you said you gave it 24 hours after the irradiation.

BERENBLUM: Five times, after each urethane injection, which, according to Kaplan's experience, seems more than adequate. We gave a great excess of bone-marrow suspension. If we had given the bone-marrow after a shorter interval than 24 hours, then the criticism would have been made that the urethane had destroyed the bone-marrow.

ALEXANDER: You say that 98% of the urethane is no longer there as urethane. To explain the multifarious activity of urethane one has to have some. . . .

BERENBLUM: When I say that 98% of urethane is gone, I mean of urethane and/or its breakdown products; because the way this was tested was by giving ^{14}C labelled urethane and then tracing the ^{14}C. The ^{14}C has disappeared from the animal to the extent of about 98% within 24 hours.

ALEXANDER: I still find it difficult to believe that all the activity of the urethane carbon is gone, because when one gives urethane for the alleviation of leukaemia in man then the activity is manifested over very long periods after the drug has been discontinued.

BERENBLUM: The effect of the damage to most cells may go on perhaps for years. I'm talking of the action itself—the action in the active sense. The consequences of urethane action depend on what kind of damage is done.

LAMERTON: You may in fact have prevented the regenerating capacity of your bone-marrow with urethane or produced conditions which don't allow the regeneration.

BERENBLUM: We have a number of experiments in progress to attempt to answer this question. In some of these, the two-stage process is being investigated in relation to thymectomy, performing the thymectomy operation at different stages in the various groups, in order to determine at which stage the removal of the thymus becomes critical. In other experiments, involving the transfer of tissues from irradiated mice and administering urethane to the recipients, the influence both of normal bone-marrow injections and of thymectomy are being tested, both on the donors and on the recipients. From all these experiments, an answer should eventually be forthcoming.

KOLLER: When do you do the thymectomy? Miller's experiments show that the thymus has done all the damage already, one week after birth. You must therefore remove the thymus one day after birth.

COTTIER: If you transfer brain, did you perfuse this brain first or did you not? What is the important thing—the blood, the blood cells or the plasma?

BERENBLUM: We used total brain mince. In the second experiment, we again used total brain tissue, irradiated *in vitro*. In the third experiment, we irradiated the animal; took out the brain; centrifuged the suspended mince at slow speed to get rid of most of the deposit and then gave four consecutive high-speed centrifugations at $7,000 \times g$ to obtain what is effectively a cell-free extract. There is a further experiment in progress which is even more critical, using filtered material. This experiment has only been going six weeks.

COTTIER: I have been wondering where the active agent would be located. You mentioned brain, lung, spleen. Now lung ordinarily contains lymphoid cells, brain to my knowledg does not, except inside the blood vessels. Now that is why I ask if you have done any experiments with perfused brain?

UPTON: I think the question of the source of the leukaemogenic virus is a very important question. Schwartz, I am told, used brain because he thought there would be the least possibility of antibodies which might bind virus present in the brain. Most virologists with whom I have discussed this are not very impressed by this argument. Prof. Berenblum's data suggest that the agent is present in every tissue although he has not perfused them. I don't know whether other lymphoma viruses have been extensively enough studied to tell us which tissues are richest in the virus. I think this is a very important question because if we are attempting to detect leukaemia viruses or tumour viruses in tumours we may be looking in precisely the wrong place. With polyoma virus, for example, the animals become viraemic shortly after inoculation, then begin to develop antibodies to the virus, and when the tumour is finally apparent the virus can often no longer be detected. So attempting to work from the tumour backwards in pathogenesis when you are dealing with a tumour virus may be a very difficult way to go at it. I should like to ask Prof. Berenblum whether a serial analysis of these activated tissues might disclose differences.

BERENBLUM: We have so far tested crude tissues, *in vitro* irradiated crude tissue, high-speed centrifuged extract and filtered extract. We have not done much beyond that. Incidentally, in all these experiments, both donors and recipients were adult animals between 6 and 8 weeks old.

UPTON: I should think it is very important to give your prospective recipients urethane before inoculation of the activated tissue to determine whether the interaction of urethane plays a role.

ALEXANDER: I have been thinking of an alternative scheme and the reason that I really want to have an alternative scheme is that I think one is loading up this poor simple formula of urethane with too many functions. We've got it as a complete carcinogen, as an initiator, as a promoter, as a cytotoxic agent, and God knows what. I just wonder whether in these particular strains of mice one would normally get small tumours as a result of irradiation with perhaps quite low doses, but in general this is prevented because repair from the bone-marrow prevents this carcinogenic stimulus. This would explain why one has to give several repeated doses in Kaplan's experiments and why bone-marrow stops it. Now, if the answer were that the urethane so damages the thymus that it is not responsive to repair by bone-marrow, perhaps only for a short period of time, then I would say that, in the absence of bone-marrow repair, 50 r is sufficient to induce a high incidence of tumours. Is it not also possible that you had a laboratory contaminant virus? Perhaps work on polyoma virus began in your laboratory at about this time?

BERENBLUM: There are two points you raised. One is the theory I mentioned earlier. I quite agree that we haven't excluded the alternative possibility of urethane acting on the thymus itself. As for the second point—the idea of a virus floating around in the animal colony—this is of course difficult to exclude with certainty. I can only say that in our animal colony of 12,000 mice, the pattern of spontaneous tumours is not what one might expect if polyoma virus were present. The spontaneous incidence of leukaemia in our C57BL mice is 0 to 1%; the incidence in urethane-treated C57BL mice is 2 to 7%. Other spontaneous tumours in this strain are extremely rare in our colony.

GRAY: Doesn't the evidence from the *in vitro* experiments suggest that this postulated virus is present almost immediately. You irradiated the tissue *in vitro* and then you inoculated it into the animal and you got your tumours. This shows that it must have been a rather immediate effect of the irradiation of these tissues.

BERENBLUM: It depends what you mean by immediate. If you irradiate a cell or a group of cells and transplant them, and if they remain alive, the postulated virus may develop later in the surviving cells within the new host. I don't see any theoretical difficulty from that point. Of course the question of the formation of new virus or provirus *de novo*, one's attitude to that depends on whether one is brought up as a chemist or a biologist. To a chemist, it is acceptable; to a bacteriologist, it appears unacceptable. Are we not starting the old Pasteur argument all over again? One could imagine that there is some factor already present which is a "pro-provirus" and that X-rays convert that entity into some product which urethane further converts into something else, and the final entity is a virus which produces leukaemia. Whether the substance right at the beginning is chromosomal material in the chemical sense or a sort of sub-living entity in the bacteriological sense—that is a very abstract sort of reasoning that hardly affects the argument at all. One can't say C57BL have no leukaemia virus. What one can say is that using the available techniques it is not normally demonstrable.

GRAY: Well, one could conceive an experiment for instance in which you made your cell-free extract from the *in vitro* irradiated material immediately. You might simply have chemically transformed some RNA or DNA for that matter.

BERENBLUM: Yes. But we have been proceeding one step at a time, and are only now approaching the stage of analysing the "transmissible factor" in terms of DNA or RNA.

A COMPARATIVE STUDY OF THE LATE EFFECTS OF CERTAIN RADIOMIMETIC DRUGS AND X-RAYS

L. NÉMETH

State Oncological Institute, Budapest, Hungary

In preliminary experiments using, first rabbits and then mice, it had been established haematologically that 25 r TBR corresponded to 5 mg of Degranol or 50 mg of Mannitol-myleran.

Groups of 50 male and 50 female mice, 2 months old and weighing 15 g, of a susceptible strain and of one not susceptible to spontaneous leukaemia, were given for 1 year, weekly intraperitoneal injections of 5 mg/kg of Degranol, 50 mg/kg of Mannitol-myleran, and physiological saline, respectively. Another group of the same number and composition was X-irradiated with 25 r. An additional 1,000 untreated and unhandled animals were earmarked for observation as controls throughout the experiment. Once a week, quantitiative and qualitative white blood cell counts were taken of every experimental animal, including those given saline, but only of random samples from among the controls. All animals were weighed systematically, initially twice and later once a week. After the third month a proportion of the mice was paired. Following the 52nd week all surviving mice were killed and subjected to histology.

RESULTS

All treatments raised the percentage incidence of lymphatic leukaemias (stem-cell leukaemias) in the mice of the susceptible strain. The increase in the rate was the lowest in the saline-treated and the highest in the Degranol-treated animals (in the latter it was to 23% from the 8% seen in the controls).

All treatments increased the incidence-rate of lymph-node tumours in the susceptible strain, to the greatest extent on treatment with Mannitol-myleran (to 53% from 19% in the controls). Only a small proportion of these tumours displayed malignancy (lymphosarcoma, reticulosarcoma, etc.).

All treatments, particularly treatment with mannitol-myleran, shortened the average life-span of the animals and impaired their resistance to infections in both susceptible and non-susceptible strains.

In addition, all treatments except that with saline produced profound histological changes, namely:

(i) hepatic lesions, cholangioma, atrophic testis, splenic lesions, giant cells, plasma-cell metaplasia of lymph nodes, etc., in both the susceptible and non-susceptible strains,

(ii) metaplasia of the Bowmann capsule, pulmonary adenoma, fibrosarcoma at abdominal sites of injection, etc., in the non-susceptible strain only.

All experimental animals of both strains, except those treated with saline, ceased to multiply and grew obese, particularly the males.

The changes in the weekly counts of the white blood cells revealed initially eosinophilia and lymphopenia after X-ray treatment and Degranol, and leucocytosis after Mannitol-myleran; later, anaemia and increases in the number of nuclear erythrocytes and young lymphoid elements followed these three types of treatment. From these pre-haemoblastic changes it was possible in some instances to predict imminent leukaemia.

Whereas the three treatments differed in their early haematopoietic effect, the late effect was invariably a lymphatic reaction.

Thanks are due to B. Kellner, K. Lapis, and Margaret Dancsházy for pathological studies.

SERIAL TRANSPLANTATION OF HAEMATOPOIETIC TISSUE IN IRRADIATED HOSTS

P. C. KOLLER AND S. M. A. DOAK

Chester Beatty Research Institute, Institute of Cancer Research, London, England

INTRODUCTION

Ionizing radiation as a cause of murine leukaemia have been demonstrated in numerous experiments, (Furth and Furth, 1936; Kaplan, 1947; Mole, 1958), in which the cells, tissues and organs involved in the leukaemogenic process have been *directly* exposed to radiation. It has also been established that non-irradiated thymuses, when grafted into irradiated thymectomized isogenic mice of certain genetic constitution, may develop into thymic lympho-sarcomas (Kaplan *et al.*, 1953, 1956; Barnes *et al.*, 1959). More recently it has been found that many leukaemias, which arose in allogenic mouse chimaeras, were of donor origin (Uphoff and Law, in press). In these instances the malignant transformation involved cells, themselves non-irradiated, but which were transplanted into irradiated hosts. From these experiments it appears that the irradiated host environment has had a leukaemogenic effect upon the non-irradiated donor cells. In both these cases, however, the possibility that the effect observed is in no way related to irradiation, but simply due to some "infective" principle, must not be ignored.

In order to clarify the issue, experiments were designed to investigate the possibility of inducing leukaemia following the serial passage of marrow cells into isogenic hosts, the latter having been exposed previously to a lethal radiation dose which destroyed the host haematopoietic tissues. It is argued, that such a procedure would cause a hyperactivity in the donated haematopoietic tissue and that this hyperactivity, if continued, might result in leukaemia.

METHODS

A group of 30 C3H/Bi mice were given 650 r (LD_{99}) and 24 hr later each mouse received an injection of bone-marrow (cell dose 4×10^6) from non-irradiated C3H donors. After 28 days the marrow from six of the primary isogenic chimaeras was used as donor tissue for injection into a second group of 30 irradiated hosts, thus making secondary chimaeras. After another 28 days, this procedure was repeated when six secondary hosts were sacrificed and used as new marrow donors. In this manner five serial transplants were made, assuming that the original iosgenic marrow persisted and functioned in the secondary, tertiary and quaternary hosts.

Analyses were made of the peripheral blood of 3-month-old untreated C3H mice and of all marrow donors prior to their sacrifice. The peripheral blood picture was also examined in the survivors of each group of 50 and 75 days post-irradiation. The number of erythrocytes/unit volume, together with the haemoglobin index, and the number of leucocytes/unit volume were determined. The relative proportions of mononuclears and polymorphs were estimated from blood smears stained with Giemsa. The non-polymorph population of white blood cells in normal control mice is mainly small lymphocytes; in the chimaeric mice, on the other hand, the cells are more generally monocytes and large lymphocytes and, occasionally, cells of the early myeloid series may be identified.

EXPERIMENTAL RESULTS

Peripheral blood analysis before and after irradiation

Five C3H mice, 3 months old, were used to obtain an average peripheral blood value as control. Two groups of 30 C3H mice were exposed to 650 r total-body radiation; one received isogenic bone-marrow (5×10^6 cells) intravenously within 4 hours of irradiation. Blood samples from each group of mice were taken on 1, 3, 6, 9, 12, 15, 21 and 28 days post-irradiation. To reduce the risk of altering the haematological picture by too frequent sampling, the mice were batch bled in groups of 6 at intervals of not less than 10 days. The results together with the control data are presented in Fig. 1.

There was no appreciable reduction in the number of circulating erythrocytes, nor a decrease in the amount of haemoglobin, until three days post-irradiation. On the other hand, the number of mononuclears in the peripheral blood dropped precipitously during the first few hours. By 3 days the polymorph population was also reduced in size, the total leucocyte count being approximately one-tenth of the control value.

Irradiated mice given no further treatment did not live after 15 days. If the mice received an injection of 5×10^6 isogenic bone-marrow cells 24 hours after the irradiation, the drop in the peripheral blood cells was arrested at about the 5th day. On the 15th day there was evidence of an over-compensatory hyperplasia of the leucocyte population. By 28 days the number of erythrocytes and the haemoglobin level, together with the total leucocyte count, were indistinguishable from the controls. However, the normal mononuclear/polymorph ratio was not restored at 28 days, the recovery of the leucocyte population being due mainly to granulopoiesis.

Serial passage of bone-marrow in irradiated isogenic hosts

Two sets of experiments were designed to study the effect of successive transfers upon the activity of bone-marrow. In the first, C3H marrow ($2\text{--}6 \times 10^6$ cells) was serially passaged into five different groups of irradiated

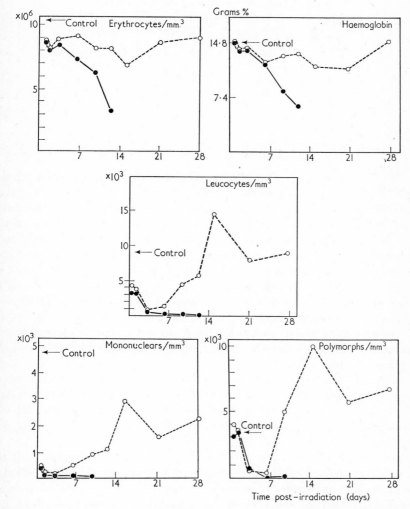

FIG. 1. Peripheral blood analysis of C3H mice before and after exposure to 650 r total-body irradiation, showing the effect of an injection of 5×10^6 isogenic bone-marrow cells given 24 hours after irradiation.

—●—●— Irradiated; — O — O — Irradiated, given bone-marrow.

C3H mice (650 r) at 28 days intervals. It was found that the number of survivors to 30 days decreased, and the decrease was proportional to the number of marrow transplants. The decrease is more striking after the third passage (Table I). Following the fifth passage, no mouse survived to 20 days.

TABLE I. *Serial transfer of isogenic marrow*

Passage No.	Percentage survival to 28 days	Blood counts at 28 days†	
		Polymorphs	Mononuclears
1	80	153	70
2	66	228	40
3	60	256	17
4	1	230	12·5
5	0	—	—

† As percentage of control value.

Marked effects were also observed on the cellular composition of the peripheral blood. The number of erythrocytes was reduced with each successive passage, the average after the fourth passage being 6·3 million/mm^3 compared to the control value of 10·5 million/mm^3. The recovery of the leucocyte population followed the pattern mentioned above, but each successive transplantation of marrow exaggerated the effect. The total leucocyte cell count was around 8–9,000/mm^3 with 80% of the cells as polyymorphs in the 4th chimaeric generation. The findings are shown in Table I. It can be seen that serial passage of non-irradiated bone-marrow into irradiated isogenic hosts destroyed the balanced leucocyte population of the peripheral blood, shown by the depletion in the number of mononuclears (affecting mostly the small lymphocytes) with an increase in the number of polymorphs.

In the second series of experiments, the peripheral blood of each group of isogenic chimaeras was analysed at 28, 50 and 75 days after the irradiation of the host to establish whether or not the cellular composition, already investigated at 28 days, represented the optimal recovery of haemato- and lymphopoiesis.

In the primary, secondary and tertiary chimaeras, the numbers of red blood cells remained relatively constant throughout the observation period of 75 days. The quaternary chimaeras on the other hand, showed a more gradual recovery of the erythrocyte population, which at 75 days was comparable with the controls.

The total leucocyte count showed a decrease in the number of cells at 28 days as the number of the serial passage increased, but with time the total number of leucocytes approached the normal control level. On the other hand the mononuclear/polymorph ratio remained consistently low; the polymorph count was two to three times greater than that of the mononuclear count. Even at 75 days after irradiation and marrow therapy the ratio was still less than 1 in all but the primary chimaeras. In the third and fourth passage chimaeras the number of mononuclears was still less than 50% of the control value.

DISCUSSION

We found that in isogenic chimaeras the composition of the peripheral blood, in terms of the relative numbers of the various cell types present, is abnormal and remains so for a considerable period. This reflects a delay in the restoration of balanced haematopoiesis in the irradiated milieu. In lethally irradiated animals the haematopoietic tissues are destroyed, consequently the rapid restoration of granulo- and erythropoiesis becomes the most urgent need. Such a demand is satisfied by the recolonization of the haematopoietic sites, effected by the relatively small number of stem-cells present in the non-irradiated donor tissue which has been injected. When the donor tissue is marrow, the haematopoietic sites will be recolonized rapidly and only after these sites are recolonized, does the restoration of the lymphopoietic organs occur (Ford et al., 1957). When the donor tissue is spleen, recolonization of the haematopoietic sites is delayed, due to the smaller number of myelo- and erythropoietic stem-cells. Thus the number of the appropriate stem-cells of the donor tissue is of primary importance to the recovery of the irradiated host.

The altered cell-composition in the blood of isogenic chimaeras observed indicates that the balance between the haematopoietic components of the donor marrow has been disturbed. When such an "unbalanced" marrow is used again as donor tissue and transferred into another irradiated animal, it might be expected that the balance in the cell composition would be further distorted and this was found to be the case in our tertiary and quaternary chimaeras. The gross disturbance in the relative proportions of the circulating blood, in particular, the almost complete disappearance of the mononuclears from the circulation might easily account for the death of the animals. Here it may be relevant to mention the recent and very important observation of Miller (1961). He found that the number of lymphocytes is extremely low in the peripheral blood of mice thymectomized at birth, and the persistence of an incompletely functioning haematopoietic system in these animals results in the "wasting syndrome" and eventual death.

Our experiment revealed that the cell population of the marrow is altered by successive transfers into irradiated milieu. The alteration consists of a distortion in the proportion of particular cell types, and with the increase in the number of the serial passage, the degree of distortion was accentuated and its duration prolonged. The relationship of these changes to the incidence of leukaemia has yet to be evaluated.

ACKNOWLEDGEMENTS

This investigation has been supported by grants to the Chester Beatty Research Institute (Institute of Cancer Research: Royal Cancer Hospital), from the Medical Research Council, the British Empire Cancer Campaign, the Anna Fuller Fund, and the National Cancer Institute of the National Institutes of Health, U.S. Public Health Service.

REFERENCES

BARNES, D. W. H., FORD, C. E., ILBERY, P. L. T., JONES, K. W., and LOUTIT, J. F. (1959). *Acta Un. int. Cancr.* **15**, 544.
FORD, C. E., HAMERTON, J. L., BARNES, D. W. H., and LOUTIT, J. F. (1957). In "Advances in Radiobiology" (G. Hevesy, A. Forssberg, and J. D. Abbatt, eds.), p. 197. Oliver and Boyd, Edinburgh.
FURTH, J., and FURTH, O. B. (1936). *Amer. J. Cancer* **28**, 54.
KAPLAN, H. S. (1947). *Cancer Res.* **7**, 141.
KAPLAN, H. S., BROWN, M. B., and PAULL, J. (1953). *Cancer Res.* **13**, 677.
KAPLAN, H. S., CARNES, W. H., BROWN, M. B., and HIRSCH, B. B. (1956). *Cancer Res.* **16**, 426.
MILLER, J. F. A. P. (1961). Lancet *ii*, 748.
MOLE, R. H. (1958). *Brit. med. Bull.* **15**, 184.

SESSION II

Chairman: H. CURTIS

LEUKAEMOGENESIS: ROLE OF VIRUSES AND CYTOLOGICAL ASPECTS

By ARTHUR C. UPTON

Chromosomes in Murine Pre-Leukaemia

By P. L. T. ILBERY, P. A. MOORE, S. M. WINN, AND C. E. FORD

The Chromosomes in Virus-induced Murine Leukaemias

By P. C. KOLLER, E. LEUCHARS, C. TALUKDAR, AND V. WALLIS

Chromosome Abnormalities and the Leukaemia Process

By J. LEJEUNE

Quantitative Aspects of Experimental Leukaemogenesis by Radiation

By R. H. MOLE

Effects of Ionizing Radiation on Cellular Components: Electron Microscopic Observations

By H. COTTIER, B. ROOS, AND S. BARANDUN

LEUKAEMOGENESIS—ROLE OF VIRUSES AND CYTOLOGICAL ASPECTS

ARTHUR C. UPTON

Biology Division, Oak Ridge National Laboratory,† Oak Ridge, Tennessee, U.S.A.

SUMMARY

1. The presence of filterable leukaemogenic agents in the tissues of irradiated mice and in those with radiogenic leukaemia implicates viral mechanisms in radiation leukaemogenesis.

2. The respective roles of radiation and virus in leukaemogenesis cannot yet be defined, but the available data suggest that murine leukaemia evolves through a multi-stage process requiring the interaction of virus, host cells, and host. In such a complex system, radiation may act in several ways, one of them being to activate a latent leukaemogenic virus. The basis for the relatively long induction period, however, in spontaneous and radiation leukaemogenesis remains an enigma.

3. Evidence points to the possibility that the same virus may induce different haematologic forms of leukaemia in irradiated and non-irradiated individuals, depending on the constitution and physiological condition of the host.

4. The occurrence of specific abnormalities of a single human chromosome (probably No. 21) in chronic myelogenous leukaemia and in Down's syndrome, which carries an increased propensity to leukaemia, suggests that cytogenetic changes may be of causal significance in human leukaemogenesis. Data on the frequency of spontaneous and radiation-induced chromosome breakage, however, argue against a simple one-break aetiologic mechanism in the pathogenesis of the disease.

5. On the basis of tracer studies, the growth of leukaemia cells is not characterized by abnormally rapid cell division. Precise details of the kinetics of leukaemic growth remain, however, to be elucidated.

Although leukaemia was one of the first neoplasms recognized as a potential hazard of ionizing irradiation (von Jagic *et al.*, 1911), it was not until recently considered possible that a small radiation dose might carry a finite leukaemogenic risk. The available data do not yet provide a clear picture of the relation between incidence of leukaemia and dose, nor of the mechanism of leukaemia induction (see Upton, 1961), but several recent advances towards an understanding of leukaemogenesis are worth noting. These include: (1) the observation that some lymphosarcomas may be induced by radiation in the mouse thymus without irradiation of the thymus itself, (2) the discovery that certain radiogenic leukaemias yield filterable leukaemogenic agents, suggesting that viruses are involved in their pathogenesis, and (3) the disclosure that specific human chromosomal abnormalities are associated with chronic myelogenous

† Operated by Union Carbide Corporation for the United States Atomic Energy Commission.

leukaemia and also with Down's syndrome in which the probability of leukaemia is increased. These developments and their implications call for reevaluation of radiation leukaemogenesis in the light of viral and cytological factors.

ROLE OF VIRUSES

The viral aetiology of fowl leukosis was established more than 50 years ago, but efforts to demonstrate this cause in mammalian leukaemia were unsuccessful until Gross proved the extraordinary susceptibility of newborn mice to cell-free leukaemogenic agents (see Gross, 1961). Since this discovery, an increasingly large and diverse series of filterable leukaemogenic agents has been found (Table I) and it has been observed that leukaemias induced by

TABLE I. *Mouse leukaemia viruses* †

Origin	Neoplastic cell	Investigators
AKR, C58 (spontaneous lymphoma)	Lymphoid	Gross (1951, 1956)
C3H (radiation lymphoma)		Gross (1958, 1959)
C57BL (radiation lymphoma)		Lieberman and Kaplan (1959)
BALB/c (sarcoma 37)		Moloney (1960)
BALB/c (C3H plasma cell neoplasm)		Breyere and Moloney (see Moloney, 1962)
Swiss, C3H (spontaneous lymphoma)	Lymphoid (sarcoma)	Schoolman et al. (1957) Schwartz et al. (1959)
Swiss (Ehrlich carcinoma)	Reticulum cell (erythroblastosis)	Friend (1957)
BALB/c (Ehrlich carcinoma)	Reticulum cell, type B	Stansley et al. (1961)
Sarcoma I, Sarcoma II, SOV 16, Ehrlich carcinoma, Sarcoma 37	Myeloid	Graffi et al. (See Graffi, 1958; Moloney, 1962)
RF/Up (radiation myeloid)	Myeloid Lymphoid (erythroblastosis)	Upton (1959) Rauscher (1962)
Friend virus	Myeloid (?)	Mirand and Grace (1962)

† Modified from Moloney (1962).

radiation in low leukaemia strain mice (C3H (Gross, 1958, 1959), C57BL (Lieberman and Kaplan, 1959), and RF (Upton, 1959; Parsons et al., 1962)) yield filterable leukaemogenic agents. So it has been inferred that this induction probably involves some form of virus activation.

Studies of radiation-induced myeloid leukaemia were carried out with mice of the RF strain, in which the incidence of this disease may be increased from the control level of about 3% by only 150 r of whole-body X-radiation (Upton, 1959). Although the presence of viral leukaemogenic agents in affected

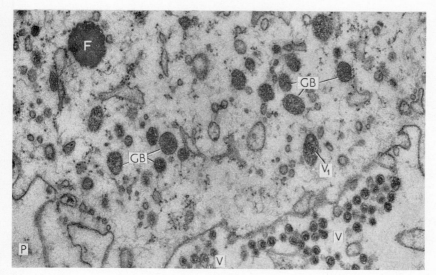

Fig. 1. Portion of a large mononuclear cell in a 14-day myeloid leukaemia spleen culture in the second passage generation. Large numbers of virus-like particles (V) are present at the outside of the cell, and an increased number of granular bodies (GB) are present in the cytoplasm, one of which contains a similar particle (V_1). (P = plasma membrane; F = fat bodies.) × 30,000. (From Parsons *et al.*, 1962.)

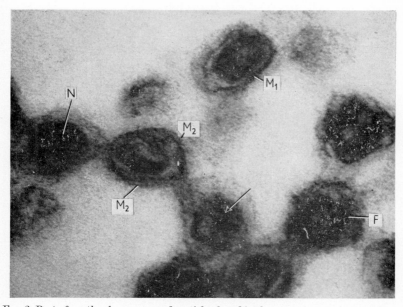

Fig. 2. Part of another large group of particles found in the same tissue culture as shown in Fig. 1. The outer membrane is 35 Å thick and appears single when sectioned normally (M_1). In other particles it appears double (M_2), but this may be due to oblique sectioning and to the thickness of section. The dense centre (N) appears to have no definite membrane and to contain filaments (F) of 40 Å diameter and fibrils of 20 Å diameter (indicated by arrow). × 240,000. (From Parsons *et al.*, 1962.)

mice was suggested by the co-leukaemogenic effects of filtrates of their tissues administered conjointly with X-rays (Upton, 1959), such effects were not consistently reproducible. Furthermore, electron micrographs disclosed few "virus particles" in such donor mice (Parsons et al., 1962). Hence, it was inferred that donors with primary leukaemia were seldom a satisfactory source of leukaemogenic virus.

To develop passage materials with greater viral activity, efforts were made to cultivate the leukaemia cells *in vitro* to promote growth of virus, as has been noted, for example, with the fowl myeloblastosis agent (Beaudreau et al., 1960) and polyoma virus (Stewart et al., 1957). In spleen cell cultures from mice with transplanted myeloid leukaemia, it was observed that the numbers of "virus particles" (Figs. 1 and 2) increased in successive passages (Table II).

TABLE II. *Evidence of viral activity in radiogenic myeloid leukaemias of mice*

Diagnosis	Frequency of virus-like particles†		Leukaemogenic activity ‡			
			Newborn		Adult	
	in vivo	*in vitro*	Myeloid	Thymic lymphoma	Myeloid	Thymic lymphoma
	No. positive / No. examined		No. positive / No. examined			
Nonleukaemic control	0/6	1/15	0/63	0/63	0/193	0/193
Primary leukaemia	5/18	1/15	0/58	0/58	0/93	0/93
Passaged leukaemia	7/18	13/28	18/274	10/274	9/279	0/279

† Electron microscopy: Sections of spleen or cultured cells; figures denote numbers of animals examined (Parsons et al., 1962).

‡ Of cell-free filtrates in non-irradiated isologous recipients: Cell-free filtrates of brain or homogenates were injected intravenously into recipients less than 24-hr old (newborn) or 10 weeks old (adult); figures denote proportion of recipients developing leukaemia during the first 6 months of life (A. C. Upton and V. K. Jenkins, unpublished data).

Furthermore, suspensions of serially passaged leukaemia cells yielded cell-free filtrates with leukaemogenic potency. The leukaemias induced by these filtrates in adult recipients were myeloid, but smaller numbers of thymic lymphomas also developed in newborn recipients (Table II). Although the number of mice developing leukaemia after injection of filtrates was relatively small, none was observed among controls of comparable age (Table II). The induction of myeloid leukaemia in newborn recipients is particularly noteworthy since the disease has not been induced by irradiation alone in the neonatal period (Fig. 3).

To explore the possibility that the same agent causing myeloid leukaemia also caused thymic lymphoma, depending on the constitution and physiological condition of the host, spontaneous and radiogenic leukaemias in

(AKR × RF)F_1 hybrids were compared with those in the two parental strains, AKR and RF. Preliminary results (Table III) show the incidence of spontaneous and induced leukaemia in the F_1 hybrids to be intermediate between that of the parental strains. The data are consistent with the possibility that the same leukaemogenic agent (1) is present in both AKR and RF strains,

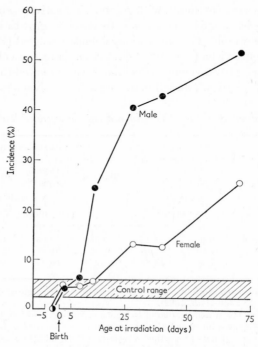

FIG. 3. Incidence of myeloid leukaemia in X-irradiated RF mice as influenced by age. (300 r, 250 kVp, whole-body X-irradiation.) The range of incidence in non-irradiated controls is shown by the cross-hatched band. (From Upton et al., 1960.)

(2) produces different leukaemias owing to strain differences in host responsiveness, and (3) is transmitted vertically to F_1 hybrids, the hybrid's response being intermediate because of genetic heterozygosity. Although such an interpretation remains to be proven, the vertical transmission of leukaemia viruses is well documented (see Gross, 1961). The extent of vertical transmission may also vary, however, depending on the virus and the host strain (Law and Moloney, 1961). Thus, the observed differences between parental strains and hybrids may conceivably reflect differences in transmission of a common agent as well as differences in susceptibility to such an agent. This possibility could be tested using viral preparations of known potency, such

as Gross's Passage A virus (see Gross, 1961) in parental and F_1 recipients. That the AKR (Gross) virus may enhance the induction of myeloid leukaemia in irradiated RF recipients has been reported by J. Furth (personal communication). The production of both lymphoid and myeloid leukaemias alternately

TABLE III. *Spontaneous and radiogenic leukaemias in RF, AKR, and F_1 hybrid mice* †

Strain	Sex	X-ray dose (r)‡	No. of mice	Median survival time (months)	Type			
					Thymic lymphoma		Myeloid leukaemia	
					%	Median age at death (months)	%	Median age at death (months)
AKR	F	0	92	10	85	8	2	18
	F	300	84	10	79	9	0	—
	M	0	79	10	69	10	0	—
	M	300	93	12	52	10	3	10
(AKR × RF)F_1	F	0	86	21	24	12	3	13
	F	300	97	12	43	8	3	16
	M	0	78	20	17	18	1	10
	M	300	88	14	24	8	16	13
(RF × AKR)F_1	F	0	94	22	27	23	0	—
	F	300	85	15	40	8	12	15
	M	0	83	24	7	19	0	—
	M	300	95	18	22	9	9	17
RF	F	0	603	19	8	17	2	19
	F	300	193	12	32	10	20	13
	M	0	512	19	5	16	3	16
	M	300	201	12	12	10	30	11

† A. C. Upton, V. K. Jenkins, and J. W. Conklin, unpublished data.
‡ Whole-body 250 kVp X-rays administered at a dose-rate of 80–90 r/min. Mice were irradiated at 10 weeks of age.

by ostensibly the same agent may occur in AKR recipients, depending on whether they are intact or thymectomized (see Gross, 1961). Also in experiments with the Moloney (1962) and Graffi (Bielka et al., 1955; Graffi and Gimmy, 1957) viruses, the production of both myeloid and lymphoid leukaemias has been observed.

The failure of irradiation to increase the incidence of lymphomas in AKR mice or to hasten their appearance conforms to earlier observations (see Upton, 1961). The significance of this failure can now only be guessed at, but viewed in the context of radiation leukaemogenesis it suggests that the "activating" role of radiation is unnecessary in the high-leukaemia AKR mouse, an interpretation consistent with the evidence that active virus is extractable from the tissues of such mice at all ages (see Gross, 1961). If this explanation is correct, the basis for the relatively long induction period of spontaneous and radiogenic leukaemia remains enigmatic.

That radiation may exert effects on host susceptibility other than of a purely "virus-activating" nature is indicated by the enhanced sensitivity of irradiated adult mice to the oncogenic effects of injected virus preparations (Graffi and Krischke, 1956; Upton, 1959; Furth, personal communication). This is also implied by the anti-leukaemogenic action of post-irradiation marrow transplantation, which is effective even when carried out after the last of four successive weekly whole-body radiation exposures (Kaplan et al., 1953). The mechanism by which marrow shielding or marrow infusion inhibits leukaemogenesis remains to be established, but several investigators (Kaplan, 1961; Axelrad and Van der Gaag, 1962) have suggested that the effect is mediated through enhanced regeneration of the irradiated thymus. According to this hypothesis, radiation-induced depopulation of haemopoietic tissues contributes toward the development of leukaemia by causing a protracted hyperplasia of undifferentiated thymic stem-cells, which are postulated to be preferentially susceptible to the oncogenic effects of leukaemia virus. How intact marrow assists thymic regeneration is not yet clear, although there is some evidence that marrow-derived stem-cells can repopulate the thymus (see Popp, 1962). Whether the counteracting effects of urethane (Berenblum et al., 1961) operate through interference with thymus cell maturation remains to be explored, as does the possible role of immunogenetic mechanisms in the anti-leukaemogenic action of intact marrow.

These observations raise puzzling questions concerning the roles of radiation, viruses, and host factors in leukaemogenesis. Although the available information is still scanty, we can make tentative generalizations.

1. Leukaemias of a wide variety of haematologic types can be induced in fowl, mice, and rats by viruses (see Gross, 1961).

2. Some radiation-induced lymphomas (Gross, 1958, 1959; Liebermann and Kaplan, 1959) and myeloid leukaemias (Upton, 1960; Parsons et al., 1962) of mice yield filterable leukaemogenic agents.

3. Induction of leukaemias by irradiation is inhibited by the presence of only a few non-irradiated heamopoietic cells (see Upton et al., 1960) and in some instances the induction has been shown to occur entirely through indirect mechanisms (Kaplan, 1959).

4. Induction of leukaemias is also promoted in some instances by administration of oestrogens (see Kaplan et al., 1954), urethane (Berenblum and Trainin, 1960), or turpentine (Upton, 1959) after irradiation.

5. Susceptibility to leukaemia induction varies markedly with age at time of irradiation (Kaplan, 1948; Upton et al., 1960) and strain (see Upton and Furth, 1957; Upton et al., 1958).

6. Removal of the thymus does not prevent induction of neoplasia in other lymphoid organs (Kirschbaum and Liebelt, 1955; Upton *et al.*, 1958) or rid the thymectomized host of leukaemogenic virus (Miller, 1960).

7. Irradiation causes an irreversible "initiating" or "priming" (Kaplan, 1961) effect in leukaemogenesis, which may be associated with the appearance of detectable leukaemogenic "initiating" activity in tissues other than the thymus (Berenblum and Trainin, 1961).

8. The evolution of malignancy in thymic lymphomas may be gradual and stepwise, as judged by serial biopsy and transplantation assays (Kaplan and Hirsch, 1956).

9. Passage materials derived from certain monomorphous mouse leukaemias contain a filterable agent, or agents, capable of eliciting neoplasia of either lymphoid cells or myeloid cells, depending on the physiological condition of the host and other factors (see Gross, 1961).

10. The earliest spontaneous and radiogenic leukaemias require a substantially longer period for their development than those induced by highly potent viral preparations.

From these generalizations, one can only speculate about possible leukaemogenic mechanisms. It would appear likely, however, that irradiation may merely initiate a multistage process of leukaemogenesis by releasing or activating a virus and that subsequent steps require the interaction of the virus with susceptible host cells and possibly of host cells one with another and with the host. The probability of completing the neoplastic transformation would depend, according to such a hypothesis, on the number of virus particles and susceptible host cells available for interaction. In such a system, potent viral preparations would be rich in mature particles, and promoting agents would increase interactions between particles and susceptible cells and favour selection of altered cells. Thoughts, however, on the possible nature of virus-host cell interactions in tumorigenesis (Latarjet, 1959; Dulbecco, 1961; and Kaplan, 1962) and on the role of radiation will remain conjectural until further quantitative information becomes available, particularly in the face of evidence that "non-oncogenic" viruses (e.g. vaccinia, adenovirus, etc.) may exert a carcinogenic or co-carcinogenic action (Duran-Reynals and Stanley, 1961; Martin *et al.*, 1961; Trentin *et al.*, 1962).

CYTOLOGICAL ASPECTS

Karyotypic changes

Since the time of Boveri (1912), oncologists have looked towards the nucleus for insight into neoplastic transformation. Although techniques for precise and detailed morphological analysis of mammalian chromosomes became available only recently, karyotypic abnormalities in blood cells are now well documented in leukaemias (Ford and Mole, 1959). The abnormalities include changes in chromosome morphology and number, although ostensibly euploid leukaemias are not unusual (Table IV).

TABLE IV. *Chromosomes in haemopoietic cells of leukaemic and non-leukaemic individuals*

Diagnosis	Aneuploid modal chromosome number	Consistent morphologic abnormality	Investigator
	(*proportion of individuals*)		
I. Man			
Leukaemia			
Blast cell	13/50	—	Ford et al. (1958); see Sandberg et al. (1961); Hungerford (1961); Kinlough and Robson (1961)
Blast cell (Down's syndrome)	7/7†	—	Tough et al. (1961); Sandberg et al. (1961)
Chronic lymphoid	0/8	—	See Sandberg et al. (1961); Kinlough and Robson (1961)
Chronic myeloid	4/42	26/34‡	See Sandberg et al. (1961); Tough et al. (1961); Nowell and Hungerford (1961); Kinlough and Robson (1961)
Non-leukaemic control	0/62	—	Sandberg et al. (1961) Hungerford (1961).
II. Mouse			
Leukaemia			
Spontaneous	7/17	—	Ford et al. (1958); Wakonig and Stich (1960)
Radiogenic	1/1	—	Ford et al. (1958)
Chemogenic	16/16	—	Stich (in press)
Graffi virus	0/4	—	Bayreuther (1960)
Gross virus	1/5	—	Bayreuther (1960)
Non-leukaemic control	0/54	—	Ford et al. (1958); Wakonig and Stich (1960); Stich (in press)

† All 7 cases trisomic for minute autosome, probably No. 21.
‡ 24 of 30 cases revealed the Ph^1 chromosome.

The significance of these abnormalities in the pathogenesis of the disease is yet to be established. However, the individuality of cytogenetic changes in certain murine leukaemias induced by radiation (Ford et al., 1958) and 9,10-dimethyl-1,2-benzanthracene (Stich, in press) suggests an on-random type of alteration which may have causal significance. Even more suggestive is the high proportion (Table IV) of human cases of chronic myelogenous leukaemia (CML) revealing the "Philadelphia" chromosome (Ph^1), a minute autosome (probably No. 21) with a terminal deletion of the long arm. It is also noteworthy that individuals with Down's syndrome, in whom the incidence of acute leukaemia is many times higher than normal (Carter et al., 1956; Krivit and Good, 1956; Stewart, 1961), are trisomic for the same autosome. These facts have prompted the speculation that chromosome-21 is important in the regulation of haemopoiesis in man (Tough et al., 1961), although no specific and consistent abnormalities of this chromosome have been noted in leukaemias other than CML or in Down's syndrome.

Exciting as these observations may be, and however strongly they may suggest a chromosomal basis for the pathogenesis of certain types of leukaemia, it must be cautioned that such an interpretation remains conjectural. It is important, for example, to determine the frequency of abnormalities of chromosome-21 in other diseases and in normal individuals and to learn whether Ph^1 may be detected in advance of the onset of CML. As yet, no abnormalities of chromosome-21 have been reported in normal individuals, although the number of the latter examined is still relatively small. It may also be inquired why trisomy for chromosome-21 in mongolism should predispose to leukaemias mainly of the acute type, whereas a terminal delection of the same chromosome should be associated with only CML in the absence of mongolism. Possibly the difference may be related to the dosage of the unbalanced chromosome (monosomy versus trisomy) or to the disparity in age and tissue distribution (post-zygotic and confined to blood cells versus pre-zygotic and present in all cells of the body). These and other questions must be answered before the cytogenetic aspects of leukaemia can be fully evaluated. The occurrence of erythrocyte mosaicism in leukaemic individuals (Tovey, 1960; Salmon et al., 1960) may be another indication of genetic instability of haemopoietic cells associated with leukaemia, as may differences in the antigenic specificity of leukaemia cells (Gorer and Amos, 1956).

Another question raised by these cytogenetic findings is the possible aetiologic influence of chromosomal aberrations in radiation leukaemogenesis. For example, can the induction of CML in irradiated human beings be attributed to radiation-induced breakage of chromosome-21 and formation of a Ph^1 through a terminal deletion? From the frequency of spontaneous and radiation-induced deletions in human blood cells (Table V), the incidence of natural and radiogenic leukaemia seems too low to be accounted for by such a one-hit mechanism alone, as argued previously (Brues, 1958, 1959). Alternatively, one might attribute leukaemogenic changes to more complex chromosomal aberrations or to multiple point mutations in the genome of the myeloblast. The latter possibilities are being examined by Burch (1962, personal communication), who argues that many tumours, including leukaemias, develop in adult humans from a generating cell with four mutant "control" genes (or their genetic equivalents) one of which is generally inherited. The Ph[1] chromosome is regarded as a deletion or translocation and equivalent to one mutation. He claims that the age-dependence and the dose-response relation for non-leukaemic malignancies induced at Hiroshima can

TABLE V. *Incidence of chronic myeloid leukaemia and frequency of spontaneous and radiogenic human chromosome aberrations* †

Parameter	Rate
Chromosome deletions	
Spontaneous	$\ll 2.4 \times 10^{-3}$ per cell (in one cell generation)
Radiogenic	$\sim 1.1 \times 10^{-3}$ per cell (in one cell generation/r)
Chromatid and isochromatid deletions	
Spontaneous	$< 2.4 \times 10^{-3}$ per cell (in one cell generation)
Radiogenic	$\sim 2.2 \times 10^{-3}$ per cell (in one cell generation/r)
Deletions of chromosome-21 per cell at risk	
Spontaneous	$\ll 3.1 \times 10^{-5}$ per cell (in one cell generation)
Radiogenic	$\sim 1.9 \times 10^{-7}$ per cell (in one cell generation/r)
Number of myeloblasts at risk per person	1.0×10^{11} (Brues, 1959)
Deletions of chromosome-21 per marrow	
Spontaneous	$\sim 1.1 \times 10^{8}$ per person (per year)
Radiogenic	$\sim 1.9 \times 10^{4}$ per person (per r)
Incidence of chronic myeloid leukaemia	
Spontaneous	$\sim 0.3\text{-}8 \times 10^{-5}$ per person (per year) (Court Brown and Doll, 1959, 1960)
Radiogenic	$> 0.3\text{-}8 \times 10^{-5}$ per person (per r)

† The rates of spontaneous and radiogenic chromosome and chromatid deletions were derived with the help of M. A. Bender from data on human nucleated blood cells irradiated *in vitro* (Bender and Gooch, 1962 and unpublished). The frequency per cell of deletions that could yield a Ph[1] chromosome was estimated by correcting the combined chromosomal deletion frequencies to adjust for that fraction of the total length of chromosomal material at risk; i.e. for the average length of the 21st and 22nd pair, of 1.4% of the total. It is recognized, however, that this correction fails to allow for the specificity and viability of the deletion within the 21st or 22nd chromosome yielding the Ph[1] chromosome; hence, the frequencies predicted are undoubtedly too high by an unknown, but probably large, margin. In estimating the radiogenic chromosomal deletion frequencies, the rates of chromosome and chromatid types were weighted to allow for the portions of the cell cycle occupied by the single- and double-stranded stages; i.e. estimated to be 75% and 25%, respectively. The frequency of such delections per person in one year was estimated by multiplying the deletion frequency per cell times the number of cells (presumably myeloblasts) at risk (1×10^{11}) times the number of cell generations per year (arbitrarily assumed to be 365).

be interpreted in detail by assuming that radiation simulates the spontaneous mutation process. However, at high dose and dose-rate, he suggests a more efficient mechanism may be responsible for the high radiogenic incidence at Hiroshima of leukaemia in adults—two specific chromosome breaks followed by a translocation and deletion could be genetically equivalent to three "point" mutations. Most of the Hiroshima victims, except for those in the highest age range, were, he suggests, carriers of an inherited mutation predisposing to a particular malignancy or group of malignancies, including leukaemia. Childhood and adolescent malignancies are said to arise from a cell with three mutant control genes, one of which is normally inherited, together with a condition of "stress". The latter condition may be either a stimulated proliferation of generating cells or an infective oncogenic virus. In all aetiologies, Burch suggests that hormonal agents or an immune reaction complete the transformation from the carcinogenic to the malignant cell. According to his theory, two loci on each of the homologous chromosomes of pair 21 must be affected concurrently to cause leukaemia. With such a premise as a working hypothesis, a variety of radiation dose-leukaemia incidence curves may be calculated, depending on the number of somatic mutations pre-existing in myeloblasts at the time of irradiation (which is presumably age-dependent) and on the conditions of exposure.

On the other hand, the indirect induction of lymphomas in non-irradiated mouse thymus cells obviously cannot be attributed to any type of radiation-induced cytogenetic changes since the cells are not themselves irradiated. The two situations, therefore, imply either different mechanisms of leukaemia development or common intermediate aetiologic pathways yet to be defined.

Cell turnover rates

The availability of isotopic tracers, especially ^{3}H-thymidine, has facilitated determination of the rate of growth of leukaemia cells. Studies of leukopoiesis to date fail to support the theory that the leukaemia cell divides more rapidly than its normal haemopoietic counterpart (see Cronkite *et al.*, 1960; Gafosto *et al.*, 1960). On the contrary, the data are consistent with a normal or even prolonged germination time of leukaemia cells. Hence, it may be inferred that leukaemic growth involves arrested maturation and unstable, quasi-exponential reproduction of non-differentiating stem-cells rather than a simple acceleration of cell division. As yet, however, precise quantitative data are too scanty to allow a precise mathematical formulation of neoplastic proliferation.

One of the most timely studies, therefore, is to characterize the kinetics of blood cell formation, survival, and elimination in leukaemias of various types and to contrast these characteristics with the normal state. It is conceivable that the elimination of leucocytes, as well as their maturation, is impaired in leukaemia, although early suggestions to this effect have not been substantiated (see Hauschka and Furth, 1959).

ACKNOWLEDGMENT

I am indebted to P. R. J. Burch for helpful discussion of the manuscript and for quotations from his unpublished hypothesis.

REFERENCES

AXELRAD, A. A., and VAN DER GAAG, H. C. (1962). *Proc. Amer. Ass. Cancer Res.* (in press).
BAYREUTHER, K. (1960). *Nature, Lond.* **186**, 6.
BEAUDREAU, G. E., BECKER, C., STIM, T., WALLBANK, A. D., and BEARD, J. W. (1960). *Nat. Cancer Inst. Monogr.* **4**, 167.
BENDER, M. A., and GOOCH, P. C. (1962). *Proc. nat. Acad. Sci., Wash.* (in press).
BERENBLUM, I., REWALD, F. E., and TRAININ, N. (1961) *J. nat. Cancer Inst.* **27**, 1361.
BERENBLUM, I., and TRAININ, N. (1960). *Science* **132**, 40.
BERENBLUM, I., and TRAININ, N. (1961). *Science* **134**, 2045.
BIELKA, H., FEY, F., and GRAFFI, A. (1955). *Naturwissenschaften* **42**, 563.
BOVERI, T. (1912). "Zur Frage der Entwicklung maligner Tumoren". Jena. (English translation by M. Boveri, "The Origin of Malignant Tumours", Baltimore, Md., Williams & Wilkins Co., 1929).
BRUES, A. M. (1958). *Science* **128**, 693.
BRUES, A. M. (1959). In "Low Level Irradiation" (A. M. Brues, ed.), p. 73. Amer Assoc. for the Advancement of Science, Washington.
CARTER, C. O., EVANS, K., and MITTWOCH, U. (1956). *Brit. med. J. ii.* 993.
COURT BROWN, W. M., and DOLL, R. (1959). *Brit. med. J. i*, 1063.
COURT BROWN, W. M., and DOLL, R. (1960). *Brit. med. J. ii*, 981.
CRONKITE, E. P., BOND, V. P., FLIEDNER, T. M., RUBINI, J. R., and KILLMANN, S. A. (1960). *Proc. IX Internat. Congr. Radiol.* **2**, 894.
DULBECCO, R. (1961) *Cancer Res.* **21**, 975.
DURAN-REYNALS, M. L., and STANLEY, B. (1961). *Science* **134**, 1984.
FORD, C. E., HAMERTON, J. L., and MOLE, R. H. (1958). *J. cell. comp. Physiol.* **52** (Suppl. 1), 235.
FORD, C. E., and MOLE, R. H. (1959). In "Progress in Nuclear Energy, Series VI, Vol. 2—Biological Sciences", pp. 11–21. London, Pergamon Press.
FRIEND, C. (1957). *J. exp. Med.* **105**, 307.
GAFOSTO, F., MARANI, F., and BLIERI, A. (1960). *Nature, Lond.* **187**, 611.
GORER, P. A., and AMOS, D. G. (1956). *Cancer Res.* **16**, 338.
GRAFFI, A. (1958). *Acta haemat.* **20**, 49.
GRAFFI, A., and GIMMY, J. (1957). *Naturwissenschaften* **44**, 518.
GRAFFI, A., and KRISCHKE, W. (1956). *Naturwissenschaften* **43**, 333.
GROSS, L. (1951). *Proc. Soc. exp. Biol., N.Y.* **76**, 27.
GROSS, L. (1956). *Cancer* **9**, 778.
GROSS, L. (1958). *Acta haemat.* **19**, 353.
GROSS, L. (1959). *Proc. Soc. exp. Biol., N.Y.* **100**, 102.
GROSS, L. (1961). "Oncogenic Viruses". Pergamon Press, New York.
HAUSCHKA, T. S., and FURTH, J. (1957). In "The Leukemias: Etiology, Pathophysiology, and Treatment" (J. W. Rebuck, F. H. Bethell, and R. W. Monto, eds.), p. 87. Academic Press, New York.
HUNGERFORD, D. A. (1961). *J. nat. Cancer Inst.* **27**, 983.

KAPLAN, H. S. (1948). *J. nat. Cancer Inst.* **9**, 55.
KAPLAN, H. S. (1959). *Cancer Res.* **19**, 791.
KAPLAN, H. S. (1960). *Proc. Amer. Ass. Cancer Res.* **3**, 123.
KAPLAN, H. S. (1961). *Cancer Res.* **21**, 981.
KAPLAN, H. S. (1962). *Fed. Proc.* **21**, 1.
KAPLAN, H. S., BROWN, M. B., and PAULL, J. (1953) *J. nat. Cancer Inst.* **14**, 303.
KAPLAN, H. S., and HIRSCH, B. B. (1956). *Proc. Amer. Ass. Cancer Res.* **2**, 123.
KAPLAN, H. S., NAGAREDA, C. S., and BROWN, M. B. (1954). In "Recent Progress in Hormone Research" (G. Pincus, ed.), Vol. 10, p. 293. Academic Press, New York.
KINLOUGH, M. A., and ROBSON, H. N. (1961). *Brit. med. J. ii*, 1052.
KIRSCHBAUM, A., and LIEBELT, A. G. (1955). *Cancer Res.* **15**, 689.
KRIVIT, W., and GOOD, R. A. (1956). *Amer. J. Dis. Child* **91**, 218.
LATARJET, R. (1959). In "Carcinogenesis: Mechanisms of Action", Ciba Foundation Symposium (G. E. W. Wolstenholme and M. O'Connor, eds.), p. 274. J. & A. Churchill, Ltd., London.
LAW, L. W., and MOLONEY, J. B. (1961). *Proc. Soc. exp Biol., N.Y.* **108**, 715.
LIEBERMAN, M., and KAPLAN, H. S. (1959). *Science* **130**, 387.
MARTIN, C. M., MAGNUSSON, S., GOSCIENSKI, P. J., and HANSEN, G. F. (1961). *Science* **134**, 1985.
MILLER, J. F. A. P. (1960). *Brit. J. Cancer* **14**, 93.
MIRAND, E. A., and GRACE, J. T. (1962). *Proc. Amer. Ass. Cancer Res.* (in press).
MOLONEY, J. B. (1960). *J. nat. Cancer Inst.* **24**, 933.
MOLONEY, J. B. (1962). *Fed. Proc.* **21**, 19.
NOWELL, P. C., and HUNGERFORD, D. A. (1961). *J. nat. Cancer Inst.* **27**, 1013.
PARSONS, R. F., UPTON, A. C., BENDER, M. A., JENKINS, V. K., NELSON, E. S., and JOHNSON, R. R. (1962). *Cancer Res.* (in press).
POPP, R. A. (1962). *Proc. Soc. exp. Biol., N.Y.* **108**, 561.
RAUSCHER, F. J. (1962). *Proc. Amer. Ass. Cancer Res.* (in press).
SALMON, C., ANDRE, R., and DREYFUSS, B. (1960). In "Proc. Seventh Congr. Europ. Soc. Haematol." (E. Neumark, ed.), p. 1171. S. Karger, New York.
SANDBERG, A. A., ISHIHARA, T., MIWA, T., and HAUSCHKA, T. S. (1961). *Cancer Res.* **21**, 678.
SCHOOLMAN, H. M., SPURRIER, W., SCHWARTZ, S. O., and SZANTO, P. B. (1957). *Blood* **12**, 694.
SCHWARTZ, S. O., SCHOOLMAN, H. M., and SPURRIER, W. (1959). *J. Lab. clin. Med.* **53**, 233.
STANSLEY, P. G., RAMSEY, D. S., and SOULE, H. D. (1961). *Proc. Amer. Ass. Cancer Res.* **3**, 270.
STEWART, A. (1961). *Brit. med. J. i*, 452.
STEWART, S. E., EDDY, B. E., GOCHENOUR, A. M., BORGESE, N. G., and GRUBBS, G. E. (1957). *Virology* **3**, 380.
STICH, H. F. (1962). (In press.)
TOUGH, I. M., COURT, BROWN W. M., BAIKIE, A. G., BUCKTON, K. E., HARNDEN, D. G., JACOBS, P. A., KING, M. J., and MCBRIDE, J. A. (1961). *Lancet i*, 411.
TOVEY, G. H. (1960). In "Proc. Seventh Congr. Europ. Soc. Haematol.", p. 1167 (E. Neumark, ed.). S. Karger, New York.
TRENTIN, J. J., YABE, Y., and TAYLOR, G. (1962). *Proc. Amer. Assoc. Cancer Res.* (in press).
UPTON, A. C. (1959). In "Carcinogenesis: Mechanisms of Action", Ciba Foundation Symposium (G. E. W. Wolstenholme and M. O'Connor, eds.), p. 249. Churchill, London.

UPTON, A. C. (1961). *Cancer Res.* **21**, 717.
UPTON, A. C., and FURTH, J. (1957). In "Proc. Third Nat. Cancer Conf.", p. 312. J. B. Lippincott Co., New York.
UPTON, A. C., ODELL, T. T., and SNIFFEN, E. P. (1960). *Proc. Soc. exp. Biol., N.Y.* **104**, 769.
UPTON, A. C., WOLFF, F. F., FURTH, J., and KIMBALL, A. W. (1958). *Cancer Res.* **18**, 842.
VON JAGIC, N., SCHWARZ, F., and VON SIEBENROCK, L. (1911). *Berlin klin. Wochschr.* **48**, 1220.
WAKONIG, R., and STICH, H. F. (1960). *J. nat. Cancer Inst.* **25**, 295.

DISCUSSION

ALEXANDER: What role do you think contamination may play in the finding of viral agents in the radiation-induced leukaemias in your laboratory? You've got the Gross virus in one animal house, and it may appear in passage materials since tumour cells are an ideal growth medium for all sorts of viruses.

UPTON: I think that experiments of the type that Dr. Berenblum mentioned, in which one irradiates an animal and then can detect in the animal filterable leukaemogenic activity, strongly suggest that such agents are in fact present in subliminal amounts in non-irradiated individuals. From the early work of Gross, and subsequent investigators, I have the impression that the amounts of virus recoverable from animals which have developed leukaemia spontaneously or as a result of irradiation are usually disappointingly small. We badly lack today a potent assay system, and I think that one of the encouraging features of Dr. Berenblum's work is that he may have shown us a better way to assay small amounts of virus. I shall admit that as cells are passaged the likelihood of contamination is appreciable. Whether success in getting a good preparation after many serial passages means that virus has entered in the course of passaging or whether an agent that was there at the outset has simply become more potent, I don't think we can yet decide with certainty.

ALEXANDER: One reason why I brought this up is that viruses were long sought in radiation-induced leukaemias and were not found. Then other viruses, among them leukaemia viruses, became popular; many laboratories adopted them, and then suddenly, many more virus-induced tumours crop up. Is Gross's radiation-induced virus distinguishable from his standard virus?

MOLE: There are three examples now of radiation-induced leukaemias associated with a virus; Dr. Upton's, Gross's and that of Lieberman and Kaplan. All three have come from laboratories where leukaemogenic viruses have in fact been in use before this demonstration was made. People don't usually report negative results so there are those who have failed to find viruses in radiation-induced leukaemia. This does seem to be a serious criticism and how you meet it and what kind of experiments or precautions you ought to take I don't quite know.

UPTON: I am not sure whether in Kaplan's laboratory there were leukaemogenic viruses in the environment before he did his work with C57BL mice, certainly in our laboratory we have raised AKR and RF mice in proximity. One of the things we need to do now to make headway with this and related problems in radiation virology is to maintain rigorously isolated animal stocks. One of the things we are hoping to do is to derive germ-free lines by caesarean section, and keep them behind absolute barriers. I think that as yet it has not been possible to culture the leukaemia viruses involved, so that many of the immunological tests which could be brought to bear to distinguish antigenic differences have yet to be applied.

BERENBLUM: Why is it, both in Gross's experience, and I believe in others, that when you try to demonstrate a leukaemia virus—when you inject newborn C3H mice, it depends on the sub-line of C3H whether the experiments will be successful or not? I don't know whether anyone has any evidence why there should be this extraordinary specificity in the recipient, except that after you've transmitted the virus many times, this peculiar specificity seems to disappear. Have you any information on this subject?

UPTON: The question of host range is a very important one. Furthermore, as concerns viral potency—what do you mean by a potent virus? Do you mean large numbers of particles per unit weight or volume of extract, or do you mean a type of particle that's more potent than another type of particle. Such quantitative aspects are still to be elucidated.

BINGELLI: Did you find that some cells of the tumour had a large number of virus particles and if so were other cytoplasmic or nuclear changes noticeable within those cells?

UPTON: It is my impression that Dr. Parsons found only one inclusion body of viral type. In most instances, isolated viral particles were found just outside cells or at the cell membrane. This was particularly true *in vitro*.

BERENBLUM: The question of multiple viruses, of course, is another bugbear. For leukaemia virology we took a few years to be certain that the polyoma virus and the leukaemia virus were separate entities that might exist together. It took a long time even to demonstrate that, although the manifestations of these two were so very striking. What is the situation with myeloid versus lymphomatous virus in a line like RF? Is there one virus or two?

UPTON: This is a question to which we are addressing ourselves now. We don't know. You may recall, however, that Gross in passaging his so-called A virus found that thymectomized recipients would occasionally develop myeloid leukaemias. When he passaged the virus from these myeloid leukaemias in thymectomized recipients back into intact recipients, he got thymic lymphomas (see Gross, 1961). Moloney also found with his virus, which characteristically induces a lymphoid tumour, that splenectomized recipients occasionally develop myeloid leukaemias; again, passage back to an intact host elicits lymphoid tumours. I believe Graffi *et al.* also noticed, on serial passage of their mouse leukaemia virus, that the characteristics of the disease would vary from one passage generation to another, i.e. they would get myeloid leukaemias interspersed with lymphomas. Whether we have two agents or one agent that can induce either kind of disease, I don't know. Certainly, however, experience with the polyoma agent would indicate that a single tumour virus need not attack a single target-cell. There can be a broad spectrum of host cell sensitivities.

DRASIL: Do you think that these leukaemogenic agents are always highly organized particles or could small pieces of changed DNA play a role. DNA was shown last year to be taken up by some haematopoietic cells.

UPTON: There is ample evidence in the literature that virus nucleic acids are infective *per se*. Hence, I would venture to suggest that, in time, there will be evidence for leukaemogenesis by nucleic acid preparations. Preliminary data have been reported by Hays *et al.* (*Nature, Lond.*, 1957, **180**, 1419) and by Latarjet (1959). It is perhaps just a question of time. We have not attempted to test nucleic acid extracts systematically in our laboratory. We have made one or two experiments with Dr. E. Volkin; in one of these we had what seemed to be enhancing effects on radiation leukaemogenesis, and in another no effect at all. I think, to reiterate, we need better assay systems than we have now. Currently, these experiments are terribly time-consuming. We don't get many myeloid leukaemias until after several months. Hence, it is very hard to make headway.

GRAY: Before we leave contamination altogether, perhaps we should remember that Hewitt could transmit mouse leukaemia with one or two cells and he has even transmitted it by rubbing a cell suspension on the snout or ear of the animal without making any apparent lesion at all, so one has to be careful even at the cellular level.

MOLE: If there was an agent that was infective naturally, and not just syringe-transmitted, there ought to be cage effects. Has anybody observed whether an animal that had been irradiated and kept with other animals could pass leukaemia on to them?

COTTIER: This question is being considered in an experiment we are doing now. Our ^{60}Co-irradiated animals are kept in cages of eight. Some cages have five animals dead from virus lymphoma—other cages have none. We wonder whether there is anything in this or not. But it is startling!

CHROMOSOMES IN MURINE PRE-LEUKAEMIA

P. L. T. ILBERY, P. A. MOORE, S. M. WINN

Department of Preventive Medicine, University of Sydney, Australia

AND C. E. FORD[†]

M.R.C. Radiobiological Research Unit, Harwell, England

SUMMARY

Chromosome anomalies are frequent in our experience in established murine leukaemias. It seems that recognizable structural changes with excess, perhaps sometimes deficiency, of total DNA (without consideration of gene mutation) are also often present in the pre-leukaemic phase. It appears these changes are associated with an early stage of leukaemia induction for transplantation of pre-leukaemic thymuses under optimum conditions, with one exception in these experiments, failed.

Abnormal chromosome combination formed the basis of Boveri's (1929) concept of carcinogenesis. It is certain that altered chromosome karyotypes, in which, parts of chromosomes or whole chromosomes are added to or subtracted from the normal complement, have since been demonstrated in a variety of established mammalian tumours. With improved techniques there has been recognition of even less pronounced deviations from normal until, within the framework of the somatic mutation theory, it is credible that a gradation of genetic changes extends beyond optical limits to mutation of genes. If this wide spectrum of genetic changes from point mutation to gross structural changes can be responsible for neoplastic transformation it is perhaps noteworthy, in the series of murine reticular neoplasms reported by the Harwell group (Ford *et al.*, 1958) and since extended, and in this present series, that there is a heavy loading to the end where abnormalities of chromosome number and/or form are visible. These cytogenetic abnormalities however may be a reflection of more subtle initial genetic changes and become manifest only at some point in time following the neoplastic transformation. This report is then concerned with an attempt to locate the development of malignancy in relation to the appearance of the altered karyotype.

MATERIALS AND TECHNIQUES

C57BL and DBA/2 mice inbred in this laboratory during the last 5 years were used in the experiments. T6/T6 mice (Carter *et al.*, 1955), homozygous

[†] Guest Worker in the Department of Preventive Medicine during 1960 aided by the Post Graduate Medical Foundation within the University of Sydney.

for a conspicuously small chromosome, were raised for us by Dr. Mary Lyon of the M.R.C. Radiobiological Research Unit, Harwell. The subsequent hybrid from the T6 cross with C57BL has been designated C6. The cytological examination was based on the hypotonic sodium citrate squash method (Ford and Hamerton, 1956).

The radiation source was a Theratron ^{60}Co Unit made available by Royal Prince Alfred Hospital, Sydney, and the method of irradiation has been previously described (Ilbery, 1960).

EXPERIMENTAL

The results of the following cytogenetic analysis of metaphase chromosomes from established radio-leukaemias illustrate the frequency of abnormal karyotypes and serve as a basis for discussion in the pre-leukaemic experiment described later.

Phenomenon of abnormal chromosome complement frequently associated with radio-leukaemia

C57BL, C6 and DBA/2 mice when 1 month old were subjected to four doses of 180 rads whole-body ^{60}Co γ-irradiation at four-daily intervals. All leukaemias were of the typical radiation-induced thymic type either localized or generalized. Either the propositus or the earliest possible passage was examined cytogenetically. The presence of a clone was considered to have been established in the leukaemic tissues when more than 5% of the cells so examined carried a novel chromosome complement. Groups of cells possessing a characteristic abnormal chromosome or a set of abnormal chromosomes are inferred to be related by mitotic descent and are therefore referred to as clones. The same inference cannot be rigorously applied to groups of cells exhibiting abnormal counts whilst lacking any morphological change. Nevertheless in this paper the term has been used in this extended sense for this class or submode of cells. Individual abnormalities of form were recorded if chromosomes that were clearly larger or shorter than any member of the normal set were identified or if a metacentric chromosome or if a chromosome with a prominent secondary constriction at an abnormal site were present. It is now usual to refer to such chromosomes as markers.

Of the 12 leukaemias observed in the propositus 8 had modes of cells containing greater than 40 chromosomes. Counts could not be obtained from the thymus in one because of technical failure but a mode of 43 was evident in the haemopoietic tissues. The remaining 3 had modes of 40 but there were classes of 42 in 2 and 39 in the third. Where spreading is good, whereas clones greater than 40 are highly significant, counts of less than 40 chromosomes may be spurious because of breakage of cells during preparation. In the case of this last leukaemia from a propositus, as well as a clone of 39 in the thymus, cells containing 39 chromosomes predominated in the bone-marrow. The validity of the 39 clone in the thymus was verified by the transplantation of the bone-marrow into a lethally irradiated C57BL. Twenty days later the resultant chimaera was sacrificed with gross leukaemic involvement of spleen and lymph nodes. Cells containing 39 chromosomes now predominated in thymus, lymph node and bone-marrow. The findings are unusual since a mouse leukaemia with a mode of 39 has not before been recorded.

In addition to the 12 propositi, 13 radiation-induced leukaemias were cytogenetically observed in the earliest possible passage. Of the total 25 leukaemic entities, 21 showed abnormalities of number and/or form. Two had modes of 40 with clones of 39 and it is preferred to consider these clones as spurious as re-examination of these leukaemias could not be made. Two more have to be considered as not showing chromosome abnormalities although bone-marrow and spleen and bone-marrow respectively in which no abnormality was detected were the only tissues examined, since technical failure was experienced with the thymic samples.

Chromosomes in pre-leukaemia

Against this background of chromosomal changes it was decided to sample mice in the pre-leukaemic phase for evidence of karyotype alteration in an attempt to elucidate which appeared first—the chromosomal anomaly or the malignant change. Pre-leukaemia for the purposes of this experiment has meant the state before macroscopic evidence of leukaemic involvement of the thymus or other organs commonly involved.

Batches of C57BL and C6 mice were subjected to four divided doses of 180 rads of ^{60}Co γ-irradiation at 4-day intervals to a total whole-body dose of 720 rads. For C57BL and C6 incidence of the thymic type of leukaemia was 70% and 68% with a mean onset expressed in days following the last irradiation of 222 and 202 respectively. All animals were examined in the period prior to the mean time to onset. The thymus was not considered macroscopically abnormal in size until a weight greater than 70 mg was recorded. Thus a thymus neoplastic to the naked eye was excluded.

Table 1 represents the cytogenetic analyses of thymus as well as bone-marrow, spleen and lymph node from pre-leukaemic mice. The first impression is the number of abnormalities to be found in this class of material. From the experience of observation of many thousands of normal cells of thymus, bone-marrow, spleen and lymph gland in scoring the presence of the T6 marker in radiation chimaerism, the presence of these abnormalities of structure and particularly number in pre-leukaemia is indeed striking. From the laboratory's experience in counting the chimaera class of normal tissue it has been somewhat arbitrarily considered that a clone is present if there are more than 5% of cells with less than 40 chromosomes or if there are more than 3% of cells with greater than 40. When more than one clone is present the clone containing the greater number of cells is listed first. Technically-satisfactory preparations were obtained from the thymuses of 19 animals in the pre-leukaemic phase. Chromosome abnormalities were detected in 12 (Table I); of these, 9 thymuses showed an increased chromosome number

TABLE I. *Irradiated pre-leukaemic mice*

Days since last irradiation	Thymic weight (mg)	Cytology			
		Thymus		Spleen, bone-marrow, lymph gland	
		Mode	Sub-mode/s Clone/s	Mode	Sub-mode/s Clone/s
113	13	40	40 S_1	40	—
119	50	40	41 S_{11}	40	—
124	70	40	42 S_{11}	40	40 S_1–S_{11} in b.m.
126	40	41	40	40	40 L_1 in b.m.
126	40	41	40, 42	40	39 S_{111} in spleen
126	22	40	41	40	—
127	13	40	40 S_1	40	40 S_1 in b.m., spleen
140	22	40	41 S_{111}	40	—
156	36	41	40, 42 S_1	40	—
168	51	40	41, 42	40	—
168	20	40 L_1–L_{11}	40	40	40 L_1–L_{11}
183	64	41	—	40	41 in b.m.

either as the mode or as the presence of clones whilst 4 of these thymic samples had chromosome complements in which a mode of greater than 40 chromosomes predominated. In the latter category 1 pre-leukaemic animal whose thymic weight was 64 mg also showed a clone of 41 in the bone-marrow evidencing very early spread. Another mouse in which a mode of 39 was observed was discounted as no cells with the normal complement of 40 were found even in bone-marrow and being a female, it was presumed to be a spontaneous XO animal.

Table II, which contains records of those mice used in the pre-leukaemic experiment found to have at sacrifice a thymic weight of greater than 70 mg, illustrates the changes typical of those we find in radiation-induced leukaemia and supports the conception of the thymus as the originator of the abnormal cell spread to other organs.

TABLE II. *Mice of thymic weight more than 70 mg*

Days since last irradiation	Thymic weight (mg)	Cytology			
		Thymus		Spleen, bone-marrow, lymph gland	
		Mode	Sub-mode/s Clone/s	Mode	Sub-mode/s Clone/s
134	187	Near tetraploid 69–96 L_1, metacentric, L_{11}, S_1			
153	318	40	39	40 in spleen 39 in b.m.	4N 40, 4N
171	172	40	41	40 in spleen 40 in b.m.	41.42 41
173	150+	42	43.41.40 (T6 missing)	42 in l g. 42 in spleen 41 in b.m.	41 43.41.40 42.43.40
223	150	42	43.41.40	84 in spleen 42 in b.m.	42 40.84.41

Transplantation of pre-leukaemic tissue

As well as the cytogenetic scoring, observations of the transplantability of various tissues from the pre-leukaemic animals were made according to the following plan.

1. Portions of the thymus were transplanted into less than 24-hour-old mice. C57BL as well as C6 thymuses were transplanted into the C6 hybrid to make use of the marker chromosome should malignancy subsequently appear.

2. Thymuses of C6 animals were also transplanted into less than 40-day-old C6 mice thymectomized within the preceding 24 hours.

3. In the case of the C6, C57BL mice were lethally irradiated and resuscitated with bone-marrow from the pre-leukaemic C6 and portions of its thymus, spleen and lymph node were transplanted into these chimaeras. In the case of the C57BL, C6 mice were lethally irradiated and resuscitated with bone-marrow from the pre-leukaemic C57BL and portions of its thymus, spleen and lymph gland were transplanted into the resultant chimaeras.

All were kept as test animals for 1 year and then sacrificed.

No animal transplanted with pre-leukaemic tissues as in 1 or 2 developed transplantation malignancy and no abnormality was detected at post-mortem. In contrast four out of the five overt leukaemias shown in Table II proved their autonomy by transplanting. Where radiation chimaeras were utilized for transplantation tests, all tests were negative with the exception of the pre-leukaemic mouse sacrificed 119 days after the last irradiation; this exception is now reported in detail.

A (C57BL × T6)F_1♀ received 4 × 180 rads γ-irradiation. One hundred and nineteen days later, this pre-leukaemic mouse was sacrificed after the administration of Colcemid (Ciba) 1 hour before death. The thymus weighed 50 mg. Of 63 cells analysed 8 from the thymus contained 40 chromosomes and 7, 41 with an Sii/iii marker chromosome present as well as T6. Forty-one cells from the bone-marrow and 7 from spleen contained 40 chromosomes. Bone-marrow from this mouse in dosage of 0·2 ml containing 2 × 10^6 cells was given to each of 5 C57BL ♀ mice irradiated with 900 rads ^{60}Co γ-rays. Two radiation chimaeras received one-quarter thymus each as a subcutaneous graft in the ear, one a graft of spleen, one a graft of lymph gland, all grafts from the pre-leukaemic mouse; one chimaera was ungrafted. One thymic grafted chimaera died from the radiation syndrome on the 14th day. The other thymic grafted mouse developed a soft whitish tumour about the ear. The remaining chimaeras survived to one year and at sacrifice at that time no abnormality was seen. The C57BL/C57BL × T6 radiation chimaera carrying the ear tumour was sacrificed following the prior administration of Colcemid 32 days from grafting. There was a tumour underlying the right ear measuring 3 cm × 2 cm × 2 cm; the right axillary node was enlarged one plus, but there was nothing else. The cytogenic data from this mouse showed sixteen cells of 42 and 14 cells of 41 in the tumour, T6 present in all and Sii/iii frequently; in the lymph gland 11 cells of 41 (one with a Sii/iii), 11 cells of 40 1 of 42, T6 present in all; and cells of 41 were seen in spleen and bone-marrow. The tumour itself had shown progression to a mode of 42. The tumour was passed by Bashford needle to 4 C57BL and 4 C6 mice. Two of the C6 grew the tumour progressively and 1 mouse showed further progression in the tumour to 14 cells of 42, 11 of 43, and 8 of 44.

This exception to the failure of all other pre-leukaemic organs to transplant is interesting because it shows that transplantation is possible before overt signs of malignancy or spread have appeared; that is the cell containing the altered karyotype is not necessarily still dependent simply because dissemination in other organs or local infiltration has not been observed. The exception involved transplantation of thymic tissue with a mode of 40 but with a clone of 30% of cells with 41. Four other pre-leukaemias with a mode of 40 and significant clones of 41 or 42 and four pre-leukaemias with modes of 41 did not transplant.

DISCUSSION

It is difficult not to associate the observed chromosome phenomena in some way with leukaemia-induction, although there *are* cases of leukaemia without observable changes. No consistent abnormality of chromosome morphology has been identifiable by other groups (Kaplan, 1959). However, if gene mutation is taken as the ultimate in carcinogenesis it is readily appreciated why cells from all lymphomatous mice do not necessarily show gross structural mutation changes. It is also conceivable that structural changes not capable of observation in this system, i.e. cryptostructural, are present. Perhaps also we have been unfortunate not to observe small clones of abnormal cells.

The observation of twelve out of nineteen pre-leukaemic thymuses containing one or other of the chromosomal structural anomalies with or without an increase in the diploid set of chromosomes approximates our experience of about 70% leukaemia-induction following irradiation. It has been estimated on microscopic evidence alone that by 50 days after irradiation virtually all thymic tumours are already apparent (Nagareda and Kaplan, 1958) although deaths due to dissemination did not begin until 120 days (110 days in our series). Combining these two sets of data one can imagine that leukaemia *in situ* or pre-leukaemia is present relatively early after irradiation but as shown by selection of cases with no macroscopic evidence of leukaemia (under 70 mg thymic weight), established leukaemia capable of transplantation occurs much later. But chromosome changes are already apparent. Importance must be attached to the observation of cells with deviations from the rigid diploid set for we have seen departure from normal chromosome complements excessively rarely (perhaps only once in a thousand cells or less) except in malignant tissues. It is implied that there must also be a departure from the normal amount of DNA. However, the significance of structural changes alone as related to the induction of leukaemia is difficult to assess for the same appearance has been seen often unassociated with neoplasia, e.g. in haemopoietic tissue in radiation chimaerism where reversion has occurred (Ford *et al.*, 1957). But all the evidence suggests that these balanced rearrangements involving neither loss nor gain of genetic material. *A priori* it is the unbalanced changes (in which there is an alteration in the amount of chromosome material through duplication or deficiency of whole chromosomes) that are the more likely to be implicated in leukaemogenesis.

In 3 of the 19 pre-leukaemic mice, clones of identical cells were disseminated throughout the tissues examined cytogenetically, i.e. thymus, spleen, and bone-marrow and yet none of these separate organs on transplantation produced leukaemia. There are a number of instances of abnormal clones also seen in haemopoietic tissue but not in thymus (Table I). Failure of the homeostatic mechanisms (Furth and Yokoro, 1960) following the heavy irradiation schedule may well allow the appearance of otherwise forbidden clones.

The majority of lymphocytic leukaemias with macroscopic involvement "take". Despite efforts to make receptive hosts by the use of immature recipients, thymectomized recipients and radiation chimaeras, pre-leukaemias, as judged by abnormal chromosome complement, in general did not take. It was thought unlikely that the failure to transplant was due to insufficient numbers of pre-leukaemic cells in the innoculum or to non-isotopic placement of cells as a very few established leukaemic cells will take when placed subcutaneously, intraperitoneally, or intravenously. Rather, as a working hypothesis, the inference was drawn that in this model, cells with altered karyotypes were present before the supervention of malignancy as judged by the inability of the changed cells to transplant.

ACKNOWLEDGMENTS

The investigation has been supported by a grant to the Department of Preventive Medicine, University of Sydney, from the New South Wales State Cancer Council and from an anonymous donor.

REFERENCES

BOVERI, T. (1929). "The Origin of Malignant Tumours". Williams and Wilkins Co., Baltimore.
CARTER, T. C., LYON, M. F., and PHILLIPS, R. J. S. (1955). *J. Genet.* **53**, 154.
FORD, C. E., and HAMMERTON, J. L. (1956). *Stain Technol.* **31**, 247.
FORD, C. E., HAMERTON, J. L., and MOLE, R. H. (1958). *J. cell comp. Physiol.* **52**, 235.
FORD, C. E., ILBERY, P. L. T., and LOUTIT, J. F. (1957). *J. cell. comp. Physiol.* **50**, 109, Suppl. 1.
FURTH, J. and YOKORO, K. (1960). "*Proc. 3rd Asian Conf. Radiobiol*". (P. L. T. Ilbery, ed.) **2**, 86.
ILBERY, P. L. T. (1960). *Aust. J. exp. Biol.* **38**, 69.
KAPLAN, H. S. (1959). Ciba Foundation Symposium on Carcinogenesis, Mechanisms of Action. (G. E. W. Wolstenholme and M. O'Connor, eds.), p. 233.
NAGAREDA, C. S., and KAPLAN, H. S. (1958). *Proc. Amer. Ass. Cancer Res.* **2**, 330.

DISCUSSION

MOLE: Radiation produces chromosomal changes. How do you know that these chromosomal changes have anything to do with the leukaemia that would have been expected to develop if the animals had been left alone?
ILBERY: The number of animals exhibiting abnormalities of chromosome number and/or form amongst the pre-leukaemias as shown in Table I represented observable cytogenetic changes in twelve out of the nineteen thymuses examined. This incidence, perhaps fortuitously, approximates the expected yield of radiation-induced leukaemia using our schedule of γ-irradiation in the C57BL strain and our C6 hybrids. In four thymuses there was a mode of 41 chromosomes present and in another 5, sub-modes of greater than 40 chromosomes—strong presumptive evidence of malignancy as I have not observed such numbers of cells containing more than the normal complement except in malignant tissues.

MOLE: Did you ever try transplanting into the region of the mediastinum? That is quite an efficient place, and it presumably provides a slightly higher environmental temperature.
ILBERY: No. although we had considered the possible advantage of so-called isotopic placement we did not use the mediastinum. The inability of the pre-leukaemic cells to transplant was being compared to a background of subcutaneous inoculation of established leukaemia.
UPTON: What proportion of primary mouse leukaemias were perfectly normal in terms of chromosome number of morphology? There has been much accumulated experience now, and obviously the mere fact that you don't find an abnormality after a good study doesn't mean that all the chromosomes, are in fact, normal. In what proportion, after an adequate examination, will you fail to find these chromosomal abnormalities?
ILBERY: A very small proportion of thymuses in primary murine radiation-induced leukaemias in this strain and its cross with the T6 would apparently seem normal. I say "apparently" advisedly as perhaps one cannot pick up in this system all the structural changes that may be present. In our experience about 85% show chromosome change in number and/or form.
KOLLER: What about spontaneous mouse leukaemias?
ILBERY: As regards the findings in spontaneous leukaemia, our results in the few we have examined are given in Table III. Except for the Ehrlich's, the tissues were from the propositi. The AK were from Metcalf at the Walter and Eliza Hall Institute in Australia, the C58 from Bar Harbour and the ascites from the Royal North Shore Hospital, Sydney.

TABLE III. *Metaphase chromosomes from mice with spontaneous leukaemia*

Mouse	Passage	Thymus		Bone marrow		Spleen		Lymph gland	
		Mode	Sub-mode(s)	Mode	Sub-(mode(s))	Mode	Sub-(mode(s))	Mode	Sub-(modes)
49 AKR	P	40	41			40	—		
50 AKR	P	40	41, 42			40	—		
51 AKR	P	41	42, 40					41	42
179 AKR	P	40	41, 42			41	40	41	42, 40
174 AKR	P	41	—	41	40, 42	42	40, 41	42	43, 41
180 Ehrlich's	* Ascites ?	43	44, 42, 86						
192 C58	P	40		40	41	40	41	40	41

MOLE: I don't think Prof. Koller will mind my saying that he has looked at quite a number of spontaneous leukaemias in AK mice which were obtained from the Chester Beatty

Research Institute. This was some years ago and they may not be the same as the ones now. He found a high proportion of modes in spontaneous leukaemia that were grossly abnormal, with stem lines of 41 and higher. Hauschka, in a paper last year, quotes Stich as finding that about half of a number of AK leukaemias showed abnormal karyotypes. So that there does seem to be quite a lot of evidence that one can get consistent abnormal karyotpes if sufficient observation is made. Even in human chronic lymphocytic leukaemia Hauschka finds that some 40% of the mitoses examined were abnormal, which is about four times higher than that for non-leukaemic individuals. It is probably not unfair to recall that Bayreuther a few years ago published a paper suggesting that there was no abnormality in structure or number in the metaphase chromosomes of a large variety of induced tumours of a large variety of species. Hauschka makes the comment that Bayreuther's results were based on looking at more than 10,000 metaphases and that his conclusion could only have been properly drawn if he'd made a detailed study of each one of these. Technical considerations come into this too—it's very easy to be overenthusiastic of course, if you aren't really a chromosome expert, and to report all sorts of abnormalities which aren't really there. To return to this point of the AK mice, Ford at any rate is quite happy that in chemically induced leukaemia, in spontaneous leukaemia and in radiation-induced leukaemia abnormal karyotypes are characteristic of something over 80% of the total.

THE CHROMOSOMES IN VIRUS-INDUCED MURINE LEUKAEMIAS

P. C. KOLLER, E. LEUCHARS, C. TALUKDAR, AND V. WALLIS

Chester Beatty Research Institute, Institute of Cancer Research, London, England

INTRODUCTION

Since the first analysis of the chromosomes in spontaneous and radiation-induced leukaemias by Ford and co-workers (1958), a considerable body of information has been added to our knowledge concerning leukaemias in man, rats and mice. It has been established that viral agents are involved in the leukaemogenic process which occurs not only in mice of AK (Gross, 1951) and C58BL strains (Gross, 1956), but also in leukaemias induced by irradiation in C57BL mice (Lieberman and Kaplan, 1959; Gross, 1959). Cytological analysis of these leukaemic cells has disclosed that the chromosome number often deviates from that found in non-leukaemic cells. Attempts have been made to find some relationship between the irregular chromosome number and the leukaemogenic process. With the aim of throwing more light on this problem, chromosome analysis has been carried out in "spontaneous" AK leukaemias and in those which were induced by the Passage-A virus (cell-free extract from AK leukaemic tissues). The data obtained from our studies are presented below.

METHODS

The preparations for chromosome studies were made according to the technique of Ford and Hamerton (1956). Colcemid was injected into mice intraperitoneally (dose: "n" ml of a 0.02% solution where "n" = 2% of the body weight in g). The animals were sacrificed 60–75 minutes later. Cell suspensions made from the thymus, spleen, bone-marrow and enlarged lymph nodes, were pre-treated in hypotonic citrate solution for 20 minutes at 37°C and fixed in acetic alcohol (1 : 3). The chromosomes were stained with 2% aceto-orcein. Before sacrificing the mice, white blood cell counts were made.

Transplantation of leukaemia was accomplished by taking $1–5 \times 10^6$ thymus or spleen cells from a leukaemic mouse and injecting them intraperitoneally into non-leukaemic hosts.

The mitotic index of various tissues has been estimated by counting the number of cells in prophase and metaphase in 1,000 cells and expressing the number of dividing cells as a percentage of the total number of cells counted. The sample used for determining the mitotic index (M.I.) was obtained from colcemid treated animals.

CHROMOSOMES IN PRIMARY LEUKAEMIAS OF AK MICE

The high leukaemic strain used for this study was the AK strain, maintained by Dr. Miller in our Institute. About 85 to 90% of these mice develop thymic lymphoid leukaemia at 9 months of age. The chromosome constitution was analysed in seven AK mice. Cells for this study were obtained from thirty sites, yielding a total number of 836 cells, out of which 72 cells (8·6%) contained more than 40 chromosomes, i.e. deviated from the normal diploid complement. The details are given in Table I.

The chromosomes of normal mouse cells and host leukaemic cells are acrocentric. Thus the extra chromosomes in cells with 41 or 42 chromosomes cannot be identified. The only alteration in chromosome structure which was observed, was in mouse 303, in which a metacentric marker chromosome was present in all the sites analysed.

A comparison of the mitotic indices of cell populations in various sites suggests that the mitotic rate was highest in the thymus, followed by the spleen and marrow. It is interesting to note that the frequency of cells with 40 + 1 chromosomes was often highest in tissues with the highest mitotic index. Thus in mouse No. 303, the M.I. of spleen was 4·7, and 11 cells out of 40 had more than 40 chromosomes; while in mouse No. 301, the M.I. of the spleen was 2·0 and only one cell out of 43 had an abnormal chromosome number.

Another interesting finding is the lack of correlation between the age of the animals and the progression of leukaemia as indicated by the peripheral blood count. The 12-month-old mouse (No. 302) had a white blood cell count (W.B.C.) of 25,000 while a six-month-old mouse (No. 250) had a W.B.C. of 41,000. Similarly no relationship is apparent between the M.I. in the marrow and the number of white cells in the peripheral blood; No. 300 had a W.B.C. of 25,800, the M.I. being 0·5 in the marrow, while No. 101 had 19,400 white cells and an M.I. 2·0.

CHROMOSOMES IN VIRUS-ACCELERATED LEUKAEMIA IN AK MICE

Cell-free extract was made from AK leukaemic tissues by Dr. Miller and injected into newborn mice. It has been demonstrated by Rudali et al. (1956) that by such treatment the long latent period of leukaemia could be considerably reduced. Up to the time of this report three such accelerated leukaemias

Serial No. of mouse	Sex	Age (months)	W.B.C.	Site	Number of cells with indicated chromosome number						Total number of cells analysed	Mitotic index	Size of organs
					38	39	40	41	42	>42			
250	♂	6	41,000	Thymus	4		34	2			40	2·5	
				Spleen			12	1	1		14	1·8	
				Bone-marrow			18	2			20	2·1	
				Axillary L.N.		1	17	2			20	2·9	
				Inguinal L.N.			14	3	1		18	1·7	
303	♀	6½	35,000	Thymus			28 (1†)	5			33	4·0	+++
				Spleen			29 (7)	8	1	2	40	4·7	+++
				Bone-marrow			35 (3)	5			40	2·8	
				Axillary L.N.			33 (1)	7			40	4·2	++
				Inguinal L.N.			35 (3)	4		1	40	2·2	
101	♀	9	19,400	Thymus			24				24	4·1	
				Spleen			2				2	0·1	
				Bone-marrow		2	20				22	2·0	
				Lymph nodes		2	36				38	2·9	++
300	♂	10	25,800	Thymus			38	2			40	1·0	+++
				Spleen			23	1	1		25	0·3	+++
				Bone-marrow			12	2		1	17	0·5	
301	♀	10	26,800	Thymus			38	1	1		40	2·18	+++
				Spleen			42			1	43	2·0	++
				Bone-marrow			36	4			40	1·02	++
302	♂	12	25,000	Thymus	1		18	1	1		21	4·03	+++
				Spleen			18				18	2·03	++
				Bone-marrow			39	1			40	2·72	++
				Axillary L.N.		1	40	3			44	1·4	++
				Inguinal			10				10	2·0	+
304	♂	13	32,100	Thymus			11	1			12	1·0	++
				Spleen			36	3		1	40	1·66	+
				Bone-marrow			15	1			16	1·1	
				Axillary L.N.			39	1			40	2·93	++
				Inguinal L.N.			39	2	1		42	2·03	++

† Number in brackets indicates numbers of cells with the marker metacentric.

Size of organs: + Slightly enlarged   ++ Enlarged   +++ Much enlarged   ++++ Very much enlarged

TABLE II.—"*Accelerated*" *leukaemia in AK mice*

Serial No. of mouse	Sex	Age (days)	W.B.C.	Site	\multicolumn{6}{c}{Number of cells with indicated chromosome number}	Total number of cells analysed	Mitotic index	Size of organs
					38	39	40	41	42	>42			
111	♀	39	42,800	Thymus							0	0.1	
				Spleen		1					1	0.1	
				Bone-marrow		1	6				7	0.3	
112	♂	41	47,400	Thymus							0	0.1	
				Spleen			7				7	0.1	
				Bone-marrow			2				2	0.1	
110	♂	48	8,800	Thymus			8	1			9	0.3	
				Spleen	2	3	36	1			42	0.4	++
				Bone-marrow			5				5	0.4	++++

occurred in our animals, and the result of cytological analysis is given in Table II. The cell samples obtained from mice No. 111 and No. 112 were not suitable for analysis—only a few cells could be studied; they are included to show the time of the occurrence of leukaemia and the mitotic index in the various sites. AK No. 110 had only 2 cells with 40 + 1 and 5 cells with less then 40 chromosomes out of 56 cells analysed.

CHROMOSOMES OF VIRUS-INDUCED LEUKAEMIAS IN C3H/Bi MICE

Primary leukaemias

Passage-A virus (cell-free extract from AK-leukaemic tissue) was injected intraperitoneally into newborn C3H/Bi mice. The first leukaemia appeared at 89 days after the injection of virus (No. 256 in Table III). In the mice both thymus and spleen were enlarged, yet the mitotic index was low at the time of sacrificing the animal. The cell population of marrow yielded 40 cells at metaphase which could be analysed; 4 hypodiploid cells were observed. In leukaemic mouse No. 251 the thymus was very much enlarged and out of the 50 cells analysed no cells were seen with the normal diploid number. The chromosome constitution in three leukaemic mice is illustrated in Fig. 1. The data obtained shows that in this AK-virus-originated leukaemia of C3H mice, the frequency of heteroploid cells is much higher than in the spontaneous primary leukaemia of the AK host. In C3H mice No. 251, 252 and 255, the incidence was 69·6% (219 out of a total of 314); while in AK mice (Table I) the frequency of heteroploid cells was 8·6%. It is particularly interesting to note, that in C3H No. 251 all the 50 cells in the thymus had irregular chromosome numbers and in C3H No. 255 both lymph nodes analysed contained about 80% heteroploid cells.

Transplanted leukaemias

Leukaemic cells of C3H mice, have been transplanted to new C3H/Bi-strain mice. A cell suspension in saline was made from the thymus or spleen of mice which had been injected at birth with Passage-A virus. In the primary leukaemias, the number of cells with 40 + 1 or 40 + 2 chromosomes was high as was shown in Table III. It was expected that by transplantation, the deviation from normal might increase and the cell population might become more heterogeneous.

The number of cells transplanted was 1×10^6 and leukaemia developed after 12 to 16 days, killing the host. The chromosome numbers in transplanted leukaemias are shown in Table IV. Two C3H mice, Nos. 205 and 206 have a much wider variation than any other leukaemias described in the present report. Both of these leukaemic mice were implanted from the primary

TABLE III. *Virus-induced leukaemia in C3H mice*

Serial No. of mouse	Sex	Age (days)	W.B.C.	Site	\multicolumn{6}{c}{Number of cells with indicated chromosome number}	Total number of cells analysed	Mitotic index	Size of organ
					38	39	40	41	42	>42			
256	♂	89	4,300	Thymus			4				4	<0·1	++
				Spleen							0	<0·1	++
				Bone-marrow	2	2	36				40	0·5	
251	♂	120	27,900	Thymus	1			44	5		50	4·4	+++
				Spleen		1	7	3			11	0·1	
				Bone-marrow	1	1	30	7	3		42	0·7	
252	♂	120	40,300	Thymus			16	4			20	0·2	+++
				Spleen		2	10	26	2		40	0·5	++
				Bone-marrow			13	3			16	0·9	
255	♀	142	26,600	Thymus		11	7	6	3		27	0·9	++
				Spleen	1	1	4	1	2		9	0·3	
				Bone-marrow			6	28	5	1	30	1·3	
				Axillary L.N.	1		1	37	1		40	1·2	
				Inguinal L.N.			2	17		1	20	1·2	

TABLE IV. *Transplanted leukaemias in C3H mice*

Serial No. of mouse	Host	Passage No.	Donor tissue	W.B.C.	Site	\multicolumn{9}{c}{Number of cells with indicated chromosome number}	Total No. of cells analysed	Mitotic index	Size of organs
						38	39	40	41	42	43	44	45	>45			
205	C3H/b	1	Thymus	99,000	Thymus			7	20	1	1			4	47	1·0	
					Spleen			8	4			4	10		12	1·0	
					Bone-marrow			1	14	1					16	0·2	++
					Lymph node			1	2		1	1	17	1	23	0·8	++++
206	C3H/b	1	Thymus	47,600	Thymus		1	31	7		1				40	1·0	++
					Spleen		1	7	20	5	1				34	1·3	++++
					Bone-marrow	3	2	10	1		6		2		23	0·4	
					Inguinal L.N.				35	3					40	1·7	
208	C3H/b	2	Spleen-Spleen	139,100	Thymus			1							1	<0·1	
					Spleen			2							3	0·2	
					Bone-marrow		2	27							29	1·4	
					Lymph node			6	1						7	0·6	
210	C3H/b	2	Spleen, Thymus	58,800	Thymus			6			1				7	0·6	
					Spleen		1	16							17	0·6	
					Bone-marrow			1							1	0·4	
					Lymph node			1							1	0·1	
222	C3H/b	6	Spleen, Thymus		Spleen	2	1	13							15	0·1	
			Spleen		Bone-marrow	4		21							25	0·7	

leukaemia. C3H No. 208 and 210 were the hosts for a second passaged leukaemia. Though the number of cells analysed is not large, the data suggests that the range of variation is reduced, and it was further narrowed in the 6th passage of C3H No. 222.

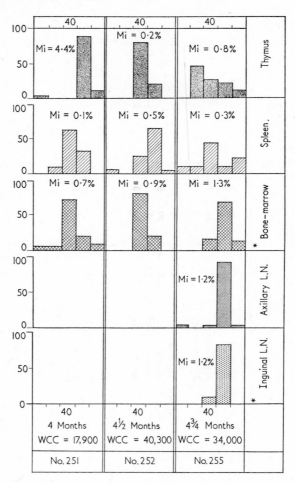

Fig. 1. Primary C3H leukaemia induced by Passage-A virus.

DISCUSSION

The presence of a viral agent in mice of the high leukaemic strains has been demonstrated first by Gross (1951) and confirmed by others (cf. Miller,

1960; Gross, 1961). All the 19 leukaemias analysed cytologically in the present investigation, were the product of the same virus, thus the causative agent being the same, uniformity in the behaviour and chromosome constitution of these leukaemias might be expected. They all display a certain degree of hereroploidy in their cell population; the frequency of heteroploid cells, however, shows considerable variation not only between different animals, but between different sites of the same animal. The difference shown by the white cell counts is a further indication that the development and progression of the various leukaemias follows an independent line, after the leukaemic transformation of normal cell by the virus has been accomplished.

The chromosome number of heteroploid cells in these leukaemias was 40 + 1 or 40 + 2; other chromosome numbers were very rare. Similar limitation of irregular chromosome number was observed by Wakonig and Stich (1960) in spontaneous AKR leukaemias, by Ford and co-workers in X-ray induced leukaemias (1958), and by Stich (1960) in chemically-induced leukaemia. The deviation from the 2n number is apparently a limited process, only cells in which the increase is not more than 1–2 remaining functional, others being eliminated by selection. Similarly the occurrence of marker chromosomes (metacentrics) and the frequency of cells carrying such chromosomes is very rare. An exception to this observation is the recent finding of Wakonig (1962) who observed a metacentric chromosome in 85 to 94% of cells in a urethane-induced mouse leukaemia.

Two transplanted lymphoid leukaemias provide further evidence in favour of the selection theory. One (C + leukaemia) originated in Balb/c, the other (EL4) in C57BL strain of mice and maintained in our Institute for over 100 transplant generations. The cytological analysis of these leukaemias, carried out by Dr. A. J. S. Davies, has shown that nearly 95% of the cell population in C+ leukaemia is composed of cells with 40 chromosomes, without evidence of structural change. The stem-line of EL4 leukaemia was found to consist of cells with 39 chromosomes, one of which was metacentric. The percentage of cells with 39 chromosomes is 93%; thus the frequency of cells with the metacentric is about the same as reported by Wakonig (1962) in her urethane-induced leukaemia (*cf.* Koller, 1961).

It is interesting to note that in cases of human leukaemias, chromosome heteroploidy is restricted to a small structural change in one chromosome of 2n complex (the Ph' chromosome in chronic myeloid leukaemia: Nowell and Hungerford, 1961) or to the presence of very few extra chromosomes found in a certain proportion of acute leukaemias so far studied (Baikie *et al.*, 1961).

The inconsistency in the chromosome composition shown by the virus-induced murine leukaemias seems to suggest that heretoploidy in the leukaemic cell population is an epiphenomenon and that the limited spectrum of variation in chromosome number is the result of selection.

ACKNOWLEDGMENTS

The authors are greatly indebted to Dr. J. F. A. P. Miller for his help and to the Medical Research Council for a grant. This investigation has been supported by grants to the Chester Beatty Research Institute (Institute of Cancer Research: Royal Cancer Hospital) from the Medical Research Council, the British Empire Cancer Campaign, the Anna Fuller Fund, and the National Cancer Institute of the National Institutes of Health, U.S. Public Health Service.

REFERENCES

BAIKIE, A. G., JACOBS, P. A., MCBRIDE, J. A., and TOUGH, I. (1961). *Brit. med. J.* i, 1564.
FORD, C. E., and HAMERTON, J. L. (1956). *Stain Technol.* 31, 247.
FORD, C. E., HAMERTON, J. L., and MOLE, R. (1958). *J. cell. comp. Physiol.* 52, Suppl. 1; 235.
GROSS, L. (1951). *Proc. Soc. exp. Biol., N.Y.* 76, 27.
GROSS, L. (1956). *Cancer* 9, 778.
GROSS, L. (1959) *Proc. Soc. exp. Biol., N.Y.* 100, 102.
GROSS, L. (1961). "Oncogenic Viruses". Pergamon Press, Oxford.
KOLLER, P. C. (1961). "Cell Physiology of Neoplasia", p. 9. University of Texas Press, Austin.
LIEBERMAN, M., and KAPLAN, H. S. (1959). *Science* 130, 387.
MILLER, J. F. A. P. (1960). *Brit. J. Cancer* 14, 83.
NOWELL, P. C., and HUNGERFORD, D. A. (1961). *J. nat. Cancer Inst.* 27, 1013.
RUDALI, G., DUPLAN, J. F., and LATARJET, R. (1956). *C.R. Acad. Sci., Paris* 242, 837.
STICH, H. F. (1960). *J. nat. Cancer Inst.* 25, 649.
WAKONIG, R. (1962). *Nature, Lond.* 193, 144.
WAKONIG, R., and STICH, H. F. (1960). *J. nat. Cancer Inst.* 25, 295.

CHROMOSOME ABNORMALITIES AND THE LEUKAEMIC PROCESS

J. LEJEUNE

Faculté de Médecine, Institut de Progénèse, Paris, France

I am taking the opportunity of adding a few words to what has already been said about the role of chromosomal changes in the occurrence of leukaemia. Very briefly it can be stated that we have three kinds of indication.

A. *Association of mongolism with leukaemia*

Even before mongolism was recognized as being determined by a trisomy, the correlation between this congenital disease and acute leukaemia had been described.

The analysis of Stewart and Hewitt (1959) shows that mongols are possibly twenty times more frequently leukaemic then normal children.

The nature of this leukaemia of mongols is also very special for in 37 cases, Stewart (1961) reports 21 blast-cell, and 16 acute lymphoblastic cases. The conclusion of Stewart being that these leukaemias are confined to one type only: the stem-cell type.

B. *Association between chronic myeloid leukaemia and partial deletion of a small acrocentric*

The discovery of Nowell and Hungerford (1960) that there were two populations of cells, one entirely normal, the other containing cells exhibiting a very small, acrocentric, chromosome has been repeatedly confirmed for chronic myeloid leukaemia.

The published data are summarized in Table I.

Hence in 28 cases the small chromosome, called Ph^1 was observed. This can hardly be considered as a pure coincidence.

The nature of the Ph^1 is not definitely established but it is most likely that it is a normal small acrocentric (21 or 22) which has lost more than half of its long arms.

C. *Association between acute myeloblastic leukaemia and deletion of a small acrocentric*

To my personal knowledge 4 cases have now been reported. One by Hungerford (1961) with 6 cells with 45 chromosomes (without determination of the lost one) among 61 examined; 1 by Fortune et al. (1962) exhibiting a clone with a quite small Ph¹ chromosome; 2 (Ruffie and Lejeune, 1962) exhibiting a clone of 45 with the loss of one of the small acrocentrics.

TABLE I

Author	Number of cases	Number of cases with Ph¹	Remarks on negatives cases
Nowell and Hungerford, 1961	10	9	1 examined in remission
Tough et al., 1961	18	13	3 are of very mild evolution, and 2 of these are probably radio-induced, the 2 others were tested in the terminal acute phase
Ohno et al., 1961	5	5	
Fitzgerald, 1962	2	1	1 possibly radio-induced
	35	28	7

Taken together we have the actual correlation:

Congenital	trisomy for a small acrocentric	mongolism Stem-cell leukaemia × 20
Clonal	partial deletion of (22–21) (Ph¹) partial or total deletion of (21–22)	chronic granulocytic leukaemia (28 cases on 35 studied) acute myeloblastic (3 cases on 3 studied)

This simple comparison excludes a chance effect and leads to the hypothesis that the same acrocentric is involved in all cases, namely the 21. It must be emphasized that this identification agrees with the microscopic observations and also with the abnormality of the nuclei of granulocytes known in mongols since 1947 (Turpin and Bernyer).

A very simple conclusion drawn from these data could be as follows: there are on chromosome 21 genes or blocks of genes which normally depress the granulocytic series. Triplication of these increases the reactivity of the lymphatic series, and conversely, loss of them permits abnormal growth of the granulocytic series.

Those reflections are, for the moment, purely theoretical, and the number of observations on which they are based is much too small. Nevertheless, it can reasonably be hoped that in the near future, accumulation of data will allow a judgment on the role of chromosomal aberrations in the production of human leukaemias.

REFERENCES

FITZGERALD, P. H. (1962). *Nature, Lond.* **194**, 393.
FORTUNE, D. W., LEWIS, F. J. W., and POULDING, R. H. (1962). *Lancet i*, 537.
HUNGERFORD, D. A. (1961). *J. nat. Cancer Inst.* **27**, 983.
NOWELL, P. C., and HUNGERFORD, D. A. (1960). *Lancet i*, 113.
NOWELL, P. C., and HUNGERFORD, D. A., (1961). *J. nat. Cancer Inst.* **27**, 1013.
OHNO, S., TRUJILLO, J. M., KAPLAN, W. D., and KINOSITA, R. (1961). *Lancet ii*, 123.
RUFFIE, J., and LEJEUNE, J. (1962). *Rev. franç Etudes Clin. Biol.* **7**, 644.
STEWART, A. (1961). *Brit. med. J. i*, 452.
STEWART, A. M., and HEWITT, D. (1959). *Brit. med. Bull.* **15**, 73.
TOUGH, I. M., COURT BROWN, W. M., BAIKIE, A. G., BUCKTON, K., HARNDEN, D. G., JACOBS, P. A., KING, M. J., and MCBRIDE, J. A. (1961). *Lancet i*, 411.
TURPIN, R., and BERNYER, G. (1947). *Rev. Hématologie* **2**, 189.

DISCUSSION

MOLE: I would draw the opposite conclusion. If, in fact each leukaemia is in any sense unique, then you wouldn't expect the chromosomes to be the same. I would expect that something that was aetiologically important should in fact be different in each leukaemic line so that the fact that each individual leukaemia appears to be unique karyotypically seems to me to be sensible. That doesn't obviate your objection—a perfectly real one—that all the chromosomal changes could be epiphenomenal. But I think it's wrong to expect that one should get a consistent change. Recent evidence confirms that the Philadelphia chromosome is not in fact a consistent finding in chronic myeloid leukaemia. Last week the Edinburgh people reported two cases of ankylosing spondylitics with chronic myeloid leukaemia without the Philadelphia chromosome in either of them. One could perhaps say that if the Philadelphia chromosome was a consistent finding in chronic myeloid leukaemia, this would be evidence that chronic myeloid leukaemia was not itself neoplastic. It only became neoplastic with the acute myeloid transformation with which it so often terminates, and then you do get karyotypic multiformity.

KOLLER: Each case must really be judged on its own. I consider that to make such a sweeping statement as "every leukaemia arises from a chromosomal abnormality" is quite illogical.

ILBERY: May I resolve this apparent conflict? Does it matter if there are cells with a mode greater than 40 present? Surely the point is that there are abnormal cells present at all. Cells with greater chromosome numbers can't be without influence on adjacent normal cells.

ALEXANDER: I have been working with a murine leukaemia which was originally produced by Law many years ago and carried in tissue culture subsequently for 5 or 6 years. Single cells are sufficient for transplantation of this tumour. Clones can be originated in tissue culture from single cells and thus one can start off with a genetically pure line. Dr. Davies has looked at these for us and he tells us that they are strictly diploid—I think something like 90% diploid in spite of the fact that they've been in tissue culture for so many years. We now have a number of mutants with different characteristics, e.g. different radiosensitivities, yet they have still 40 chromosomes. Certainly Prof. Koller's conclusion that one cannot say that leukaemia is invariably associated with changes in chromosome number seems to be very well exemplified here.

LAMERTON: May I ask whether with trisomics you have much variation in the morphology of one or more of the chromosomes? If you don't, it's a bit difficult to see why you only get the leukaemia in a few of the cases. In the trisomics, do the three 21's look the same always?

LEJEUNE: Yes they are. There is no detectable change in the karyotype of bone-marrow or blood.

LAMERTON: You don't get any variation, and yet you still only get an increase in the leukaemia by a factor of 10. Many people who have those three apparently identical chromosomes don't get leukaemia.

LEJEUNE: That is not so simple, because we don't know at all whether they get it or not. When we are dealing with the mongols, we have to say that we are dealing with a new population—because they are now allowed to live their lives with antibiotics but previously they were dying much earlier. We have a small proportion of leukaemia which was 20 times higher than in the normal population, but they have been at risk for a relatively short period. It is not known, but likely that mongols will possibly get a lot more leukaemia, when they reach 20 or 30 years old, which is quite unobserved at the moment.

LAMERTON: Is not this figure of twenty times very misleading?

LEJEUNE: It is very misleading. Roughly it compares 6-year-old children—mongols—with the normal 6-year-old population—that does not give an estimate of the general risk during the whole life-time.

UPTON: I've been very much interested in discussing this problem with P. R. J. Burch who is spending a year at Oak Ridge. I think it's his belief that the Philadelphia chromosome acts merely as the equivalent to mutations in two loci. He visualizes that one must have homologous loci on both chromosomes, in other words a four-hit situation for chronic myelogenous leukaemia. With the Philadelphia chromosome you have two to four hits.

ALEXANDER: Dr. Lejeune, you had 2 cases of these acute leukaemias. Does this mean that only 2 cases have ever been looked at?

LEJEUNE: So far as I know, for acute myeloblastic leukaemia; yes, just last week a third one was published; in that case there was quite a big deletion of the 21, not a complete disappearance.

KEPES: What is the frequency of this chromosome anomaly?

LEJEUNE: In both cases there are 2 populations of cells in these individuals. One is entirely normal, and the other shows the abnormality. In the cases we have checked there was one boy and one girl—the girl died too early, but we could also do the skin of the boy—to show that he has quite normal cells of the skin. As to the frequency of the abnormal cells it is of the order of one-half or more—but the actual numbers are possibly greatly disturbed by differential survival *in vitro*.

BINGELLI: Have you done any electron microscopy on this problem?

LEJEUNE: Electron microscopy of human chromosomes is technically very difficult. They are too big and too thick.

BINGELLI: I thought you might have done serial sections.

LEJEUNE: Yes, but when you do that you are on a very different scale so you do not know what you are seeing. The only published attempt has given an enlargement which is not greater than that obtained with the optical microscope; so the electron microscope for the moment is of no use at all. The only way in our opinion is for the cytogeneticists to become as intelligent as *Drosophila* and to become able to produce polytene chromosomes *in vitro*.

QUANTITATIVE ASPECTS OF EXPERIMENTAL LEUKAEMOGENESIS BY RADIATION

R. H. MOLE

M.R.C. Radiobiological Research Unit, Harwell, England

SUMMARY

When considering possible mechanisms for leukaemogenesis by radiation the following experimental observations need to be taken into account. Irradiation need not increase leukaemia incidence, it may decrease it. Continuing irradiation need not lead to a continuing increase in leukaemia. The effect of fractionation depends on the size of the total dose. Dose-rate may be significant of itself.

Because of the curvilinear relation between leukaemia incidence and dose, experiments on physiological and other factors affecting the induction of leukaemia by radiation cannot be easily interpreted unless dose-response curves in the different experimental situations are compared.

A good deal of information exists on the quantitative aspects of experimental leukaemogenesis by radiation and again the awkward observations seem to be the more interesting ones.

When mice are exposed to single doses of whole-body irradiation the leukaemia incidence need not necessarily increase as the dose increases even when corrections are made for life-shortening by other processes and the consequent reduction in the population at risk (Upton et al., 1958). With increasing radiation-dose in fact one form of leukaemia may decrease in incidence while another increases (Upton et al., 1960). The effect of a wide range of dose-rate does not seem to have been examined.

With multiple doses of irradiation, also, leukaemia incidence in mice need not necessarily increase as the dose increases (Mole, 1959a, Fig. 1) and the failure of the incidence to rise with dose was not due to any differential loss of population at risk. With a fixed total dose of radiation the details of the way in which the irradiation is given may be of over-riding importance. With a given kind of fractionation, daily exposure 5 days a week for 4 weeks, leukaemia incidence varied from 5 to 40% depending on the dose-rate of the individual exposures (Mole, 1959b).

Fractionation was shown some years ago to be of similar quantitative importance (Kaplan and Brown, 1952) but the experiment was defective in one respect. The over-all exposure-time varied with the fractionation over the range 4 days to 3 weeks so that the different groups of mice also varied in the ages at which they were irradiated. It had already been shown that the leukaemic response of mice of the strain involved varied fairly sharply with age (Kaplan, 1948). Therefore fractionation experiments have been repeated covering a much wider range of fractionation of dose and keeping constant the over-all exposure-time and therefore the age of the mice during irradiation.

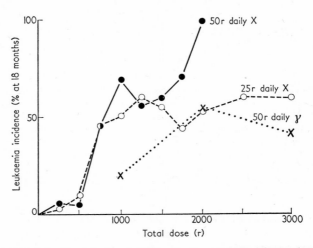

FIG. 1. Leukaemia incidence in female CBA mice given daily irradiation by X- or γ-rays for limited periods. X-Rays 5 days a week at about 1 r/sec for 1–24 weeks (Mole, 1959a). ^{60}Co γ-rays 5 nights a week at about 3 r/hr for 4–12 weeks. (Unpublished data.)

With a total dose of 1,000 r the optimum fractionation was found to be very similar to that reported earlier by Kaplan and Brown, the effect of dose-rate being evident as well (Fig. 2). However with a total dose of 2,000 r given in just the same over-all exposure-time and in just the same way by the same radiation source, the optimum fractionation was clearly quite different (Fig. 3). It is worth emphasizing that when a dose was given every hour for 672 hours (28 days) 1,000 r produced very little leukaemia. However when each individual fraction was doubled in length from 5 minutes to 10 minutes, there was roughly a 50% incidence of leukaemia. Thus the first half of each fraction had little effect, whereas the second half of each fraction did a great deal. Conversely when the dose was given once a day, 5 days a week, for 4 weeks, the incidence with 2,000 r was about the same as with 1,000 r, so that with this system of fractionation the second half of each fraction had no additional effect at all. Clearly the leukaemogenic effect of the γ-rays depended far more on the circumstances of the irradiation than on the actual exposure dose.

Although it is interesting to note the similarity of the optimal fractionation of 1,000 r in mice (Fig. 2) to the standard radiotherapeutic practice of treating ankylosing spondylitis by several exposures weekly at high dose-rate for a few weeks, it will be readily admitted that these experimental dose-response curves have little, if any, direct application to man. They have,

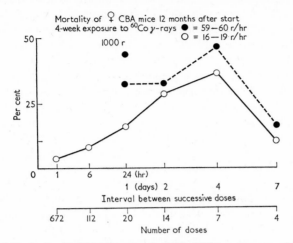

FIG. 2. Mortality in female CBA mice given 1,000 r of ^{60}Co γ-rays over a 4-week period at different dose-rates and with different schedules of fractionation. At 12 months practically all deaths were due to leukaemia.

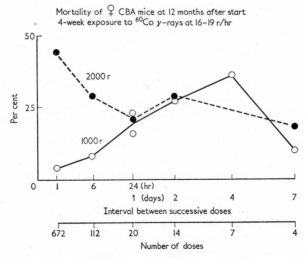

FIG. 3. Mortality in female CBA mice given 1,000 r or 2,000 r of ^{60}Co γ-rays over a 4-week period at a fixed dose-rate but with different schedules of fractionation. At 12 months practically all deaths were due to leukaemia.

however, a direct relevance to the interpretation of experiments on factors which alter the incidence of leukaemia in irradiated mice. Such experiments are commonly carried out by choosing some particular course of radiation exposure which reproducibly induces leukaemia in high frequency and then scoring the effects of experimental intervention in terms of leukaemia incidence. This is the way in which the importance of hormones and of various organs like the spleen and thymus have been demonstrated. However, with dose-response curves which are markedly curvilinear or dependent on fractionation, scoring in terms of leukaemia incidence may not, in fact, mean very much. The problem of interpretation seems to be closely analogous to the problem of interpreting the activity of substances which reduce acute mortality from radiation when administered before exposure. It is easy with a lot of substances to show reduction in mortality from say 80 to 10% but due to the curvilinearity of the dose-response curve this is equivalent to altering the lethal effect of a roentgen by only a few per cent. Experiments on physiological factors affecting radiation leukaemogenesis do not ever seem to have been carried out with more than one schedule of radiation exposure so it is impossible to know whether the change in the leukaemogenic efficiency of a roentgen is large and important or small and trivial. Thus it would seem that judgment should be reserved on the quantitative importance of the different physiological factors which have been shown experimentally to change the incidence of radiation-induced leukaemia.

One logical problem is always going to face the experimenter. Experiments are concerned with tens, sometimes with hundreds, rarely, if ever, thousands of animals. Whatever mechanism can be demonstrated at the incidence levels which are measurable in groups of these sizes, say 100 down to 10 or even 1%, it is always possible that some other mechanism will be the important one at the lower levels of incidence which are of practical concern to human populations. The only way round this seems to be to produce leukaemic cells by *in vitro* treatment, for by varying the number of cells inoculated, it should then be possible to quantitate rare cellular events with probabilities of down to 10^{-8} or so.

REFERENCES

KAPLAN, H. S. (1948). *J. nat. Cancer Inst.* **9**, 55.
KAPLAN, H. S., and BROWN, M. B. (1952). *J. nat. Cancer Inst.* **13**, 185.
MOLE, R. H. (1959a). United Nations Peaceful Uses of Atomic Energy Proc. Second International Conference, Geneva, 1958, **22**, 145.
MOLE, R. H. (1959b). *Brit. J. Radiol.* **32**, 497.
UPTON, A. C., WOLFF, F. F., FURTH, J., and KIMBALL, A. W. (1958). *Cancer Res.* **18**, 842.
UPTON, A. C., KIMBALL, A. W., FURTH, J., CHRISTENBERRY, K. W., and BENEDICT, W. H. (1960). *Cancer Res.* **20**, part 2, 1.

DISCUSSION

GRAY: Well, is it fair to say that in the thymic cases the depression of the bone-marrow is playing a very important part? I got the impression from the differences in doses, dose-rate and incident that, on the whole for these tumours, the more damage you were doing (the more effective the radiation in the killing sense) the more tumours you obtain; whereas in the other, the non-thymic ones, where the incidence decreases with increasing dose, if those were not involved, the bone-marrow depression would fall at the dose level you use. This is, I think, something one would almost excpet if an initiating event were in competition with cell destruction, because at these large doses you will be killing all or a considerable number of cells. Is this a general way of looking at this correctly, or can this be knocked down completely by the fact that I have taken a wrong interpretation of bone- marrow damage?

MOLE: The first table was of Upton's results not mine, but I would say that Berenblum's experiences do not agree. He stepped up his dose to 400 r in order to get sufficient initiation, and the kind of range that was covered in the table was 200–400 r.

EFFECTS OF IONIZING RADIATION ON CELLULAR COMPONENTS: ELECTRON MICROSCOPIC OBSERVATIONS[†]

H. COTTIER[‡], B. ROOS

Institute of Pathology, University of Bern, Switzerland

AND S. BARANDUN

Medical Department, Tiefenauspital, Bern, Switzerland

SUMMARY

Lymphatic, bone-marrow and liver tissues of adult mice and rats given short term whole-body X-irradiation with 600, 700 and 1,000 r were examined by electron microscopy at early intervals after exposure. The first detectable changes appeared in the nuclei of radiosensitive cells (chromatin derangement with focal condensation, margination and oedema) at approximately 15 minutes post-irradiation. Apparent membrane fading (real or due to tangential sectioning) was more often seen in irradiated than in non-irradiated cells, but no precise evaluation of this finding is possible without serial sectioning. No morphologic evidence is found for intracentriolar damage as the cause of radiation-induced mitotic delay.

It is conceivable that electron microscopic studies of the ultra-structure of cells following irradiation may provide some clue as to the vulnerable element(s) which are important for altered function and/or cell death. Such observations, if possible, may well contribute to a better understanding of the nature of the basic and initial effects of radiation upon the cell at a chemical level. However physiological variations and artifacts of normal tissue being studied by electron microscopy are still insufficiently known to permit a precise characterization of radiation-induced *early* and *discrete* abnormalities. Accordingly the observations to be presented must be interpreted with caution. Some ultrastructural changes in irradiated cells have already been described; but most of these reports deal with cell types that do not undergo early death after exposure to radiation doses such as used in our studies (rat liver cells—Glauser, 1956; mouse Paneth cells—Hampton and Quastler, 1958; *Paramecium*, Schneider, 1961).

This preliminary report will include electron microscopic observations made on components of highly *sensitive* as well as more resistant cells from

[†] Research supported by the Swiss National Foundation for Scientific Research (Commission for Atomic Science).

[‡] Present address: Brookhaven National Laboratory, Medical Research Center, Upton, L.I., N.Y.

whole-body X-irradiated mice and rats during the first 24 hours after exposure.

MATERIAL AND METHODS

Six-week-old Swiss albino mice and rats (inbred strains, Institute of Radiology, University of Bern) were given 600, 700 and 1,000 r air doses of X-ray delivered by a 250 kVp X-ray machine with the following conditions: 250 kVp X-rays, filter Thoraeus I, HLV 1·52 mm Cu, 15 mA, t.s.d. 60 cm, average dose-rate 24 r/min. The dose-rate was calibrated by Victoreen dosimeter.

Lymphatic organs (thymus, spleen, lymph nodes, Peyers patches), bone-marrow (mice) and liver tissue (rats) were taken 1, 5, 10, 15, 26, 34 and 60 minutes, 2, 4, 6, 12 and 24 hours after exposure. The material was fixed immediately in 2% buffered osmic acid and embedded in butyl methyl methacrylate according to Bernard's (personal communication) procedure. The blocks were sectioned with glass or diamond knives on a Porter-Blum microtome, placed on copper grids covered with a formvar film and examined in a Siemens Elmiscope I electron microscope.

RESULTS AND DISCUSSION OF RESULTS

Nuclear chromatin

The first detectable nuclear change in small lymphocytes consists in a derangement of the normally rather diffusely arranged and finely granular or fibrillar chromatin (diameter of granules in intact cells approximately 100 Å) into larger and denser aggregates, leaving osmiophobic spaces between the latter (Figs. 1(a), (b), 2(a)–(d)). This may be seen throughout the nucleus, without a noticeable predeliction for chromatin located near the nuclear membrane. With the doses and dose-rate used in these experiments some very discrete alterations of the same quality may show immediately after the end of exposure (e.g. 24 to 40 minutes after the start of irradiation); but on blind examination of the electron micrographs it was often impossible to differentiate between irradiated and non-irradiated lymphatic tissue taken before 15 minutes after radiation (= 39 to 55 minutes after start of exposure; cf. the independent studies of Bauer and Stodtmeister, 1961). The difficulty in detecting very early ultrastructural changes may in part be due to slight variations in the electron microscopic aspect of non-irradiated lymphocytes as they occur even under standardized conditions of fixation. Later there is, in many cells, a marked progression of this peculiar chromatin condensation resulting in larger aggregates and/or clumping in some parts and oedema or vacuolation in other pasts of the nucleus (Fig. 2(d)), with formation of perinuclear halos (Figs. 1(b), 2(c)) and margination of the chromatin

(Fig. 2(d)). When focal or complete nuclear pyknosis occurs, the corresponding part of the perinuclear space widens and/or a clear zone or vacuole often forms in the adjoining cytoplasm, suggestive of transfer of fluid into the latter (Fig. 3(a)).

Nucleolus

Swollen, but morphologically intact cells may show within 1 hour after the end of exposure unusually large nucleolar areas (Fig. 2(a)). But with osmic acid fixation alone distinction between an increase in size of the nucleolus and a condensation of nucleolus-associated chromatin may be very difficult.

Nuclear membrane

A lack of sharpness, or discontinuities in the osmiophilic inner or outer nuclear membrane has been observed (Figs. 1(a), 2(a), (b), (d)) but must be interpreted with great caution; images looking like membrane defects are often due to tangential sectioning and folding of the nuclear surface (cf. review by Bernhard, 1958). To overcome part of this difficulty a great number of serial ultra-thin sections would be necessary; unfortunately the technical possibilities for such a procedure have been unsatisfactory until lately, and we have not yet sufficient data in this respect. It may be mentioned, however, that an early loss of distinct membrane structures was more often seen in irradiated than in non-irradiated cells; if all these findings were due to tangential sectioning, the frequency would be expected to be equal. With these restrictions in mind we may conclude from our present findings that:

1. No *definite* and clearly detectable radiation damage to the nuclear membrane was observed on single sections *prior* to the appearance of chromatin derangement.

2. Wherever, apart from "pores", discontinuities in the nuclear membrane are found within minutes after exposure, these single observations as such cannot be acknowledged as proof of membrane injury before serial sectioning has shown them to be real defects.

3. It remains to be seen whether circumscribed cytoplasmic and/or nuclear oedema in spatial connection with an apparent partial lack or "fading" of the nuclear membrane (Figs. 1(a), 2(d)) can be regarded as circumstantial evidence for membrane damage.

4. No morphologic alteration of the so-called pores of the nuclear membranes is observed prior to radiation-induced chromatin derangement.

Centrosome and spindle apparatus

Morphologically intact centrioles are found throughout the first hour after exposure (Fig. 1(a)), e.g. the period of time it takes the nuclear damage to become clearly visible. This is consistent with observations made after partial-cell irradiation which do not seem to indicate that centrioles are especially radiosensitive structures (Bloom, et al., 1955; Zirkle et al., 1955, 1960). It may be mentioned that we several times observed non-symmetrical partition ("budding") of centrioles in irradiated cells (Fig. 3(b)). Doubling of centrioles in protozoa (Cleveland, 1957) and higher invertebrates (Mazia, 1960, 1961) has been described to take place by birth of a tiny "infant centriole" from a mother centriole. But until now only one such asymmetric pair of centrioles connected by a strand has been reported in electron micrographs (de Harven and Dustin, 1960). It remains to be seen whether the relative frequency with which this peculiar picture was seen in our studies on irradiated cells is not coincidental. Since the spindle apparatus most often is not readily demonstrated in ultra-thin sections (reviewed by Bernhard, 1958), we do not know whether radiation-induced derangements of the spindle fibres have anything to do with these findings. Such a process is considered as one of the possible causes of mitotic delay, a very sensitive and early biochemical reaction of irradiated cells (reviewd by Bacq and Alexander, 1961). In view of Schrader's (1953) hypothesis that kinetochores may become transformed into centrioles, this question deserves further attention. Wendt (1961) reported that doses of 600–1,000 r delivered as short-term irradiation to cultured chick fibroblasts produced within 3 to 5 hours a reversible vacuolation and milky change of the centroplasm detectable by phase-contrast microscopy. In our studies we were unable to see more than an occasional and questionable halo around the centrioles.

Mitochondria

Swollen, disrupted or partly dissolved mitochondria with focal loss of visible inner cristae are readily found in irradiated cells showing advanced nuclear damage (Fig. 3(a)); but we could not establish beyond doubt that discrete changes of the same quality occur very early after exposure and in absence of detectable nuclear injury. The marked reduction in size and increased over-all electron density of rat liver cell mitochondria 4 hours after whole-body X-irradiation with 1,000 r (Fig. 4(a), (b)) does not necessarily, and exclusively, reflect direct radiation damage since local exposure to the same dose does not produce an equal degree of alteration. The difficulties in evaluating eventual damage to mitochondrial membranes (Figs. 1(a), 2(a)–(d)) are the same as those discussed above for the nuclear membrane.

Endoplasmic reticulum, ergastoplasm and annulate lamellae

Except for an occasional widening or collapse of the vesicular, canalicular and spatial system of the endoplasmic reticulum and ergastoplasmic cisternae in several irradiated cells within 1 hour after exposure, no definite early

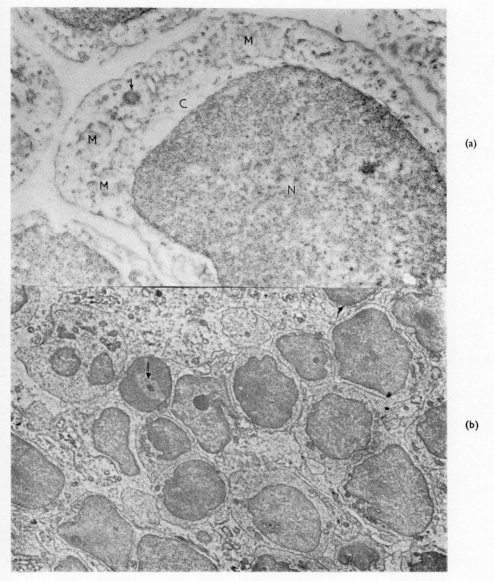

Fig. 1. Mouse thymus. (a) Lymphocyte 34 minutes after whole-body X-irradiation with 600 r—N, nucleus with slight derangement (patchy condensation) of the chromatin pattern; C, clear zone in cytoplasm; ↓, centriole; M, mitochondria. × 20,000.

(b) Thymic medulla, 1 hour after whole-body X-irradiation with 600 r— ↓ , perinucleolar halo. Various discrete stages of nuclear chromatin derangement and margination. × 5,000.

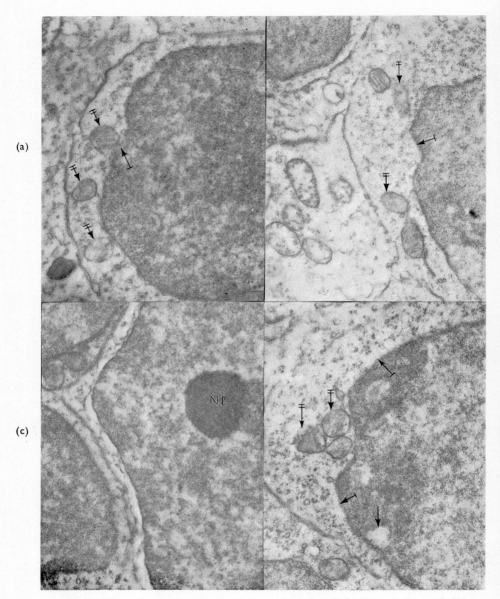

FIG. 2. Mouse thymus lymphocytes 1 hour after whole-body X-irradiation with 600 r. Slight 2(a)–(c) to marked (d) chromatin derangement with focal condensation, nuclear oedema and vacuolation ($\downarrow$). On the basis of single sections alone it cannot be decided whether apparent membrane "fading" is merely due to tangential sectioning or whether it may in part represent real membrane defects ($\bar{\downarrow}$, nuclear membranes; $\stackrel{\scriptscriptstyle\mathrm{I}}{\downarrow}$, mitochondrial membranes). Nl, large nucleolus with thin halo. × 30,000.

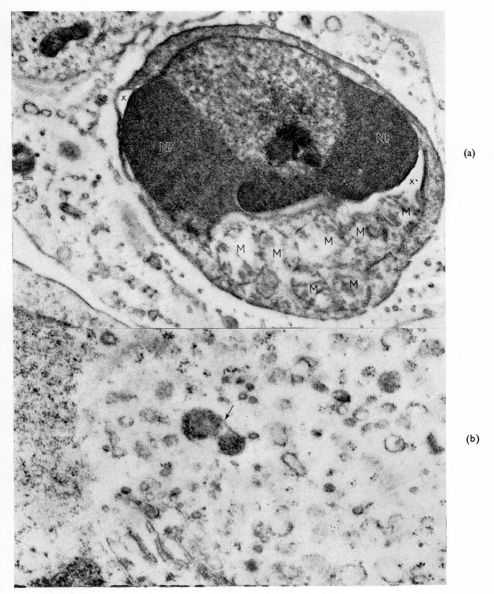

Fig. 3. Mouse bone-marrow. (a) Undifferentiated small round cell 6 hours after whole-body X-irradiation with 1,000 r: partial nuclear pycnosis (NP), widening of the perinuclear space (x) and marked swelling and/or disruption of mitochondria (M). × 30,000.

(b) Myelocyte 24 hours after whole-body X-irradiation with 600 r. Asymmetrical partition of a centriole (↓). × 47,000.

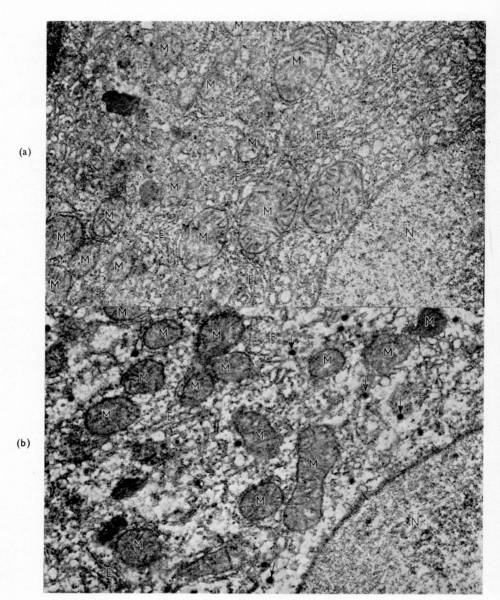

Fig. 4. Rat liver cells. (a) Unirradiated animal; (b) 4 hours after whole-body X-irradiation with 1,000 r. Condensation of the mitochondria (M), derangement and partial vacuolation of the ergastoplasm (E) and agranular endoplasmic reticulum. There is a reduced number of clearly visible inner cytoplasmic membranes. N, nucleus; ↓, small lipid droplets. × 20,000.

change was observed in these structures. Again possible membrane defects were extremely difficult to judge. The derangement of the ergastoplasmic pattern in rat liver cells seen within 4 hours after whole-body X-irradiation with 1,000 r (Fig. 4(a), (b)) was less obvious after local irradiation. No definite early changes were seen in annulate lamellae.

Cytoplasmic matrix and outer cell membrane

As early as 1 hour after exposure, when nuclear damage is definitely visible in many lymphocytes, several cells show cytoplasmic swelling. Reliable morphologic evidence for cell membrane defects developing prior to nuclear damage has not as yet been found.

CONCLUSIONS

The earliest changes in mouse lymphoid and bone-marrow cells after whole-body X-irradiation with 600 to 1,000 r, detected in this study by single-section electron microscopy, are found in the *nucleus* (chromatin derangement with focal condensation and nuclear oedema). No comment is as yet possible on eventual earlier *membrane* defects (nuclear membrane, mitochondrial membranes, endoplasmic emmbranes, outer cell membrane). Therefore, and since radiation-induced membrane damage may occur within seconds, we cannot—on the basis of our findings—evaluate the so-called "enzyme-release hypothesis" of radiation injury. Further studies, including the application of very short exposure times and serial ultra-thin sections, are necessary to elucidate this problem. No morphological evidence is seen to suggest that mitotic delay is due to radiation damage within *centrioles*.

REFERENCES

BACQ, Z. M., and ALEXANDER, P. (1961). "Fundamentals of Radiobiology". Pergamon Press, Oxford.
BAUER, W., and STODTMEISTER, R. (1961). Unpublished data.
BERNHARD, W., (1958). *Exp. Cell Res. Suppl.* **6**, 17.
BLOOM, W. ZIRKLE, R. E., and URETZ, R. B. (1955). *Ann. N.Y. Acad. Sci.* **59**, 503.
CLEVELAND, L. R. (1957). *J. Protozool.* **4**, 230.
DE HARVEN, E., and DUSTIN, P., Jr. (1960). C.N.R.S. (Paris). Colloque No. **88**, 189.
GLAUSER, O. (1956). *Schweiz. Z. Path. Bact.* **19**, 150.
HAMPTON, I. C., and QUASTLER, H. (1958). Proc. 4th Int. Congr. Electron Microscopy, Berlin, Vol. II, p. 480.
MAZIA, D. (1960). *Ann. N.Y. Acad. Sci.* **90**, 455.
MAZIA, D. (1961). In "The Cell, Biochemistry, Physiology, Morphology" (Brachet, J., and Mirsky, A. E., eds.), p. 77.
SCHNEIDER, L. (1961). *Protoplasma* **53**, 530.
SCHRADER, F. (1953). "Mitosis." 2nd edn. Columbia University Press, New York.
WENDT, E. (1961). *Z. Zellforsch.* **53**, 172.

ZIRKLE, R. E., BLOOM, W., and URETZ, R. B. (1955). *Proc. Int. Conf. Peaceful Uses of Atomic Energy Geneva* **11**, 173.

ZIRKLE, R. E., URETZ, R. B., and HAYNES, R. H. (1960). *Ann. N.Y. Acad. Sci.* **90**, 435.

DISCUSSION

ALEXANDER: Dr. Cottier whetted our appetites by saying that he'd seen membrane changes and then that he wasn't so sure whether they were real. This was exactly the position which we reached at this Institute, when Drs. Birbeck and Mercer examined a whole lot of irradiated tissues and were at first very enthusiastic about changes in the membranes. However, after looking at a really large number of sections of control material, one just couldn't be sure. Do you think from your studies that one could make this electron microscopic morphology quantitative so that one could say that the uumber of membrane abnormalities is increased as a result of irradiation? It seems to me that at the present time electron microscopy can only given us a positive or negative answer, and if the radiation change is something similar to the sort of change that may happen in the handling of the cell, then one can't be sure unless one can make it quantitative. Do you think it should be possible to do this?

COTTIER: It seems possible that by serial sectioning and constructing a three-dimensional model of the cell, one should be able to differentiate between actual membrane defects and membrane folding. For *in vivo* studies of this kind the thymic cortex may be particularly suitable since it is composed mainly of the radiosensitive small lymphocytes and does not show the same degree of spontaneous cell death as is seen in spleen and lymph nodes.

ALEXANDER: We tried to examine this in what we thought would be an even more simple system, namely with cells in tissue culture. The interesting thing with these cells is that one doesn't even get any gross abnormality until after about 24–36 hours. It's only when one irradiates the same cells *in vivo* that one gets—as you showed with your liver cells—these gross abnormalities, so these must be due to an abscopal effect. The disappointing thing really is that not only does the electron microscope not pick up anything immediately with certainty—it doesn't even pick anything up with certainty after 24 hours.

UPTON: May I comment briefly on some work which Dr. Parsons carried out at Oak Ridge with the electron microscope, in which, following up the observations of Burke and Russell on the very high radiosensitivity of the early oocytes in the immature mouse, he chose to look at oocytes within a few minutes of low doses of the order of 10 to 15 r. After going to great pains to eliminate artifacts he was able to find quite dramatic general changes in mitochondria within just a few minutes after doses of less than 10 r. This of course is the LD_{50} for oocytes, a big dose for this type of cell, but a small dose in terms of whole-body irradiation. These changes occurred very early and the cells that survived appeared to be normal. He wasn't able to determine whether the mitochondrial changes preceded the earliest detectable nuclear changes. He was anxious to try to see whether nuclear changes could be seen earlier, but this point could not be resolved with the techniques that were available.

COTTIER: Mitochondrial swelling of a variable degree is often seen in lymphocytes. On the basis of our findings we could not decide whether it precedes or follows nuclear damage.

SESSION III

Chairman: A. HADDOW

CARCINOGENESIS
By A. GLÜCKSMANN

The Effect of X-Irradiation Compared to an Apparently Specific Early Effect of Skin Carcinogens
By F. DEVIK

Long-term Consequences of ^{90}Sr in Rats and the Problem of Carcinogenesis
By K. SUNDARAM

The Influence of Radiation on the Survival of Mice Injected with Ascites Tumour Cells
By V. DRÁŠIL

Radiation-induced Bone Tumours—Fractionation Studies
By J. P. M. BENSTED, N. M. BLACKETT, V. D. COURTENAY, AND L. F. LAMERTON

Carcinogenesis as the Result of Two Independent Rare Events
By R. H. MOLE

Preliminary Studies on Late Somatic Effects of Radiomimetic Chemicals
By A. C. UPTON, J. W. CONKLIN, T. P. MCDONALD, AND K. W. CHRISTENBERRY

CARCINOGENESIS

A. GLÜCKSMANN†

Strangeways Research Laboratory, Cambridge, England

The first case of X-ray cancer in man was reported just sixty years ago (Frieben, 1902) and eight years later experimental proof of the carcinogenic action of external radiation was provided by Marie *et al.* (1910). The short interval between incidental observation and experimental proof contrasts strongly with the long interval between the description of the chimney sweep cancers by Pott in 1775 and the first experimental induction of animal tumours by chemical agents by Yamagiwa and Ishikawa in 1914. Similarly the observation of bone tumours in dial painters after ingestion of radioactive compounds (Martland, 1931) was followed quickly by the experimental production of osteosarcomas in rabbits injected with radium and mesothorium (Sabin *et al.*, 1932). Finally the carcinogenic effect of small doses of whole-body radiation was discovered experimentally (Krebs *et al.*, 1930) before similar experiences in man due to the explosion of the first atomic bombs. For all three main forms of carcinogenic action of radiation experimentation has followed quickly, or even preceded, the observation of clinical cases. Nevertheless the assessment of the carcinogenic risk of radiation still presents difficulties.

For the induction of cancers by localized external or internal radiation fairly large doses are required, while doses as low as 50 r given to the whole-body of mice are sufficient to induce ovarian tumours in 70% of the animals (Deringer *et al.*, 1955). The carcinogenic risk varies not only with the total dose, but with time and dose fractionation, with the type of organ and volume of tissue exposed, and with the species and strain of animal. For whole-body radiation the role of hormonal mediation and of other systemic factors suggest an "indirect" action of radiation. Thus oestradiol injection into irradiated mice prevents (Gardner, 1950), while the testosterone injection promotes formation of ovarian tumours (Gardner, 1950, 1953). If only one ovary is irradiated, tumours do not develop as long as the other ovary functions, but develop when the functioning ovary is removed (Lick *et al.*, 1949). Irradiated ovaries grafted into spayed or irradiated mice produced tumours, but failed to do so in untreated females with functioning ovaries (Kaplan, 1950). On the other hand, unirradiated ovaries grafted into irradiated mice failed to produce tumours. Thus direct and systemic effects are necessary

† Supported by a grant from the British Empire Cancer Campaign.

for tumour production. On the other hand, an unirradiated thymus grafted into an irradiated thymectomized mouse may produce tumours (Kaplan and Brown, 1954). The relation of radiation effects to other systemic actions in these carcinogenic events are far from being understood. It might be thought that the understanding of carcinogenesis by direct local action will be less difficult.

Radiation is known to cause mutations in cells and the somatic mutation theory is one of the popular concepts of cancer. The assumption that radiation causes cell mutation and thus cancers is obvious and might be considered the simplest theory of direct cancer induction: cells with suitable mutations proliferate into visible tumours. An alternative hypothesis is the possibility that radiation causes excessive changes in the irradiated tissues and that these conditions cause secondarily cancerous changes in irradiated or even in unexposed cells. I propose to devote most of my time to radiation changes induced by localized radiation and to compare the process of radiation-induced carcinogenesis in the skin with that following the application of chemical carcinogens.

For chemical induction of skin cancers we painted mice once weekly with a 1% solution of benzpyrene in acetone, and rats, also once weekly, with a 1% solution of 9, 10-dimethyl-1.2-benzanthracene (DMBA) in acetone. The irradiation experiments were carried out in collaboration with Dr. J. W. Boag and made use of an electron beam generated by a van de Graaff linear accelerator. This arrangement is very suitable for irradiating the skin of mice and rats, since the depth and area of the irradiated tissue can be defined very clearly and the total radiation energy is absorbed in the skin. The irradiation was given through the intact hair coat, the field defined by a saddle-like lead shield and the depth regulated by the voltage at which the irradiation was given. Figure 1 illustrates the definition in depth and area of the beam by means of paper dosimetry. On the right is the image of the electron beam passing at right angles through photographic paper in the centre of the field. For A the field size in mice is a 1 cm circle and the depth of blackening related to 0·7 MeV. For rats a circular region of the dorsal skin of 2·5 cm diameter was exposed at 1·0 MeV. B shows the image of the beam in the photographic paper in a similar position to A, while in C the top edge of the photographic paper is shaped to resemble the curvature of the dorsal region of the rat. On the left hand side of the picture a number of sheets of photographic paper were clamped together and exposed to an electron beam generated at 1·0 MeV. The definition of the area and the sharp cut off of the radiation energy at depth are clearly evident.

We shall start with a comparison of the histogenesis and then discuss tumour yield and duration of the induction period for chemical and radiation carcinogenesis, first in mice and then in rats. The application of benzpyrene

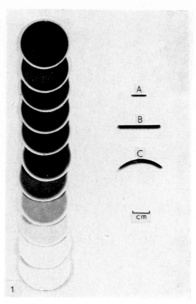

Fig. 1. Blackening of photographic paper by an electron beam generated by a van de Graaff generator under the conditions used for experiments in mice (A) and rats (B, C). The beam enters from the top of the picture through a saddle-like lead shield and penetrates a number of sheets of photographic paper clamped together (on left). The circular field is 2·5 cm in diameter and well-defined at the periphery. The radiation falls off sharply between the 9th and 10th sheets of paper. On the right a single sheet of photographic paper was exposed in a slot of a wax phantom at right angles to the beam under similar conditions: (A) over a field of 1 cm diameter to a beam of 0·7 MeV, (B) and (C) over a field of 2·5 cm diameter to a beam at 1·0 MeV. In (C) the upper edge of the paper is shaped to simulate the dorsal contour of a rat. (Courtesy of Dr. J. W. Boag.)

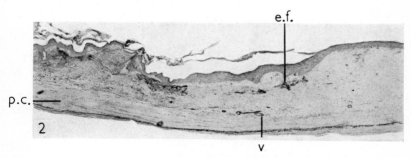

Fig. 2. Section through a lesion in the dorsal skin of a C57BL mouse exposed over a 1 cm field to a dose of 8,000 rads given in 30 seconds at 0·7 MeV, 419 days before fixation. The hyalinized central part of the lesion is seen on the right (cf. Fig. 3). Subepidermal blistering extends peripherally over a dermal region in which the elastic fibres (e.f.) appear as distinct clumps, and coincides in position with the blood vessels (v) above the panniculus carnosus (p.c.). Weigert's Elastica. × 40.

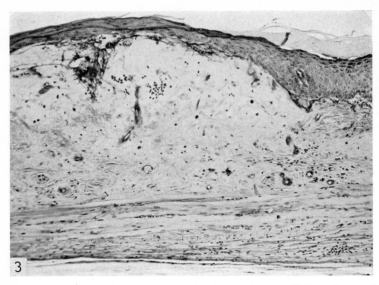

Fig. 3. The central part of the same lesion showing an almost acellular and avascular scar which is hyalinized and undergoing lysis. Note the disappearance of the basal epidermal layer, the blistering below it in the centre and the folding of the basement membrane at the periphery. Periodic acid–Schiff. × 100.

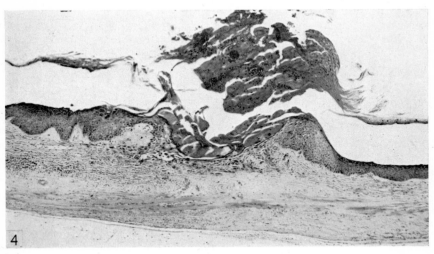

Fig. 4. A peripheral section of the same lesion showing a small hyperkeratotic lesion and, at its base, the invasion of the dermis by malignant epidermal cells. Haematoxylin–Eosin. × 65.

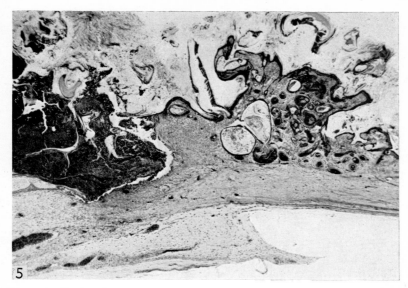

Fig. 5. A radiation ulcer in a rat with a papilloma at its edge 978 days after exposure to 11,000 rads given in 20 seconds at 1·0 MeV. Haematoxylin–Eosin. × 25.

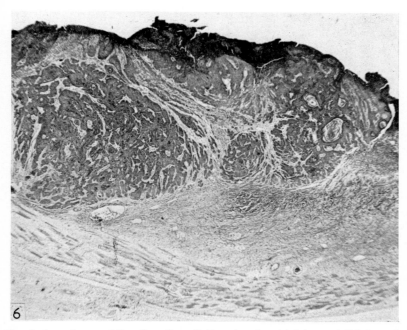

Fig. 6. A carcinoma at the edge of a radiation ulcer (on right) in a rat 494 days after 2 doses of 2,300 rads each given at an interval of two months, at 1·0 MeV. Haematoxylin–Eosin. × 25.

to the mouse skin induces almost immediately a marked epidermal hyperplasia which involves also the hair follicles—depending on their stage in the hair cycle and also causes a squamous metaplasia of the sebaceous glands. The dermal and vascular changes are at first very slight, though later on a mild inflammatory reaction is found and also a replacement of the dense dermal fibres particularly in the superficial regions by a thin-fibred, cellular connective tissue. With continued painting the hyperplasia shows quantitative variations coinciding with the hair cycle, and leads after about two months to the formation of papillomas. These are formed by epidermal proliferations as well as by abortive hair follicles in which proliferation replaces the normal production and differentiation of hairs. These changes occur over a wide field and lead to the confluence of neighbouring hyperplastic and papillomatous foci. After about eighty-four days, invasive tumours are found which infiltrate and penetrate the panniculus carnosus. Thus in chemical carcinogenesis a cellular proliferation with delayed maturation of the potential dividing cells of the epidermis and the hair follicles is induced and this change is followed subsequently by the acquisition of new characters of the cells consisting of their assumption of morphological and functional anaplasia. The morphological anaplasia is indicated by increased proliferative activity, decreased ability for differentiation, by loss of adhesiveness of cells, loss of polarity and ability to invade other structures. The functional anaplasia is seen in the ability of the tumour tissue to grow at other sites in the same host—i.e. in metastatic localizations—and in other hosts, i.e. the acquisition of transplantability. These changes occur at irregular intervals, often as separate steps but always in the descendants of the originally treated cells.

In mice given a single dose of 8,000 rads from an electron beam generated at 0·7 MeV over a circular area of 1 cm diameter the histological changes are as follows: mitotic inhibition in the epidermis and in the hair follicles is followed by a hyperkeratosis and, with the absence of replacing basal cells, the keratinized epidermal layers and the hair follicles are shed. At the same time the dense dermis shows vascular and cellular changes and is sloughed down to the dermal fat layer after about fourteen days. The panniculus carnosus tends to be perfectly normal, though in response to the developing superficial ulcer there is a deep dermal vascular reaction with fibre formation. The ulcer heals after about thirty days but the scar is thin-fibred, cellular and served by dilated and insufficient blood vessels. These changes lead to hyalinization of the scar which in turn is shed and this time the panniculus carnosus may be partly shed too. Subsequent scarring again leads to the formation of an ultimately acellular, avascular hyalinizing scar which lyses and breaks down. After repeated cycles of ulceration and regeneration carcinomatous changes are seen at the periphery of the lesion among the immigrating epidermal cells. Most, if not all, of the exposed cells of the epidermis and hair follicles are shed during the ulcerative phases and the regeneration of the epidermis is due to the immigration of unexposed neighbouring epidermal cells. None of the hair follicles are left in the exposed areas. The dermal scar is formed largely from the connective tissue accompanying the blood vessels and originally situated below the panniculus carnosus. Figure 2 shows a central section through a lesion 419 days after exposure to radiation: on the right the scar tissue appears hyalinized, almost avascular and acellular and begins to lyse. At the bottom of the picture the elastic fibre layer below the panniculus carnosus can be seen and at the periphery of the scar two vessels are seen just above the panniculus carnosus. The subepidermal elastic fibres are clumped and this change has spread beyond the limits of the original lesion. Above these clumps the epidermis is thickened while the epidermis above the hyalinized scar is keratotic and about to be shed. The peripheral epidermis at the other side of the lesion is also thickened and lifted from its basement membrane as Fig. 3 shows. In these regions the absence of cells and lysis in the scar is also clearly seen as well as the thickening of the capillary walls. Still further at the periphery of the lesion marked epidermal thickening is seen (Fig. 4) and the irregular invasion of the peripheral dermal scar by obviously malignant epidermal cells.

Thus the malignant change occurs in the periphery of the lesion in cells concerned with the repair of the progressive injury caused by radiation and probably in cells whose ancestors were outside the field of radiation. The malignant change occurs after repeated cycles of ulceration and regeneration in the regenerating epidermal cells which are subjected to the adverse conditions caused by the radiation-induced vascular damage. This damage is progressive; it is at first confined only to the immediately exposed superficial vessels but spreads in depth as well as laterally to the deeper vascular plexus which was outside the range of the electron beam. Though this extent and amount of vascular damage may not be the only necessary condition for carcinoma induction, it should be mentioned that with smaller doses of radiation and with a less penetrating beam, i.e. of only 0·3 MeV, we failed to induce tumours.

In rats the histogenesis of chemically induced tumours resembles that of mice in that the descendants of treated cells ultimately become malignant. There are some differences in details: the process takes longer in the rat; the spacing of hair follicles is greater and they change more independently from the epidermis than in mice and tend to produce basal cell carcinomas; sarcomas occur quite frequently in rats either as separate tumours of starting as sarcomatous changes in the stroma of carcinomas, while sarcomas are rare in mice.

Irradiation of the dorsal skin of rats is also followed, as in mice, by mitotic inhibition, hyperkeratosis and ulceration. The epidermis as well as the hair follicles are shed and epithelial healing is due to the immigration of neighbouring epidermal cells. The dermal changes resemble those in the mice in that an unstable scar is produced, but differ in that the changes are between keloidal and thin-fibred states instead of between hyalinization, lysis and thin-fibred scars. In either case the keloidal or hyalinized tissue is replaced by a cellular tissue which in turn becomes keloidal or hyalinized until in the rat this tissue undergoes a sarcomatous change. The instability of the scar tissue can be traced in both species to progressive vascular injury manifested by endarteritic and teleangiectatic vessels of even the deeper plexus and causing late ulcers. At the edge of these ulcers papillomas (Fig. 5) and carcinomas (Fig. 6) occur. As in the mouse these malignant changes are likely to occur in cells whose ancestors were not exposed to radiation. The majority of tumours in the rat are sarcomas which arise in or close to the panniculus carnosus which was outside the range of the electron beam and which as in the mouse becomes involved in the lesion only very much later.

TABLE I. *Induction of papillomas, carcinomas and sarcomas in the dorsal skin of rats by DMBA, by localized irradiation and by irradiation followed after fifteen months by DMBA*

Treatment	Number of rats	Percentage incidence of		
		Papillomas	Carcinomas	Sarcomas
DMBA	68	15	84	37
Radiation	95	3	8	32
DMBA + radiation	15	0	20	40

Thus in mice and rats radiation-induced cancers arise in regenerating tissue rather than in the descendants of treated cells as in the case for chemically induced tumours. This "indirect" action of radiation as compared with the more "direct" effect of chemical carcinogens is reflected in the duration of the induction period and the tumour yield.

In the dorsal skin of mice a dose of 8,000 rads given at 0·7 MeV over a circular field of 1 cm diameter produced only four carcinomas in 27 mice surviving for 400 days. The weekly application of benzpyrene to 48 mice resulted in 48% carcinomas and an additional 38% of papillomas in a period of three to five months (Fig. 7). With smaller single doses of 4,000 rads or with an electron beam generated at 0·3 MeV no tumours were found. A single dose of 4,000 rads at 0·7 MeV followed by weekly applications of croton oil failed to increase the incidence of papillomas over that obtained with the croton oil painting alone.

In rats the tumour yield following irradiation, or the application of DMBA to the dorsal skin, is given in Table I. In the group of irradiated animals are included only those which survived for the induction period of the first tumours in the following experiments: a single dose of 11,000 rads at 1·0 MeV; two doses of 2,300 rads each at an interval of two months and eight doses totalling 6,800 rads given over a period of six months at monthly intervals. While the yield of sarcomas is not very different in the DMBA-painted and the irradiated group, the incidence of carcinomas is ten times greater after the DMBA than in the irradiated animals. The difference in the duration of the

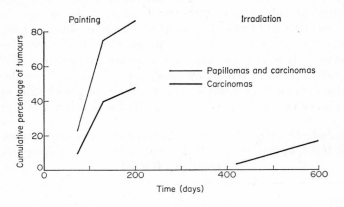

Fig. 7. The induction of skin tumours in mice by weekly applications of benzpyrene and by exposure to a single dose of 8,000 rads at 0·7 MeV. Mice surviving at least seventy days for painting and 400 days for the irradiation are considered "at risk".

induction time is very striking indeed as Fig. 8 shows. The induction period for the same incidence of sarcomas is about 3 times greater for radiation than for painting while for carcinomas the difference is even greater. Animals given a single dose of 2,300 rads of an electron beam generated at 0·7 MeV— which itself failed to induce any tumours—followed 15 months later by weekly applications of DMBA, show a shortening of the induction period for sarcomas as compared with the painting of the unirradiated rats and a lengthening of the induction period and reduction in tumour yield for carcinomas. These effects might be due to the destruction of hair follicles and the reduction in number of epidermal cells by previous irradiation and the increase of cell population in the unstable dermal scars following irradiation.

As regards the induction of carcinomas in mice and rats the histogenetic investigation suggests that with carcinogenic hydrocarbons descendants of treated cells produce cancers, while with radiation most if not all the exposed cells are killed and cancers are formed in the regenerating tissue involved with repair. The direct action of the chemicals as compared with radiation is also reflected in the greatly shortened induction period for the hydrocarbons and the significantly greater tumour yield. For sarcomas in rats the same holds true as regards histogenesis and duration of the induction period, but the yield of sarcomas is about the same for the two carcinogenic agents. The essential condition for tumour formation in the case of heavy local irradiation appears to be the induction of specific and progressive vascular changes in the form of endarteritis, periphlebitis and teleangiectasies which in turn cause unstable scars. These vascular changes in rats differ from those induced by repeated thermal burns which in our hands failed to induce tumours locally.

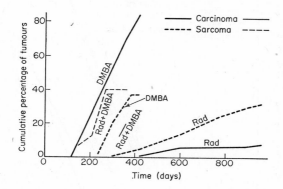

Fig. 8. The induction of skin tumours in rats by (a) weekly paintings with DMBA, (b) exposure to an electron beam at doses varying from 2 × 2,300 rads to single doses of 11,000 rads, (c) a single dose of 2,300 rads followed fifteen months later by weekly paintings with DMBA.

These observations do not support the hypothesis that in local tumour formation it is the mutagenic effect of radiation directly on exposed cells which is responsible for carcinogenesis, but suggest that this process is rather indirect. With chemicals, carcinogenesis is more direct in that treated cells or rather their offspring form the tumours. Even here, however, the process involves a passage through various stages and is dependent on systemic factors which may promote or inhibit the progress to malignancy. To illustrate this Fig. 9 shows the results of DMBA-paintings of the vulva and vagina of intact and castrated rats. By this means carcinomas are induced in the vulva and sarcomas in the vagina. The rate at which sarcomas and carcinomas occur is about equal in intact rats, but in castrated rats the vulval carcinomas occur in the same percentage though slightly later, while the sarcomas are not only delayed but very significantly reduced in incidence. In order to test whether, by lowering the immune reactions in the rats, we could speed up the carcinogenesis, we irradiated such rats treated with DMBA either by whole-body exposure to X-rays or by exposure of the pelvic region only. The results for both types of exposure were essentially the same and only those for pelvic exposure are given. In intact animals the pelvic and whole-body radiation shortened the induction period for vulval carcinomas, but decreased the incidence of sarcomas to about the same level as for castration. In castrate rats, on the other hand, pelvic like whole-body exposure to X-rays strikingly shortened the induction period for vulval carcinomas and for vaginal sarcomas

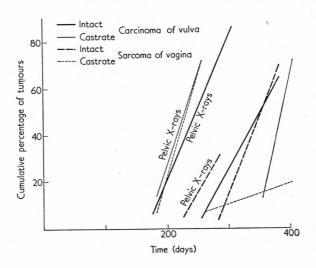

FIG. 9. The effect of irradiation on the induction of vulval carcinomas and vaginal sarcomas by weekly DMBA paintings in castrate and intact rats. The irradiation was given at 200 kVp in whole-body exposures (five times 400 rads in sixteen weeks) or pelvic exposures with shielding of the body (six times 400 rads in twenty weeks).

and also significantly increased the yield of vaginal sarcomas in castrate animals. Since the effects of whole-body and pelvic radiation are the same in this instance too, it is unlikely that a mere lowering of the immune response is responsible for rapid tumour formation and the fact that the vulva and vagina of intact rats behave differently under radiation supports this view. Thus hormonal and other, as yet undefined, systemic factors enter even into the local induction of tumours by the application of carcinogens.

This example brings us back to the intricate problem of the interaction of hormonal and other factors in the induction of tumours by whole-body irradiation. We quoted the example of the induction of ovarian tumours in mice in the beginning and pointed out that irradiation of the ovaries is one of the necessary conditions for carcinogenesis, but that the presence of one functioning ovary or treatment with oestrogens can prevent, or at least delay, the tumour formation while for the thymoma the irradiation of the thymus does not appear to be one of the necessary conditions. The point I wanted to make, is that even in the induction of tumours by locally acting carcinogens systemic factors play a major role and that the process of carcinogenesis due to local radiation may be an indirect one.

REFERENCES

DERINGER, M. K., LORENZ, E., and UPHOFF, D. E. (1955). *J. nat. Cancer Inst.* **15**, 931.
FRIEBEN (1902). *Fortschr. Röntgenstr.* **6**, 106.
GARDNER, W. U. (1950). *Proc. Soc. exp. Biol. N.Y.* **75**, 434.
GARDNER, W. U. (1953). "Advances in Cancer Research", Vol. 1, p. 173. (A. Haddow, and J. P. Greenstein, eds.), Academic Press, New York.
KAPLAN, H. S. (1950). *J. nat. Cancer Inst.* **11**, 125.
KAPLAN, H. S., and BROWN, M. B. (1954). *Science* **119**, 439.
KREBS, C., RASK-NIELSEN, H. C., and WAGNER, A. (1930). *Acta Radiol.* Suppl. X.
LICK, L., KIRSCHBAUM, A., and MIXER, H. (1949). *Cancer Res.* **9**, 35.
MARIE, P., CLUNET, J., and RAULOT-LAPIONTE, G. (1910). *Bull. Ass. franç. cancer* **3**, 404.
MARTLAND, H. S. (1931). *Amer. J. Cancer* **15**, 2435.
SABIN, F. R., DOAN, C. A., and FORKNER, C. E. (1932). *J. exp. Med.* **56**, 267.

DISCUSSION

ALEXANDER: In producing skin cancers by irradiation, do you inevitably have to produce this chain of events of shedding of the irradiated tissue and scar formation? If you rearrange your dosage schedule of irradiation in such a way as to avoid it, do you then never get skin cancers or can you still get cancers sometimes?
GLÜCKSMANN: I have never been able to get it without such previous local damage. There are some experiments with α-particles of very low penetration which do not cause ulceration and which Professor Lamerton has used. Did you get any cancers?
LAMERTON: We did not get any tumours in these experiments.
BERENBLUM: I did not quite understand the irradiation procedure. Was the dose given as a single exposure?
GLÜCKSMANN: In most of the experiments the dose was given as a single dose, but we have given two equal doses at intervals of two months in rats and also supplied X-rays in eight doses at monthly intervals over a period of six months. Henshaw *et al.* have given daily fractions of β-rays to mice and rats at various dose levels and also single doses and found that single exposures were more efficient in producing skin tumours than the fractionated régime.
CASARETT: Were the doses of radiation which reduced the yield of chemically induced carcinomas and delayed the time of their onset large doses? Have you any idea of what the mechanism of this delay or inhibition might be?
GLÜCKSMANN: The doses were of the order of 2,300 rads in the skin and almost the same for the genital tract. I do not know what the mechanism of the delay or inhibition of carcinogenesis may be. Under certain conditions, in the vulva for instance, we get a

shortening of the induction period with irradiation. For chemical carcinogenesis we have found that continuous presence of chemicals in the form of injected deposits can shorten the induction period as compared with weekly applications. For both forms of application we found that a change in the hormonal pattern of the experimental animal can shorten or prolong the induction period. In radiation carcinogenesis the latent period is always long and this agrees with almost all other observations. Similarly the doses are fairly large, except if there are some predisposing factors such as the injection of silicates by Lacassagne and by Burrows *et al.* Unless you change the biological system, you cannot cut down the dose of local radiation nor shorten the induction period.

CURTIS: Yesterday we talked a great deal about the two-stage hypothesis of tumour induction in the case of leukaemia. Fractionation of the dose into two, three or four doses of radiation separated by one or two week intervals proved much more effective than a single dose and this with other facts discussed yesterday, indicates that apparently this is a two-stage phenomenon. Have you done something like this in the skin and is there any suggestion of a similar two-stage phenomenon in the skin?

GLÜCKSMANN: I did the classical experiment of following a single dose of radiation from an electron beam with weekly applications of croton oil to the skin of mice. I failed to produce more papillomas in irradiated than in control mice and never got a fully malignant tumour, i.e. a carcinoma or sarcoma. Shubik *et al.* reported a similar experiment some time ago and got more tumours after previous β-ray exposure with croton oil. Unfortunately they did not do any histology and we do not know whether these tumours were only papillomas or malignant. They used a smaller dose of radiation than we did. In our experiments cancer formation seems to occur in regenerating cells rather than in the offspring of directly-treated cells and we have thus conditions very different from the chemical carcinogenesis of the skin where the two-stage phenomenon applies.

GRAY: In relation to Shubik's experiment: though he got papillomas and no sarcomas or carcinomas, his results are quite clear-cut. He had three groups of mice: those given a dose of 800 rads of β-rays had no papillomas, those treated with croton oil alone had few papillomas while those given 800 rads and subsequently painted with croton oil had a great many papillomas. I was very interested in this experiment because if there is a specific effect of irradiation in inducing tumours, I think perhaps the use of low doses of radiation which do not cause cell destruction, followed by a promoting agent like croton oil might be a good way of finding out about it.

There is some evidence in the literature of some kind of specific initiating act in tumour formation in the experiments of Bond and Cronkite with breast tumours in rats. There is a hormonal factor involved. But they have shown that the natural incidence of breast tumours assayed at one year is extremely low, that there is an increase in tumour yield with doses from 25 r upwards and that the tumours occur in irradiated zones only. When they segmented the breast for irradiation, the tumours appeared in the irradiated zones only. Whether this is an initiation of tumour induction or a hastening of some other process, it is necessary to irradiate the cells in which the tumour arose and in some cases the doses were as low as 25 r, though the dose range was from 25 to 800 r.

GLÜCKSMANN: The situation for the breast is very similar to that for the ovary; irradiation is necessary but hormonal factors come in at some stage all the time.

COTTIER: You showed pictures of the initial necrosis of skin after heavy radiation. Were you able to follow whether the regenerating epithelium originates from within the irradiated field or perhaps also from an outer irradiated zone or were there some follicles or single cells left?

GLÜCKSMANN: All the follicles and the whole epidermis were killed and the vascular changes extended beyond the irradiated region. Thus even the first regenerating tissue was subsequently shed and the repair could be brought about only from outside the irradiated zone, i.e. in the case of the epidermis from neighbouring unirradiated epidermis.

GRAY: But if you are giving 12,000 rads in the irradiated zone, the periphery surely might have had 100 rads.

GLÜCKSMANN: Boag's paper dosimetry (Fig. 1) shows a very sharp cut-off of the radiation in depth and laterally. Even if some peripheral cells survived irradiation and participated initially in the regeneration, the progressive vascular change with the repeated cycles of regeneration and sloughing would effectively remove most, if not all, of the peripheral cells.

MARTINOVITCH: Have you tried to graft irradiated epidermis to unirradiated dermis or alternatively grafted unirradiated epidermis on the irradiated region?

GLÜCKSMANN: No, we have not done this.

CHASE: We have not done carcinogenic experiments as such, but have some incidental pertinent observations indicating that you have to get an ulcer before you obtain tumour formation. We exposed skin to slightly lower doses in the 800 to 1,000 r range and repeated the doses several times. Though accumulating a considerable dosage, we did not get ulceration and no tumours. We then gave a dose of 1,800 to 2,000 r, got ulceration and after repeating the dose we obtained papillomas and then malignant tumours.

GLÜCKSMANN: I should mention one other thing: you can get necrosis of dermal tissue and its resorption and replacement by regenerating tissue without apparent ulceration and a similar replacement can take place also in the epidermis. Fundamentally this is still cellular death and repair whether it takes place at the surface or underneath.

BERENBLUM: In the earlier days of skin carcinogenesis I was always impressed by the fact that the more potent carcinogens all produced hyperplasia and it was therefore assumed that the hyperplasia was secondary to some sort of damage. Everything fitted very well and when one extended this work to give a single painting of a carcinogenic hydrocarbon and then croton oil, there appeared to be a similarity between the degree of damage produced by the carcinogen and its effectiveness as an initiator. But this changed when Salaman introduced urethane—perhaps the most potent initiator of all—which produced no hyperplasia, and no damage to the epithelium. What is more, one can inject urethane subcutaneously and then apply croton oil to the skin, and still get tumours. We have now become very cautious about assuming that the component which has anything to do with damage is that responsible for carcinogenesis. Now, I realize that with radiation it is very difficult to set up a group of experiments at the same time. But if you have to give a single dose, it has to be big enough to produce a certain damage. On the other hand, if you give multiple doses, it will complicate the experiment to such an extent that you cannot do it. I am just wondering whether the damage produced is really as critical as it appears from your experiments, or whether, if one could design a suitable type of experiment where the radiation per unit time were so low, one could not get carcinogenesis without damage to the epithelium.

GLÜCKSMANN: These experiments have been done to a certain extent with β-rays by Henshaw et al. The answer, I think, was that up to 25 reps daily he did not get any visible damage or tumours. With doses above 50 reps per day he gets damage and ultimately tumours. For local irradiation it seems that you have to have some degree of damage before getting carcinomas or sarcomas.

BRINKMAN: You could protect the dermis chemically against radiation damage without protecting the epidermis. Thiosulphate protects the dermis without protecting the epidermis and this might help in differentiating your problem.

GLÜCKSMANN: Thank you, this is a very good suggestion.

MAYNEORD: Since the experiments I did some time ago with Burrows were quoted, I should like to mention that we were quite convinced that the tumours arose immediately outside the field of irradiation. It would, of course, be quite impossible to say what dose had been given by scatter. But I remember we gave single doses of the order of 250 r or so and I think it is very improbable that the dose was more than 20 r in what seemed to be the site of origin of the tumour.

CASARETT: I would like to suggest that the promoting factor in carcinogenesis may be a secondary change in the cells of origin. This may be due to small doses of radiation or to some tissue disorder at the site of origin which would require bigger doses of radiation. We may be dealing with two different dose problems and two mechanisms of carcinogenesis. In other words, we may have potential cancer cells existing before irradiation and to produce a second change, more radiation must be added. A high dose of radiation causing tissue disorder may be required as the promoting factor so that the full potential of the existing pre-malignant cells can be realized. This sort of duplication of mechanism may account for cancers arising outside the field of irradiation in an area which is seriously damaged with regard to the supporting tissue.

ALEXANDER: I would like to ask two questions: How strong is the evidence for saying that there must be at least two quite different pathways: the one following whole-body irradiation, where the tumour induction is highly dependent on the genetic make-up of the animal and where there does not seem to be any *a priori* demand for severe or readily detectable tissue injury, and the second one from localized radiation where as far as one can see genetic make-up plays a relatively minor role and where tissue injuries form the key point. I wonder whether this is a reasonable summary of the present situation?

My second question: when you gave doses of radiation which reduced the carcinogenicity of the hydrocarbons, did those doses produce detectable damage or not?

GLÜCKSMANN: Your summary is very reasonable. The doses of radiation we gave to the rat skin before DMBA caused depilation.

ALEXANDER: I think Dr. Hochman at Jerusalem reported that doses of 700 or 800 r of X-rays reduced the effect of DMBA and this dose induced no obvious skin damage.

MOLE: I want to object to Dr. Glücksmann's conclusion: if he uses only doses of radiation that produce severe damage, he cannot conclude that severe damage is necessary.

GLÜCKSMANN: That is obviously a good point, but we did try smaller doses and did not get any effect.

GRAY: It seems to me that it may not be correct to sum up this morning's discussion by saying that in order to produce tumours you must produce gross damage and that the tumour cells need not necessarily have been irradiated. I think Dr. Glücksmann said that the surest way of producing tumours is by way of gross damage and that very likely, the tumours arise from cells which have not been irradiated. But I would like to know what Glücksmann thinks about the type of experiment done by Bond and his collaborators, because I think we certainly ought to bear in mind a certain possible mechanism which might be distinct from the one you have been talking about. With the experiments of Shubik *et al.* and with those of Bond *et al.* we have definite evidence that you can get tumours at low doses and that they arise in irradiated cells. Just to keep the balance, I wonder if you would comment on the experiment. Am I wrong in my interpretation?

GLÜCKSMANN: I quite agree with you that there are different mechanisms of carcinogenesis by irradiation. The breast cancers, like the ovarian cancers, can be induced under certain hormonal and other systemic conditions by small doses of irradiation, and they arise in irradiated cells. These systemic conditions are difficult to analyse and I have concentrated on the mechanism of local effects of radiation in carcinogenesis. Here the conditions are different in that the primary carcinogenic effect of radiation is on the blood vessels. This is a somewhat specific effect of radiation as regards the progressive nature of the vascular injury and this cannot be induced even with repeated thermal burns, for instance. Secondary to this radiation-induced tissue disorder carcinogenesis may supervene in the regenerating tissue.

REFERENCES IN DISCUSSION

BURROWS, H., MAYNEORD, W. V., and ROBERTS, J. E. (1937). *Proc. roy. Soc. B.* **123**, 213.
HENSHAW, P. S., RILEY, E. F., and STAPLETON, G. E. (1947). *Radiology* **49**, 349.
HENSHAW, P. S., SNIDER, R. S., and RILEY, E. F. (1949). *Radiology* **52**, 401.
LACASSAGNE, A., and VINCENT, R. (1929). *C.R. Soc. Biol., Paris* **100**, 249.
LACASSAGNE, A., and VINCENT, R. (1929). *C.R. Soc. Biol., Paris* **100**, 249.
SHUBIK, P., GOLDFARB, A. R., RITCHIE, A. C., and LISCO, H. (1953). *Nature, Lond.* **171**, 934.

THE EFFECT OF X-IRRADIATION COMPARED TO AN APPARENTLY SPECIFIC EARLY EFFECT OF SKIN CARCINOGENS†

F. DEVIK

Statens Radiologisk-Cysiske Laboratorium, Montebello, Oslo, Norway

An early test for possible skin carcinogens has recently been described by Iversen (1961). He found good correlation between carcinogenic activity and the type of reaction that was recorded after one day by means of a tetrazolium reduction method (Iversen, 1959). In a blind test of twenty different compounds, the method singled out six carcinogens (e.g. 1.2,5.6-dibenzanthracene, 3.4-benzpyrene, and others) from the fourteen chemically-related but non-carcinogenic compounds. With the carcinogenic potentialities of X-irradiation in mind, it is of interest to compare the effect of X-irradiation with that of chemical carcinogens recorded by this method.

The tetrazolium method as applied by Iversen was used to estimate changes in the rate of formazan deposition in the epidermis of hairless mice, following application of different skin irritants. The skin was taken immediately after sacrificing the mice, and incubated for one hour at 37°C in a tetrazolium solution, with no substrate added. The reduction of tetrazolium salts by living cells is considered to be an indicator of dehydrogenase activities of the cells. The formazan formed is insoluble and is deposited in the cells at the site of reduction. The formazan is coloured, and the amount in the epidermis is measured by colorimetric methods.

The carcinogenic compounds produced an initial rise in formazan deposition on the first day, whereas the non-carcinogenic compounds showed a more or less pronounced depression at the same time (Fig. 1). Iversen has discussed the significance of the test, and considers that the increased deposition of formazan signifies a disturbance in the function of the mitochondria, with release of enzymes (Iversen and Evensen, 1962).

In collaborative work with Iversen, the effect of local X-irradiation on the epidermis of the same strain of mice has been tested in the same way as for the carcinogens, using X-ray doses ranging from 500 to 2,700 r (Iversen and Devik, 1962). A gradual increase in formazan deposition was observed, which was more pronounced the higher the dose, but more delayed in time when

† Work performed at Statens radiologisk-fysiske laboratorium, Montebello, Oslo, and Institutt for Generell og Eksperimentell Patologi, Rikshospitalet, Oslo.

following the application of the carcinogens (Fig. 2). On the other hand, there was no clear indication of any initial decrease, as with the non-carcinogenic compounds.

One main difference between X-radiation and toxic or pharmaceutical substances is the distribution in space: the former acts very uniformly at the

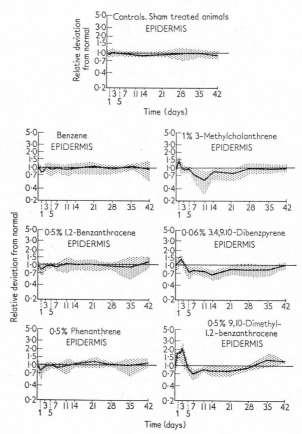

FIG. 1. Formazan deposition in the epidermis after application of three non-carcinogenic compounds (to the left), and three carcinogenic compounds (to the right), relative to the deposition in normal epidermis. Reproduced by kind permission of Dr. O. H. Iversen (Iversen and Evensen, 1962).

cellular level, whereas the latter are expected to show differences in concentration both at the cellular, and the subcellular, level. For this reason one should not expect too close a parallelism neither with respect to the acute nor to the late reactions and effects.

Without bringing out any similarity which is immediately striking, the experiments do indicate that a functional disturbance of mitochondria may be common to chemical carcinogens and to X-irradiation. Whether the disturbance is closely related to the mechanism of carcinogenesis remains to be determined.

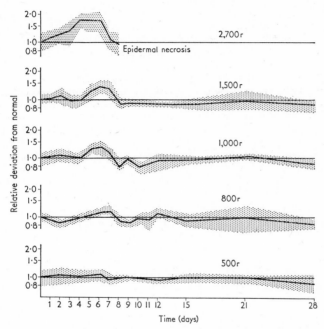

FIG. 2. Relative amounts of formazan deposited in epidermis after irradiation. (Reproduced by kind permission of the Editor of *International Journal of Radiation Biology*.)

REFERENCES

IVERSEN, O. H. (1959). *Acta path. microbiol. scand.* **47**, 216.
IVERSEN, O. H. (1961). *Nature, Lond.* **192**, 273.
IVERSEN, O. H., and DEVIK, F. (1962). In press, submitted to *Int. J. Rad. Biol.*
IVERSEN, O. H., and EVENSEN, A. (1962). *Acta path. microbiol. scand.* Suppl. 156.

DISCUSSION

GLÜCKSMANN: May I ask one question? In both cases you would get hyperkeratosis in the skin, from the carcinogens as well as from the radiation injury. Does your reaction simply mean that you have different types of cells present in the skin?

DEVIK: This I do not know, but this reaction is recorded very early, well before you get any keratosis. We can hardly differentiate histologically in our experiments between irradiated and non-irradiated skin at an early age.

BACQ: I wish to point out to Dr. Devik that Tabashnik from the U.S.A. has recently published a paper showing the liberation of RNase in the skin very early after irradiation of guinea-pigs with β-rays of ^{90}Sr. He points out that his test is very good because he has no changing population, he has no dead cells and no reproduction, so that he is constantly dealing with the same population of cells.

UPTON: May I ask, how sharply localized the change was? Did you find any diffusion of activity along the margin of the irradiated area?

DEVIK: The irradiation was very sharply localized. We only sampled the epidermis well within the irradiated field, we did not examine the margin.

UPTON: Due to the finding that tumour formation tends to occur at the edges, it would be be interesting to find out how sharply localized this change might be in relation to the margin of the irradiated area.

DEVIK: We have excluded so far the periphery of the field.

KOLLER: I would like to mention our experiments with skin autografts after total-body irradiation. In a control experiment we also transferred skin from an irradiated animal to these mice. And we were very impressed that within twenty-four hours we had such a thickening of the skin after irradiation. The effect is very dramatic and very quick.

DEVIK: What dose did you use?

KOLLER: The LD_{99}.

LONG-TERM CONSEQUENCES OF ^{90}Sr IN RATS AND THE PROBLEM OF CARCINOGENESIS[†]

K. SUNDARAM
Medical Division, Atomic Energy Establishment, Trombay, India

Among the fission products released by nuclear explosion, ^{90}Sr predominates over other long-lived radionuclides from the point of view of biological hazards, since it has the combined characteristics of a high fission yield, long physical and biological half-lives, a long residence time in the bones and a low maximum permissible level. ^{90}Sr being a moderately powerful β-emitter, is capable of irradiating a larger volume of bone tissue as well as the haematopoietic system. However, a factor of discrimination against strontium with respect to calcium is operative during the passage along the ecological series from soil to bone via the food chain.

In the present paper, the data reported by Casarett *et al.* (1961) on the long-term effects of ^{90}Sr in rats have been analysed both with regard to the fraction retained and the consequent biological effects. Table I summarizes the

TABLE I

Group	Number of animals	Age at start of experiment in days	Average age at death	Average days at risk	Total dose (μc)	Skeletal burden at 5 months (μc)
A. ^{90}Sr	64	*425*	805	380	330	1
Control	64	(425)	(747)			
B. ^{90}Sr	80	*346*	718	372	650	2
Control	80	(346)	(674)			
C. ^{90}Sr	80	*117*	471	254	790	11
Control	80		(718)			
D. ^{90}Sr	40	*40*	106	106	464	33
Control	40	(40)	(740)			

results with respect to the administered dose and the amount retained in the skeleton, five months after ten to thirty days of feeding of ^{90}Sr in the drinking water.

[†] Communicated by A. R. Gopal-Ayengar.

From the above it is seen that the skeletal burden of ^{90}Sr is a function of age of the animal at the beginning of the experiment. The animals in the youngest group retain thirty-three times more ^{90}Sr as compared to the oldest group. This is in conformity with the earlier findings of Finkel et al. (1960). In addition, the specific activity and the site of deposition of ^{90}Sr are dependent upon the metabolic state of the skeleton.

It is thus justifiable to assume a significant difference in both the dose-rate and the total dose received by the organ systems under consideration. These differences would be of the same order of magnitude as the skeletal burdens in the four age groups.

The histological changes in the haematopoietic tissue contained in the femur of these animals further substantiate these assumptions. The bone-marrow cells in the youngest group of animals with a skeletal burden of 33 μc showed marked atrophy (Fig. 1). In animals belonging to group C with a skeletal burden of 11 μc, the damage was moderate (Fig. 2). In the oldest group of animals with skeletal burdens of 2 or 1 μc of ^{90}Sr, the cellularity of the bone-marrow did not show any significant alteration. These variations in the degree of damage are reflected in the frequency of leukaemia, or osteogenic sarcoma, in the four groups. Table II gives the incidence of leukaemia and osteogenic sarcoma in these various groups.

TABLE II

Experiment	Skeletal burden (μc)	Period of observation (weeks)	Number of rats	Osteogenic sarcoma per rat week	Leukaemia per rat week
A	1	54·3	64	0	0
B	2	53·1	80	0	0·0007
C	11	36·3	80	0·0076	0·0017
D	33	15·1	40	0·023	0

The lowest skeletal burden of ^{90}Sr to cause leukaemia, is observed in group B, suggesting that the dose-rate and the total dose received by the bone-marrow in this group may be regarded as the minimum effective dose for the induction of leukaemia in these experiments. In the group A with 1 μc of ^{90}Sr as the skeletal burden, the absence of leukaemia during the entire life-span of the animals, which incidentally was the same as the controls, would indicate that the potential hazard is insignificant if the organism is sufficiently old at the beginning of exposure. In group C with a higher skeletal burden (11 μc) and a moderate reduction in the life-span, the incidence of leukaemia per rat week of observation shows an approximately 2·5 times increase over that of group B. With the highest skeletal burden (33 μc) in group D and with marked reduction in life-span, no leukaemia was observed.

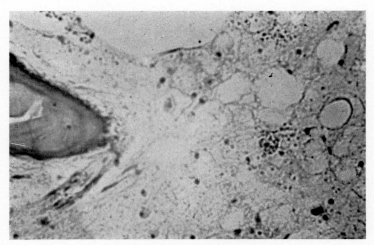

FIG. 1. Transverse section through the upper end of femur from a rat with 33 μc of ^{90}Sr in the skeleton. The bone-marrow shows marked degree of atrophy.

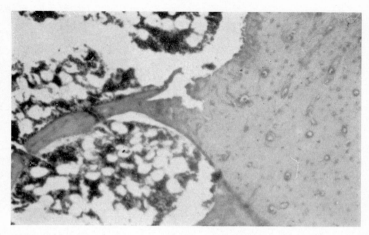

Fig. 2. Similar section from a rat with 11 μc of ^{90}Sr in the skeleton. The bone-marrow shows moderate cellularity.

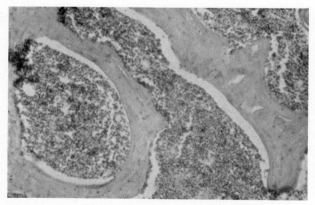

Fig. 3. Section from a rat with 2 μc of ^{90}Sr in the skeleton. The bone-marrow is essentially normal.

Among the delayed effects of ionizing radiations on mammals the induction of leukaemia is well recognized. While the precise processes leading to leukaemogenesis are still not clear, we may recognize two components in the action of ionizing radiations at the cellular level; the probability of tumour induction and a cytotoxic effect. The latter effect may be mediated by primary injury intrinsic to the cell or by extrinsic factors such as the impairment of cell nutrition. The production of a neoplasm may then be regarded as a balance between these two components. There are indications that the probability of induction of leukaemia is dose-dependent, to a limit which is determined by the preponderance of cytotoxic effect. In other words the chances of inducing leukaemia depend upon the extent of injury to the haematopoietic system and the percentage of such altered cells surviving and capable of multiplication. It would appear from the data that such a situation does exist, inasmuch as the incidence of leukaemia per rat week of observation shows a 2·5 times increase between groups B and C. In both these groups the histological study of cytotoxic effect on the bone-marrow indicated only mild or moderate effects. However, in group D the bone-marrow was completely atrophic, suggesting a profound cytotoxic effect with hardly any haematopoietic cells capable of survival or multiplication to cause the appearance of leukaemia.

In contrast to the induction of leukaemia the incidence of osteogeneic sarcoma shows some interesting variations. This may be explained by comparing the relative radiation sensitivity of the osteoblasts and the cells of the haematopoietic system. It is well recognized that the osteoblasts are relatively less radiosensitive than the cells of the blood-forming organs. Table II shows that the lowest level of skeletal burden of ^{90}Sr to cause osteogenic sarcoma is in group C. In group D the incidence rate is increased by about 2·7 times. This indicates that the minimum dose to cause osteogenic sarcoma is different from that required for leukaemogenesis and the limiting dose for the osteogenic sarcoma is not attained even in animals with as high a skeletal burden of 33 μc of ^{90}Sr. This presumably suggests that those cell types which are less radiosensitive require a higher dose for the induction of neoplastic change and the limiting dose is significantly increased, since the latter is determined by the dose which has the highest cytotoxic effect on the particular cell system under consideration.

By comparing the incidence rate of leukaemia and osteogenic sarcoma in groups B, C and D, there is an indication that at low levels of skeletal burden of ^{90}Sr, the primary delayed somatic effect would be leukaemia, at intermediate levels both leukaemia and osteogenic sarcoma and at high levels only osteogenic sarcoma provided the animals survive a sufficient length of time before pancytopenia or other causes of death supervene. It is however necessary to exercise caution in regard to these observations, since the number of animals in each group is relatively too small to warrant statistical evaluation. Again the non-uniformity of distribution of ^{90}Sr in the bone, makes it equally difficult to compute the dose-rate or the total dose received by the cells under consideration. It would be worthwhile to undertake further investigations to validate these conclusions, by determining the type of neoplastic changes in amimals of identical age groups with different levels of skeletal ^{90}Sr.

ACKNOWLEDGMENT

I wish to acknowledge my sincere thanks to Dr. George Casarett, Radiation Biology, Atomic Energy Project, University of Rochester, Rochester, N.Y. for making available to me the histological material and other facilities for this report.

REFERENCES

CASARETT, G. W., TUTTLE, L. W., and BAXTER, R. C. (1961). University of Rochester, AEC Document UR-597.
FINKEL, M. P., BERGSTRAND, B. S., and BISKIS, B. O. (1960). *Radiology* **74**, 458.

DISCUSSION

ALEXANDER: Do you think that these conclusions have any significance from the statistical point of view? By rough arithmetic you get in 80 rats by fifty weeks two to three cases of leukaemia in the whole experiment. Is two to three significantly different from zero?
GOPAL-AYENGAR: Perhaps Dr. Casarett might be able to answer the question of actual number of cases.
ALEXANDER: Well, the numbers involved must have been so small that none of the figures can be significant. Is there any difference between 0·0017 and 0·0007?
CASARETT: As I recall there were between five and ten cases of leukaemias spread between the B and C groups and leukaemia is a very rare disease in rats more so than in man. Our experience is that there is a significant increase in the incidence of rat leukaemia.
MOLE: One ought to pursue this point because even if there were five to ten cases of leukaemia you still cannot say that there is a bigger incidence in one group or another, because it can still occur by chance.
CASARETT: I am sure that you are correct Dr. Mole in the statistical sense, unless one takes into account the rarity in the larger number of control animals studied in these experiments. A, B, C and D are the groups in each of the treated and control groups. For osteogenic sarcoma both in the treated and controls in A and B groups it was 0%, 27·5% in the treated C group with a body burden of 11 μc and 17·9% in the D group. Skin carcinomas of the face occurred at the same dose level as that which produced an increased incidence of leukaemia—7·5% in the B group and 11·25% in the C group and none in the A and D groups. The incidence of leukaemia was 0% in both the treated and

controls in the A group, 3·75% in the treated B group and 6·25% in the treated C group. The body burdens of ^{90}Sr in A, B, C and D groups were 1 μc, 2 μc, 11 μc and 33 μc respectively.

UPTON: What sort of leukaemia was this?

CASARETT: Lymphatic.

MOLE: So this is really three cases of leukaemia in group B and five cases of leukaemia in group C. Three out of eighty or five out of eighty are different. Three out of eighty and nothing out of eighty are not different. Do you think that groups B and C are really different?

CASARETT: I do not insist on that myself. I do feel convinced that leukaemia was produced by the treatment but the difference between B and C to me is questionable.

POCHIN: Would you consider A, B, C and D are different when you have χ^2 of 8·5?

CASARETT: As I indicated, my conviction that this treatment has produced leukaemia is based not solely on the number of animals in this experiment but on knowledge that in a careful study, haematologically and pathologically, of all animals that died in our colony, we found leukaemia very rarely, extremely rarely.

POCHIN: I was not questioning the evidence but merely whether it is different between the four-dose levels.

CASARETT: This I do not yet know. These are four experiments of a large group in which different age, different feeding doses and different body burdens are being investigated.

LAMERTON: I take issue with Dr. Mole when he says that there is no difference between zero out of eighty and three out of eighty; there is in fact quite a difference. I think we have to recognize particularly in carcinogenesis when one does not use large numbers of animals that one has sometimes to accept results which are statistically significant at the 5% level. We are far too inclined to say that a thing is not different because it is not a 20 to 1 chance. May I make a second point about something Dr. Gopal-Ayengar said. He was comparing bone tumour and leukaemias incidence. In point of fact there is probably quite a difference in the radiation dose received by some of the osteogenic cells and by the bone-marrow cells. Even with ^{90}Sr in the mouse there may be a difference of 3 to 1 or 5 to 1 if, for example one considers the osteogenic cells at the face of the plate, you are in fact giving osteogenic cells quite a large dose in some cases. And secondly, you said that the osteoblast was much more radioresistant than the blood-forming cells. What do you really mean by this?

GOPAL-AYENGAR: I mean that the ostegoenic cells are relatively less sensitive than the blood-forming cells.

UPTON: This experiment has raised several interesting points. First of all, squamous cell carcinoma. I would like to enquire whether there was ulceration of the skin or any noticeable epithelial breakdown in the skin. Then the other point is that in a variety of experiments, with internal emitters, not necessarily with ^{32}P but ^{89}Sr or ^{90}Sr leukaemias have appeared in increased numbers in rats and in mice, despite the observation that whole-body radiation is much more effective, if not actually necessary, to produce this kind of leukaemia. In view of the dose-rate dependency, admittedly complex as Dr. Mole said yesterday, I think this is a very interesting experimental finding. In the work we did with RF mice we thought that we were much more likely to induce myeloid leukaemias with internally administered ^{89}Sr or ^{90}Sr. To our great surprise, we did not affect the myeloid leukaemia incidence at all with the doses used, but we did increase the incidence of thymic lymphoid leukemias.

MOLE: I would like to pursue something Dr. Upton said. Although the animals were grouped according to the terminal body burden, they were in fact given a great deal more strontium in their drinking water, 10 or 100 times more, and presumably a good deal of that got into the body and irradiated it before being lost in the urine and in the faeces. So these animals were getting quite a large dose of whole-body radiation if they were given half a millicurie or something like that. Further because it was in the food and because the turn-over time of strontium is very rapid, they were, in fact given a set of pulsed doses of whole-body radiation. The doses varied during the day so that it could easily be that one could have a kind of leukaemogenic whole-body irradiation from giving animals ^{90}Sr.

CASARETT: There was ample opportunity for lymphatic tissue to be irradiated. In answer to your question, Dr. Upton, there was ulceration either before or after tumour development in these face tumours. As to the figures in the table, the survival time at the high-dose level was generally much shorter than the average induction time for leukaemia, following the treatment with strontium. I think, in these cases when one speaks of increased incidence at the intermediate level and relative reduction at the high level, we have consistently a picture where the survival time is too short for the observed induction time of the tumours involved. At the lowest level apparently the dose is insufficient to produce an increased incidence of malignancy.

THE INFLUENCE OF RADIATION ON THE SURVIVAL OF MICE INJECTED WITH ASCITES TUMOUR CELLS

V. DRÁŠIL

Institute of Biophysics, Czechoslovak Academy of Sciences, Brno, Czechoslovakia

If Ehrlich ascites tumour cells (EAT) are injected otherwise than intraperitoneally, there arise solid tumours, the localization of which depends on the manner of injection. After intravenous injections of EAT, the tumours develop mostly in the lungs, less frequently in the mesentery, liver and other organs.

In this communication we present the results of preliminary experiments aimed at finding an answer to the question of whether the irradiation of the animal prior to injecting it with EAT can influence the resulting formation and growth of tumours.

The diploid form of EAT grows in mice of all strains hitherto tested in this respect (CBA, C3H, C57Bl, ASW). Non-irradiated mice (three-month-old females) die after an injection of 10 million cells, their mean survival-time being about thirty days. The mean survival-time depends on the number of injected cells; when 1 million cells are injected, it amounts to forty-five days. Tumours are found uniformly in all experimental animals after an intraperitoneal injection. It appears that the injected cells produce a very weak homologous reaction. The EAT for our experiments with intravenous injections was obtained from the peritoneal cavity of mice through which the tumour is passed. The suspension for intravenous injections was diluted at least ten times and injected very slowly in an amount of 0·8 ml.

The experimental mice were divided into three groups, viz. non-irradiated, irradiated one day prior to injection with EAT, and irradiated thirty days before being injected. The mice were given doses of 20 r, 60 r, and 180 r (180 kV, 0·5 mm Cu, 1·0 mm Al). The results of one typical experimental series are shown in Fig. 1. From these results it follows that, after an injection of EAT, irradiated mice die much faster than non-irradiated ones. Upon repeating the experiments we have found a good reproduceability after doses of 60 r and 180 r. In mice irradiated with 20 r, the reduction of the mean survival time after an injection of EAT lies on the borderline of statistical significance.

From the results of these experiments it follows that even small doses of radiation can exert such an influence on mice that from the EAT injected into them intravenously tumours are formed at a faster rate, and consequently, the mice die after a shorter period. It is rather surprising in this case that the influence of radiation can be demonstrated even on mice irradiated thirty days prior to being injected with EAT.

For the time being, only hypotheses may be expressed about the mechanism of the influence of irradiation described above. In the first place we might consider the reduction of the immunity reaction as a result of irradiation. However, from the work of Ilbery (Ilbery et al., 1958) and those of other authors dealing with immunity reactions against tumour cells after irradiation

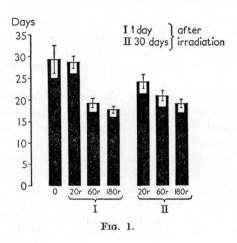

Fig. 1.

it follows that the ability to form antibodies reverts to normal one month following irradiation even after doses of radiation higher than 200 r. This must occur more quickly after doses of 180 r, 60 r or 20 r. Consequently, we presume that the suppression of the immunological activity will not be the reason why irradiated mice die faster after being injected with EAT.

When "metastases" are formed after an intravenous injection of EAT, the state of the capillary vessels, in which the cells are intercepted and the walls of which they destroy in the course of their further growth, most probably also plays an important role. It is likely that irradiation may result in some damage in the wall of the capillaries, or that it may produce in the adjoining connective tissue some changes that promote the formation of the tumour, and its further growth.

REFERENCE

ILBERY, P. L. T., KOLLER, P. C., and LOUTIT, J. F. (1958). *J. nat. Cancer Inst.* **20** 1051.

DISCUSSION

SCOTT: Have you performed some quantitative immunological tests?

DRÁŠIL: These are preliminary results. Quantitative tests for DNA synthesis and histological tests are in progress. We have not yet performed any immunological tests.

SCOTT: Could I suggest that you might try a very simple test between these two systems? That is pre-immunizing the animals with irradiated cells and then doing quantitative titration of the number of viable cells, which would produce a tumour in a pre-immunized animal. It is a very simple test and you can get a clear-cut answer.

DRÁŠIL: Yes, I agree, we will perform this test in further experiments.

KOLLER: One comment on the same question as Dr. Scott has asked about testing quantitatively. Now first of all, what about the strain of mice you used? Did you reduce the dose and give intravenously 10^6 cells, and did you try and give much less than this, to come down and see at what level the mice react and do not give a tumour at all? In this way I think you can really prove that you are dealing with the suppression of an immune mechanism when you give total-body irradiation.

DRÁŠIL: We have performed experiments with different cell numbers in non-irradiated mice only. We studied the dependence of survival on the number of cells injected. We found that when these cells were injected intravenously, after 10^6 cells the survival time increased about 50% and after 10^5 cells injected intravenously the survival-time was greater than 120 days. When the cells are injected intraperitoneally, about 10^2 to 5×10^2 cells are enough to kill the animal. As to the strain of mice, we used our laboratory strain H and C57BL. The tumour grows also in C3H and CBA and other strains of mice.

POCHIN: Was it possible to obtain any evidence as to whether the earlier death was actually due to more rapid tumour growth or could it have been due to other causes?

DRÁŠIL: This is a question for further quantitative tests but the post-mortem examinations showed that after death these irradiated mice had about the same organ distribution of tumours as the non-irradiated ones.* The size of tumours is at the time of death the same but it seems that tumours in irradiated mice develop faster, and in a greater number, then in non-irradiated mice.

Note added in proof: Further, more detailed experiments showed an increased percentage of liver metastases in irradiated animals.

RADIATION-INDUCED BONE TUMOURS—FRACTIONATION STUDIES

J. P. M. BENSTED, N. M. BLACKETT, V. D. COURTENAY, AND
L. F. LAMERTON*

Physics Department, Institute of Cancer Research, Royal Cancer Hospital, London, England

INTRODUCTION

The finding that the incidence of radiation-induced thymic lymphoma in mice was dependent on the fractionation of the exposure, for the same total dose, opened up a fruitful field for the study of the mechanisms of lymphoma induction. If a similar dependence on fractionation or protraction of exposure could be shown for other types of radiation-induced tumour, a much more direct attack on mechanisms would be possible than has hitherto been provided from studies of dose-incidence relationships. It is probably true to say that, apart from a few special cases (e.g. when linearity is demonstrated), a dose-incidence curve without additional data can give little information about the mechanisms involved.

Only a few fractionation studies have been reported for radiation-induced bone tumours and most of these have been made with bone-seeking isotopes of relatively long half-life and with short intervals between successive injections. Under these circumstances the radiation exposure conditions in the bone may not be very different from those with single injections. This could explain why little effect of fractionation was found by Finkel (1959) using repeated ^{90}Sr injections in mice, or by Kuzma and Zander (1957), using daily injections of ^{45}Ca. If bone-seeking isotopes are to be used for fractionation studies, the interval between injections should be of the same order as, or greater than, the half-life of the isotope in order to produce sufficient change in the pattern of radiation dose and dose-rate to the bone. For this purpose a suitable bone-seeking isotope would be ^{32}P although, as in all studies with bone-seeking isotopes, it must be recognized that growth and remodelling of the bone will complicate the radiation dose pattern within the bone from successive injections.

In this paper a brief resumé will be given of the effect of fractionation on bone-tumour production in young rats using (a) ^{32}P, a high energy β-emitter, with a relatively short half-life (14 days), (b) ^{239}Pu, an α-emitter with a long half-life ($>$ 20,000 years), and (c) localized external radiation with X-rays.

* Read by L. F. Lamerton.

MATERIALS AND METHODS

In these studies male F1 hybrid rats (aged 6 to 8 weeks) of the inbred August and Marshall strains have been used. The ^{32}P was given as intraperitoneal injections of sodium phosphate in saline with a specific activity between 1 and 4 µc per µg P. In the external radiation experiments one hind limb was irradiated, using 140 kVp radiation, half value layer 0·3 mm Cu, at a doserate of about 50 r/min, and one experiment where both hind limbs were irradiated at 230 kVp, half value layer 1·5 mm Cu, at 20 r/min. In the ^{239}Pu experiments the injection was of $Pu(NO_3)_4$ in nitric acid, which was brought to about pH 5 with sodium carbonate solution immediately before injection.

No spontaneous bone tumours have, so far, been observed in unirradiated rats of the hybrid or parent strains.

In the data to be presented there was, in some groups, an appreciable mortality following irradiation from causes other than bone tumours. Since the numbers of animals used are not sufficient to make a meaningful correction for this, both the time of appearance of bone tumours and mortality from other causes (including non-skeletal tumours) are given.

^{32}P EXPERIMENTS

The ^{32}P experiments have already been reported in some detail (Bensted et al., 1961; Blackett, 1962). A very marked difference in the time of appearance of tumours was found when repeated injections of ^{32}P were given. Figure 1 shows the mortality, with time of appearance of bone tumours (in groups of about twenty animals) for a single injection of 3·0 µc per g body weight, and also for a total injection of 2·8 µc/g given as a single injection of 1·0 µc/g followed by three injections of 0·6 µc/g at two-weekly intervals. It can be seen that with the single injection the first bone tumour appeared at eight months after injection, the other five tumours in this group appearing between eleven and thirteen months. With repeated injections the first tumour appeared six months after the first injection and a further fifteen tumours by ten months, without any deaths from other causes during this period.

Earlier work (Blackett, 1962) had suggested that the time of appearance of tumours was dependent on the precise timing of the repeated injections, and in particular on whether the interval between the first and second injection was two, four or six weeks. Repetition of these experiments has given inconsistent results and the effect of the timing must now be regarded as in doubt. However, a marked difference between the effect of single and repeated injections has been consistently demonstrated.

The bone tumours have been found most frequently in the tibia and femur and less frequently in the spine and humerus, and only very occasionally elsewhere in the skeleton.

In growing bone, the epiphyseal plate moves away from the deposit of ^{32}P as the bone lengthens, with the result that the isotope deposit and concomitant bone damage appears to move down through the metaphysis and eventually becomes a part of the diaphysis. With repeated injections the continued growth of bone leads to a new band of maximum uptake in the newly-formed bone just below the epiphyseal plate, after each injection. The volume of bone irradiated is therefore greater than with a single injection.

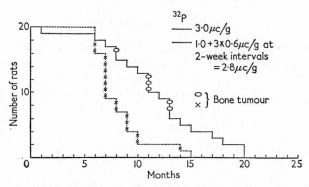

FIG. 1. Comparison of bone-tumour incidence with single (3·0 μc/g) and repeated (1·0 + 3 × 0·6 μc/g) injections of ^{32}P.

Histological studies show that, with repeated injections, obvious damage does extend over a greater area and is of a somewhat different character. In particular, there is a greater degree of peritrabecular fibrosis, which in the past has been suggested by some workers as an important factor in bone-tumour production.

Another factor which might increase the relative carcinogenic efficiency of the repeated injections is that the mean radiation dose to the cells in the irradiated volume is less than with a single injection, permitting a greater survival of potentially malignant cells. On the other hand there may be some groups of cells which receive a *higher* radiation dose with repeated injections. In support of this suggestion there is evidence of a migration of cells up the trabeculae, following the retreating epiphyseal plate (Kember, 1960), and such cells could be in the region of maximum radiation dose for each of the repeated injections. For 3·0 μc/g total injection, radiation dose measurements in bone (Blackett, 1962) show that such cells would receive an accumulated dose of about 11,000 r with repeated injections, compared with 6,000 r for a single injection. On the other hand cells just below the plate at the first injection and not subsequently moving up would receive an accumulated dose of 9,000 r for the repeated injections and 10,000 r for the single injection.

There are still other factors which may be of importance in determining the difference in response between single and repeated injections, for instance the change in radiation dose-rate and the change in type and number of cells irradiated due to the changing cell population in the bone. Because of the considerable difference in dosimetry between repeated and single injections it is bound to be difficult to pin-point the factor or factors responsible for the difference in tumour response.

^{239}Pu EXPERIMENTS

A long lived bone-seeking α-emitting isotope such as ^{239}Pu will give a very different pattern of radiation dose distribution in bone from that given by ^{32}P (Taylor et al., 1961). It is of interest to study the effect of fractionation of the injection and Fig. 2 shows a comparison of mortality and bone-tumour

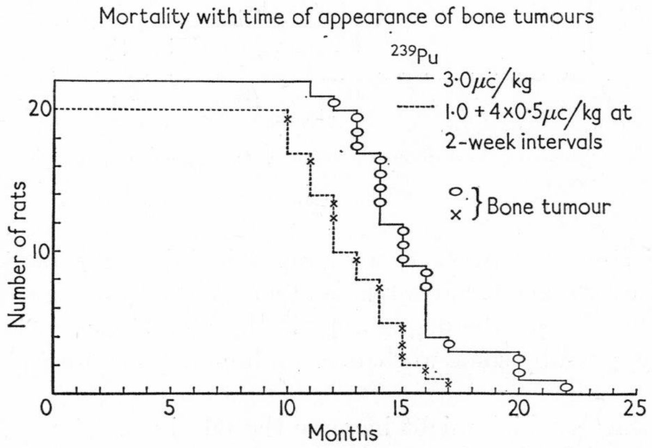

Fig. 2. Comparison of bone-tumour incidence with single (3·0 μc/kg) and repeated (1·0 + 4 × 0·5 μc/kg) injections of ^{239}Pu.

incidence for a single dose of 2·9 μc/kg and for an injection of 1 μc/kg followed by four two-weekly injections of 0·5 μc/kg. About twenty animals were used in each experiment and it can be seen that the final incidence of bone tumours was high in both groups. With a single injection, 17 out of 22 animals developed bone tumours between twelve and twenty-two months. With repeated injections 11 out of 20 animals developed tumours between ten and seventeen months. However, the greater mortality from other causes in the repeated injection group prevents one from drawing firm conclusions about a difference in response, although there is some indication that the bone tumours appear slightly earlier with the repeated injections.

Fibrosis was not a constant finding in the bones of these rats and, when observed, it consisted of no more than a localized area of rather loose connective tissue in the marrow associated with some surrounding marrow aplasia. There was little evidence of the dense fibrotic change described by some other workers (Lisco, 1956; Jee et al., 1957) and sometimes assumed to be an important factor in bone tumour production by α-emitters. There was no apparent difference in the incidence and distribution of the fibrosis in the rats given either single or repeated doses of ^{239}Pu.

With the α-emitting bone-seekers, tumours are frequently found in sites other than the femur and tibia. In the present experiments, of the twenty tumours (in 11 tumour-bearing rats) following repeated injections, twelve were in the femur and tibia and three in the spine, with five in other bones. Following single injections, of the twenty-five tumours (in 17 tumour-bearing rats) eleven were in the femur and tibia, seven in the spine and seven in other bones.

With regard to non-skeletal tumours there was one case of myeloid leukaemia following the single injection and three cases following the repeated injections. Spontaneous cases of myeloid leukaemia have not been observed hitherto in our rat colony.

EXTERNAL RADIATION EXPERIMENTS

Experiments with localized external radiation involve fewer problems of dosimetry than those with the bone-seeking isotopes, since the dose distribution is more uniform. Figure 3 compares the mortality and bone-tumour incidence for (a) a single dose of 3,000 r to one hind limb and (b) three doses of 1,000 r at two-weekly intervals to one hind limb. 140 kVp radiation was used in both cases.

In the single dose experiment there were a number of early deaths, but of the thirty-four animals surviving six months, 12 developed bone tumours between six and twenty months. In the repeated dose experiment only one bone tumour developed (at seventeen months) from 15 animals surviving six months.

In this experiment a considerable difference was observed in the degree of histological damage between the two groups. With a single dose of 3,000 r about one-half of the bones examined showed obvious peritrabecular fibrosis, especially in the early post-irradiation period, and more than three-quarters showed considerable abnormality in the epiphyseal plate. With the repeated 3 × 1,000 r treatment there was little evidence of peri-trabecular fibrosis and only about one-half of the animals showed epiphyseal plate abnormality.

This finding suggests that gross histological damage exemplified by fibrosis and epiphyseal plate damage could be an important factor in the production of bone tumours by radiation. However, some other experiments

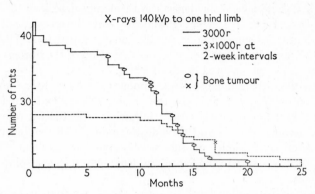

FIG. 3. Comparison of bone-tumour incidence with single (3,000 r) and repeated (3 × 1,000 r) doses of 140 kVp X-radiation to one hind limb.

have not confirmed this suggestion. Figure 4 shows results for localized irradiation with 230 kVp radiation to both hind limbs where six doses of 500 r were given at 0, 1, 4, 5, 6 and 7 weeks. Here 7 out of 17 animals (that is,

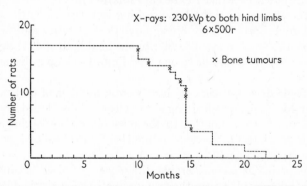

FIG. 4. Bone-tumour incidence with repeated (6 × 500 r) doses of 230 kVp X-radiation to one hind limb.

7 out of 34 irradiated hind limbs) developed bone tumours between ten and fifteen months. This incidence is more than half that observed after a single dose of 3,000 r at 140 kVp, yet there was very little evidence indeed, in the

6 × 500 r experiment, of peri-trabecular fibrosis and only a slight incidence of obvious plate abnormality.

The different quality of radiation in these two latter experiments could considerably affect the pattern of radiation dose distribution in the bone so that, apart from the small numbers of animals used and the consequent large statistical errors, it is difficult to draw conclusions from these data about the effect of fractionation.

However, in the results of Figs. 3 or in the comparisons of Figs. 3 and 4 there is no evidence that, under the conditions of these experiments, fractionation of the dose increased the tumour incidence. The present evidence suggests, in fact, a decrease.

DISCUSSION

With ^{32}P there is clear evidence of an increase in bone-tumour incidence and shortening of tumour development time with fractionation of the dose. With ^{239}Pu no very marked difference was observed, though there was some indication of earlier development of tumours with fractionation. With localized external radiation to the hind limb there was certainly no increase in tumour incidence with fractionation, but possibly a decrease.

These experiments do not therefore establish any common pattern of response which could be used to elucidate mechanisms. In the case of ^{32}P it must be recognized that the radiation dosage pattern within the bone is complex and one or more of a number of factors might be involved in the difference of response between single and fractionated treatments.

The suggestion that the greater tumour incidence with the ^{32}P fractionation treatment was because fractionation permitted a greater survival of potentially malignant cells does not appear to be supported by the X-ray results. However, the dose of 1,000 r used in the X-ray fractionation experiments will produce much cell death and it would be of great interest to study bone-tumour production with external radiation using a larger number of smaller fractions.

With ^{239}Pu the effect of fractionation of treatment on cell survival would be expected to be less than with low LET radiation because of the severe damage caused by the passage of a single α-particle through a cell.

Another possible explanation for the high carcinogenic efficiency of repeated ^{32}P treatments and of ^{239}Pu, both single and repeated, is that the radiation exposure is protracted enough to involve the irradiation of regenerating tissue. In the fractionated treatment with external radiation (3 × 1,000 r) the interval between first and last treatments may not have been sufficiently long. It is possible that the high incidence of bone tumours at 6 × 500 r is the result of a lengthened period of exposure. Again, external radiation studies with a larger number of smaller fractions, together with histological studies of regeneration, would be useful.

The present experiments have not been of a large enough size to determine whether there is any significant difference in favoured sites of tumour production in bone between the single and fractionated treatments with the same radiation source. Anatomical and metabolic differences between different sites are probably of importance in determining the carcinogenic efficiency of a given radiation dose distribution and ideally comparisons should only be made for tumours arising at the same site. If subsequent work were to show significant differences in tumour sites between single and fractionated treatments the conclusions drawn here regarding the effects of fractionation might have to be revised.

Work with bone tumours has the advantage that, using the bone-seeking isotopes, high tumour yields can be obtained but the heterogeneous nature of bone and the changes with age of histological and metabolic pattern greatly complicate the interpretation of experimental data. It may be possible to design experiments on the mechanisms of bone-tumour production by radiation which will avoid many of these difficulties. Besides fractionation, the quality of the radiation is a factor that can be altered. A study of the variation of the comparative effectiveness of an α- and a β-emitting bone-seeker over a wide range of injected amounts could be a very useful approach. Some comparative data are already available from the work of Finkel (1959) and studies are in progress in other laboratories, but for the purpose of studying mechanisms it would be necessary to extend to low levels of isotope concentration, which would require the use of very large groups of animals.

ACKNOWLEDGMENTS

The authors wish to express their appreciation to many colleagues for help and advice and particularly to Dr. D. M. Taylor, to Miss K. Adams, Mr. J. Blackmore, Mr. E. J. Perry, Miss M. Shreeve and Mr. L. Smith and to Professor W. V. Mayneord, Director of the Physics Department, Institute of Cancer Research, Royal Cancer Hospital.

REFERENCES

BENSTED, J. P. M., BLACKETT, N. M., and LAMERTON, L. F. (1961). *Brit. J. Radiol.* **34**, 399.
BLACKETT, N. M. (1962). Council for International Organisation of Sciences Symposium "Radioactive Isotopes and Bone" 1960. Blackwell Scientific Publications (in press).
FINKEL, M. P. (1959). "Radiation Biology and Cancer", p. 322. University of Texas Press.
JEE, W. S. S., OTTOSEN, P., MICAL, R., and LOWE, M. (1957). Annual Progress Report Radiobiology Laboratory, University of Utah College of Medicine, Salt Lake City. Utah. p. 86.

KEMBER, N. F. (1960). *J. Bone Joint Surg.* **42B**, 824.
KUZMA, J. F., and ZANDER, G. E. (1957). *Arch. Path.* **63**, 198.
LISCO, H. (1956). CIBA Foundation Symposium on Bone Structure and Metabolism, p. 272.
TAYLOR, D. M., SOWBY, F. D., and KEMBER, N. F. (1961). *Phys. Med. Biol.* **6**, 73.

DISCUSSION

BENSTED: I should very much like to support these ideas but also suggest that we pay more attention to the idea that some of these results may be explained in terms of the greater volume of bone tissue which is being irradiated as a result of fractionating a dose of ^{32}P. I feel that some further investigations on these lines might be fruitful.

CURTIS: I want to enquire about the experiments with ^{32}P. You mentioned that as you gave a single dose, the metaphyseal region is irradiated and with the next dose you irradiate a different region and so on. But is it not true that as the plate moves up, it does so by the division of cells which are lower down? So that the second dose I would say, is being given to the progeny of cells which have originally been irradiated and so on. Is this right?

LAMERTON: This may indeed be true. In fact, it is very difficult to talk about the actual radiation dose given to any population of cells since there is migration of cells within the tissue certainly as far as bone is concerned.

CURTIS: So that the two-event phenomenon we spoke of yesterday would certainly hold for the ^{32}P experiments if not for the X-ray experiments?

LAMERTON: Yes, this is one explanation of the ^{32}P experiments. But we should have liked to have found the difference in response with the fractionation of the external radiation but this did not turn out to be so.

GLÜCKSMANN: With regard to the significance of damage, I think you ought to distinguish between the types of damage present. Have you looked for vascular damage? This type of damage may be crucial.

BENSTED: We have thought about vascular damage as a strong possibility in the phosphorus experiments but we haven't done the appropriate experiments. In this context I might perhaps quote the experiments of Jee and his co-workers at Utah on dogs with thorium, plutonium and radium burdens. They made a study of the micro-circulation in the bones of these dogs by the technique of India Ink injection. As a result they were able to show that there was impairment of the circulation to the Haversian system. They have some rather pretty pictures of the trabecular surface of the bone separated from the capillary by quite a thick layer of fibrosis. They concluded that the death of osteocytes was certainly due to the disturbance in the vascular supply. This disturbance in the local environment might well also influence the behaviour of the connective tissue in this region.

MULLER: To add to what Dr. Curtis had to say, I think one might raise another question in connection with fractionation and that is whether the repeated dosing would not result in a wider distribution of the damage, that is to say the cells at the end would be cells which had received the whole dose of repeated applications either in their ancestral cells or in themselves, whereas they would have left behind a trail of other cells deeper down which had also received a good deal more than which the cells in the corresponding position would have received as a result of a single dose.

LAMERTON: I think this is quite possible, Dr. Muller. The trouble is, we know so little about the migration of cells in a growing bone. Dr. Kember has done some work with tritiated thymidine which shows it's not a simple problem. There are some cells which definitely crawl up the trabeculae as the plate retreats and these may receive nearly the whole dose from repeated injections.

BERENBLUM: In the case of the repeated dose experiments with ^{32}P what was the interval between the three doses?

LAMERTON: The interval between doses was two weeks so that the animals given these doses were subjected to irradiation for at least four weeks longer than those given a single dose.

BERENBLUM: Well, what I had in mind was, supposing that the length of time necessary before you got your irreversible change was, for the sake of argument, twelve weeks, after which time the carcinogenic process was more or less committed, then in that case a period of four weeks would be quite a significant proportion of the twelve weeks. I don't know whether that might or might not influence your interpretation.

LAMERTON: One would have expected to have obtained a similar difference with repeated doses of X-rays if such a mechanism were operative.

COTTIER: Speaking of bone tumours induced by ionizing radiation, I should like to draw your attention to the possible importance of some endocrine factors. In our strain of mice we observed after whole-body irradiation, at the age of 3 months with 450 to 600 r, osteogenic sarcomas only in females and the important thing was that this strain also developed a high percentage of ovarian tumours after the irradiation. They also developed an ossification of the bone-marrow space, internal hyperostosis and only in such cases did we find osteogenic sarcomas in a significantly different percentage, that is 9 in the females and none in the males. Now I should like to ask if anybody working with bone tumours has observed sex differences such as this?

LAMERTON: Our own work has been primarily with males but there were some observations of Pybus and Miller some time ago when their Newcastle strain of mice suddenly developed a high spontaneous incidence of bone tumours. These began to appear at about 6 months of age and if I remember rightly, the incidence in females was something like 65% and in males about 25%. So in this case spontaneous bone tumours showed a considerable sex difference. Miriam Finkel has worked both with male and female animals, I believe, in her experiments with the bone-seeking isotopes but she has not found any considerable difference between them as far as I know.

ALEXANDER: My comment is really rather similar to Dr. Cottier's. I wondered whether the difference between ^{32}P and ^{90}Sr couldn't be due to an abscopal effect? With ^{32}P the whole-body will be irradiated and if, in fact, you give fractionated doses you are giving continuous whole-body radiation. If continuous whole-body irradiation is producing a continual hormonal stimulus this might then result in the much higher incidence of tumours. Whereas with ^{90}Sr the fractionation would not give you a uniform whole-body irradiation and you wouldn't get this continual stimulation even by fractionating it.

LAMERTON: You would get some degree of whole-body irradiation in a mouse with ^{90}Sr because a mouse is small.

MOLE: I only want to comment on Dr. Cottier's comment. What seems to me to be unusual is the occurrence of osteogenic sarcoma. I think everybody has probably observed a high incidence of ovarian tumours as well as internal hyperostosis in various strains of mice given single doses of radiation but I think that this is the only report I know of with a reasonable incidence of osteogenic sarcoma in animals given a single dose of external radiation.

ZELENY: Do you have any estimate of the absorbed radiation dose in your plutonium experiments and could you compare it with the ^{32}P dose?

LAMERTON: This is a very good question and we would very much like to be able to compare the plutonium and ^{32}P and X-ray experiments on a basis of dose. We have done a certain amount of thick-section autoradiography in order to look at some aspects of this problem but we have to recognize that we don't really know what we mean by dose with alpha emitters. With alpha radiation you are getting relatively few cells given a very high dose whereas with X-rays and β-rays you are getting a very large majority of cells given a much smaller dose. This is just one of the problems. The other is of course that the range of the alpha emitters is only 30–40 μ, and in which cells are we to calculate the dose?

CARCINOGENESIS AS THE RESULT OF TWO INDEPENDENT RARE EVENTS

R. H. MOLE

M.R.C. Radiobiological Research Unit, Harwell, England

SUMMARY

The human evidence on the increase in cancer incidence in survivors of the atom bomb explosion at Hiroshima (which varies with the age of the individual at the time of exposure) and the animal evidence relating quantity of bone-seeking isotope and bone-cancer incidence are both compatible with the general hypotheses that cancer arises as a result of two independent rare events.

The intention of this note is not to report new results but to illustrate a new way of looking at old results. The basic ideas of Armitage and Doll (1957) already adumbrated (pp. 13,15) can provide a common explanation of two

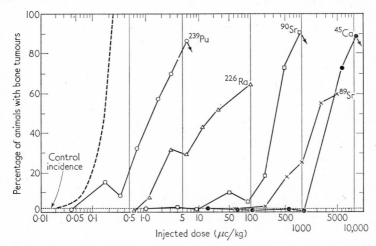

FIG. 1. Incidence of bone tumours in CAF_1 mice given single injections of different bone-seeking isotopes (data of Finkel and her colleagues as plotted by Lamerton, 1958). The dashed curve gives the shape of the theoretical incidence curve if bone-tumour production was proportional to the square of the injected dose, life-shortening and excretion of the dose being disregarded (cf. Mole, 1962).

apparently unrelated phenomena, (a) the markedly curvilinear dose-response in experiments on bone carcinogenesis by a variety of bone-seeking isotopes in the mouse (Finkel, 1959; Fig. 1) and in the dog (Dougherty, 1962), and (b) the pattern of cancer induction in Japanese exposed at Hiroshima where the induced cancer rate appears to increase with the sixth power of the age just as the natural cancer rate does (Harada and Ishida, 1960; Fig. 2).

The basic hypothesis is that cancer follows the occurrence of two independent cellular events which each have a constant probability of occurrence with time if environmental conditions are constant. The first event is followed by clonal growth of cells all carrying this first change, the second event occurs in any of the cells carrying the first change and is inevitably followed by the development of overt cancer. The kinds of event are not further specified.

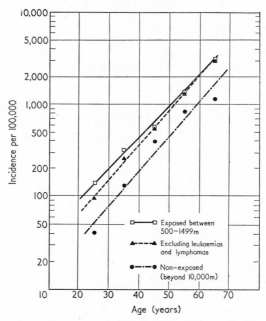

FIG. 2. Age-specific incidence rate of all malignant neoplasms in Japanese at Hiroshima (from Harada and Ishida, 1960). The upper continuous line gives the incidence rate of all malignant neoplasms, the lower broken line the incidence rate excluding leukaemia, in those exposed between 500–1,400 metres from the hypo-centre. The interrupted line gives the incidence rate of all malignant neoplasms in the non-exposed (beyond 10,000 metres).

If the first kind of event is produced by radiation (as well as occurring spontaneously) then irradiated individuals will carry a larger number of different clones than unirradiated people of the same age and will show the cancer incidence to be expected of older unirradiated individuals with the same number of clones. The age-specific cancer rate of the particular group of irradiated Japanese recorded by Harada and Ishida (1960) is in fact equivalent at all ages to that of unirradiated Japanese 7 to 8 years older (Fig. 2). Since there is a logarithmic increase with age in mortality from natural cancer there will be a similar logarithmic increase with age in radiation-induced cancer and a given radiation exposure will produce more cases of cancer the older the individuals at the time of exposure.

This radiation-induced logarithmic increase above natural age-specific cancer rates is to be expected if the probability of the first kind of event but not the second kind is appreciably increased by radiation. In the experimental situation where the induced cancer rate is very much higher than the spontaneous rate, as in the experimental induction of bone cancer by bone-seeking isotopes, it may be supposed that radiation can cause both kinds of event with probabilities so much larger than the spontaneous probability that control incidence and spontaneous probabilities may be neglected. In this case whatever the relative probabilities of induction by radiation of the first and second kinds of event, the incidence of bone tumours at any fixed time after administration should depend on their combined probabilities. With radioactive materials emitting α- or β-particles there is no problem in deciding what parameter of radiation dose to use because clearly it must be the particles themselves which are responsible for the (hypothetical) cellular events. Thus the incidence of bone tumours at any fixed time should depend on the square of the number of radioactive distintegrations since the administration of the isotope. Figure 1 gives the observed over-all tumour incidence in a series of experiments in which mice were given a single injection of radioactive material and a range of amounts of different bone-seeking isotopes were used. If life-span is not shortened by the induction of other lesions then over-all tumour incidence should depend on the square of the administered dose and it can be seen that the data agree reasonably well with this expectation.

The hypothesis also requires that the rate of development of tumours with time after the injection of an isotope should be proportional to the square of the number of radioactive events in the time interval concerned. An experiment in which the body burden of radioactivity is maintained at a constant level by monthly injections should provide a test of this prediction but it should be noted that what is scored in such an experiment is not the time of origin of a tumour but the time when there is a gross bone-tumour, often a tumour large enough to kill. When the data of Brues (1949) on ^{89}Sr were used in Fig. 3, a fixed time $\theta = 150$ days was subtracted from the time at which the bone tumour was recorded, this period of time being considered to be the average development time of a tumour, the time interval from its start to the moment when it was big enough to kill or be recorded. When this is done it can be seen that there is a good fit of the observed data to the prediction that the rate of tumour development, dN/dt, scored over successive fifty-day periods, should be proportional to the square of the number of preceding radioactive events. The abscissa is the square of time so that the rates of tumour development with different body burdens should be proportional to the square of the body burden as they seem in fact to be.

When a single injection of any radioactive material is given there is a progressive fall in the amount retained so that the number of radioactive disintegrations per unit of time is variable not constant. When allowance is made for this progressive reduction in body burden with time, the data from

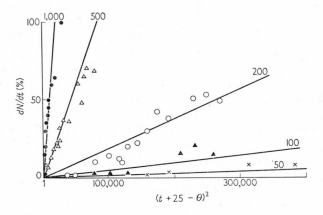

FIG. 3. Age-specific incidence rate of bone tumours in mice given monthly injections of ^{89}Sr (data of Brues, 1949). The bone-tumour rates are plotted against the square of the number of radioactive disintegrations in a manner detailed by Mole (1962). The straight lines are drawn by eye through the origin so that their slopes shall be proportional to the squares of the monthly doses (shown in μc/kg against each line).

an experiment of Finkel and her colleagues utilizing a single injection of ^{90}Sr also seem to fit the prediction that the rate of tumour development with different body burdens should be proportional to the square of the number of preceding radioactive events (Fig. 4).

Unfortunately without knowing more about the exact site in bone where the carcinogenic change begins it is not possible to utilize the hypothesis to compare the carcinogenic potency of radioactive materials of different physical chemical and metabolic properties. It is also the case that the hypothesis has not yet been checked against a number of important experiments in which the method of administration of the radioactive isotope has been varied as by fractionation of injections or by continuous feeding in the diet (Finkel et al., 1960a, b). The full results of these experiments have not yet been reported. Nevertheless one general conclusion may be drawn. If it is true that overt cancer begins only after the conjunction of two rare events then in a group of animals with just 100% incidence of tumours this conjunction will have occurred just once in each animal. This means that it will be very unlikely indeed that this conjunction or start of a tumour can be recognized morphologically simply because of the problem of sampling. In fact whatever the mechanism of carcinogenesis, if the origin of a tumour is a rare event then any morphological change which is seen more than once in a whole series of microscopical sections is *ipso facto* not the relevant change.

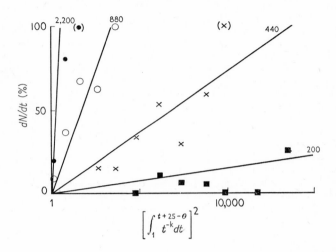

Fig. 4. Age-specific incidence rate of bone tumours in mice given a single injection of ^{90}Sr (data of Finkel et al., 1960). The bone-tumour rates are plotted against the square of the number of radioactive disintegrations in a manner detailed by Mole (1962). The straight lines are drawn by eye through the origin so that their slopes shall be proportional to the squares of the injected doses (shown in μc/kg against each line).

REFERENCES

Armitage, P., and Doll, R. (1957). *Brit. J. Cancer* **11**, 161.
Brues, A. M. (1949). *J. clin. Invest.* **28**, 1286.
Dougherty, T. F. (1962). In "Some Aspects of Internal Irradation" (T. F. Dougherty, ed.). Pergamon Press, Oxford.
Finkel, M. P. (1959). *Radiation Res.* Suppl. 1, 265.
Finkel, M. P., Bergstrand, P. J., and Biskis, B. O. (1960a). *Radiology* **74**, 548.
Finkel, M. P., Biskis, B. O., and Bergstrand, P. J. (1960b). In "Radioisotopes in the Biosphere" (R. S. Caldecott and L. A. Snyder, eds.). University of Minnesota Press, Minneapolis.
Harada, T., and Ishida, M. (1960). *J. nat. Cancer Inst.* **25**, 1253.
Lamerton, L. F. (1958). *Brit. J. Radiol.* **31**, 229.
Mole, R. H. (1962). In "Some Aspects of Internal Irradation" (T. F. Dougherty, ed.). Pergamon Press, Oxford.

DISCUSSION

LAMERTON: Dr. Mole had to put Dr. Brue's data through a good many manipulations to demonstrate linearity on a square law basis. I think I have a good reason for feeling distrustful of this sort of analysis because what, in fact, has been done is to compare body burden essentially with incidence. This is fine if you've got a uniform dose among the cells at risk but not if there is a non-uniform dose. You can't prove a square law with a non-uniform irradiation unless you weight the radiation in your irradiated individual appropriately. This, I think, was also a snag in the curve that Dr. Mole showed the other day of Dr. Burch's analysis of the spondylitic data using the mean dose to fit incidence with the square law. But we do know for the spondylitics that there wasn't a uniform exposure. Even if you only take the exposure to the spine there were separate fields and non-uniform irradiation. Under those conditions you can't even guess what the weighted dose is. This, in fact, is a criticism of the whole Court-Brown and Doll analysis which I think they recognize. They didn't show that there was linearity, they merely set out to discover whether the results were consistent with linearity, and to do this they made the assumption that the relation was linear when they took a mean dose over the spine or over the person examined.

MOLE: I do not really think that that's a logical objection to the analysis of the bone-tumour data, because, as long as the distribution of doses of different sizes is the same, then you could apply a weighting factor, whatever it is, to each dose and just go straight ahead. If the distribution isn't the same with doses of different sizes than you are quite right, but if it is the same I don't think it makes any difference.

LAMERTON: In the case of spondylitis the distribution wasn't the same because there were lots of different fields.

MOLE: As far as I know the distribution of strontium is largely independent of the dose; this has been looked at.

LAMERTON: I'm not convinced that if the distribution is the same you can still use mean dose as your parameter.

UPTON: I may prove to be mistaken on this point but I think if one adopts the thesis that the carcinogenic transformation consists of two successive stages, then would it not follow that, given an individual with one step completed, one would get a one hit type of response curve. Looking at the data from Japan, you have a hodge-podge of different kinds of dose response curves which, in the aggregate, simulate the effects of a certain number of years of ageing. Burch has looked at the linear dose response curve mentioned earlier that Bond and Cronkite published on Sprague–Dawley female rats on the assumption that this strain inherits already one of the mutations required and radiation simply adds another. The fact that they have a high natural incidence means that they don't have to live very long before they acquire the necessary additional mutation or whatever event one postulates for the second genetic step.

BERENBLUM: I take it that the suggestion you made is based on the assumption that the two events are both two single sudden events. Now would the same sort of reasoning apply if the first event were a sudden one but the second depended on a long continuing action?

MOLE: I don't think it would make any difference as long as the continuing action is proportional to the number of radioactive events per unit of time, but it's much easier with particles to think of each particle as doing something.

CASARETT: I am concerned about the elimination of latent periods and this may have something to do with the nature of the second event. I don't think that we need assume the second event has to be an event in the cell of origin of the tumour, it can be a progressive pathologic change involving endarteritis of fibrosis, or whatever it may be, or even the changes of ageing which simulate such conditions, in which case one cannot eliminate latent period if this is indeed a secondary event in the situations being considered. This applies to the whole question of linearity and threshold also. If you consider that a heterogeneous population, such as the human population, is made up of individuals with a wide range of dose thresholds for any particular late radiation effect, all the way from almost no threshold to a very high threshold, then it's easy to see why one might show that a small dose of radiation would produce an increase in incidence in some individuals of such a population. This would mean an increase in absolute incidence as far as the population is concerned, by a small dose of radiation. It's also not surprising when we think that experiments usually use relatively homogeneous groups of young adult animals, that we get curvilinear relationships because the range of dose thresholds here is very narrow and similar, although not identical. In other words, there's a difference in amount of non-radiation induced causes for each of these non-specific events which determines the threshold of an individual. When you said the other day that the threshold dose problem was theoretical then I certainly agree.

DANIELLI: If you do come up with an answer that requires two events there must be some biological explanation and, however dubious the data that you're working on may be, nevertheless an explanation is required.

MOLE: I would agree with you that it badly needs an explanation in biological terms. It is very easy to think of chromosomal changes in view of what we learned yesterday in a review of chromosomal changes in neoplastic cells. Perhaps it's too easy to make that jump, and I'd be very reluctant to do it just now. The precision of the data, the number of bone tumours that you find depends entirely on how intensively you look. All I've done is to assume the same intensity of observation throughout one experiment. If, however, you do compare the two experiments with repeated doses and with single doses, which were done many years apart, you come up with the same probability of biological event per radioactive particle, within less than a factor of two. The data do seem to form a consistent whole, in spite of the fact that experiments were done over a long time, and that there were differences in the survival times, in animal hygiene and this kind of thing. But the actual errors I wouldn't have any idea about at all.

ROTBLAT: I would like to ask Dr. Mole about this application of the square ratio between incidence and dose, to the two events which we discussed yesterday. If you have a square ratio between incidence and dose, doesn't this imply that the two events must come very close after each other, when you have given the dose?

MOLE: Not necessarily, no; because what I have done you see is to subtract 150 days from the time of the occurrence of death. This is what I have called θ or development time which includes not only the time from the beginning of the tumour to the point at which it kills the animal, but also any other time required for an interval between the two events. I am trying to get still further and see in fact if I can get any sort of idea of what the intervals between the two events should be. But I don't think with relatively long-lived radioactive materials, like ^{89}Sr and ^{90}Sr one can get anywhere.

MULLER: Did I understand you to say that you had also tried a fit with a one hit and a three hit event, but that they didn't go as well?

MOLE: I didn't actually say that, but they certainly don't fit a one hit idea, because Brues tried that and he had to put in a "latent period" (which has an uncertain meaning) which varied with the dose. I think that Finkel herself might like to feel that the three hit event might fit some of the data better.

MULLER: Didn't you put in a "latent period"?

MOLE: I put in this fixed development time, which. . . .

MULLER: What's the difference between that and a "latent period"?

MOLE: Well, I don't normally use the phrase "latent period" which has all sorts of meanings to different biologists. I have said that there is a certain probability of inducing the tumour which increases from zero time. And I have just subtracted from the survival time the time required for the growth of the tumour assessing that on the average this is the same for all tumours. And so I tried to avoid this idea of latency.

MULLER: May I go back to the question which I asked originally? If you make a suitable assumption, neglecting its plausibility, can you make these data fit either a one hit, a two hit or a three hit hypothesis? Because if so, the reliability of any of these hypotheses seems open to question.

MOLE: Well it depends on what you mean by differences in assumptions. Brues had to assume that tumour incidence depended on the number of radioactive events, and also on a latent period, unspecified as to meaning and varying with dose. This further quantitative assumption was only made in order to draw straight lines. I thought I had reduced the number of assumptions by at least one anyway!

BERENBLUM: The fact that it is difficult to define latent period doesn't mean that it doesn't exist. The point is, if you adopt another term, and then assume that it is a fixed period and deduct it, in a sense you may be introducing an error In other words to assume that this latent period of whatever you call it, is a fixed time may be wrong, it may be proportional to the total time in which case would it not induce errors in your calculations?

MOLE: Oh well, all I was saying was that if you don't assume a fixed value for it, the data will not fit the proper lines. That's all I'm saying.

UPTON: I would like to speak again about the question of the Japanese data and parallel curves. To me this is an extremely revealing situation, in that, I am told that the more recent data from Japan on tumour incidence are consistent even though the numbers are still very small. But what experimental evidences we have are consistent with the idea that the log of the death rate for tumours plotted against age is displaced by giving doses of radiation. From the raw data it doesn't seem to matter how old the individual is, you seem to get about the same degree of displacement. As Dr. Mole emphasized since this is a log-scale, doubling the tumour incidence at an early age represents a far smaller net number of tumours induced, than doubling the tumour incidence at a later age. In this sense radiation is in fact adding something akin to time, if you simply look at actual data.

ALEXANDER: Would the situation be, Dr. Mole, that with regard to the bone-tumour data, they can be fitted to a one hit law, when one assumes a latent period which is proportional to dose, and can be fitted to a two hit law if you assume a latent period independent of dose, and therefore, we can take our choice as to which we think is more probable—the latent period independent of dose or latent period dependent on dose. So really you haven't thrown one assumption away, we have still got two alternatives, and if we once know about the latent period, then we can decide whether we want a one hit or a two hit event.

MOLE: If you like to think of latent period as being anything more than a mystical assumption.

MAYNEORD: I have a suspicion that something of the same kind occurs if you look at the normal incidence. Dr. Turner and I have recently been very interested in the question of the incidence of cancer of the stomach in relative to radioactivity. In fact, we find very little, but this has led us to plot, let us say the incidence of carcinoma of the stomach against age. And you get varying degrees below, I think about a 5·8 or 6th power. Incidentally, I still don't understand how you fit a square law to that! What does intrigue me, and I wish Prof. Berenblum would tell me about it, is that if you take the data of incidence for different countries, you will find that they are extroardinarily parallel. The fascinating thing which I don't understand, is that if you take the curves for the different countries, Japan, Israel, Wales, Scotland, etc., and simply displace them along the time axis, then they all fit together. In other words, the Japanese seem to be behaving exactly as if they were about 70 years older than the Israelis. Is this a latent period?

UPTON: That's not a latent period, that's original sin!

MOLE: Well, I think if you assume that there is some exogenous factor, which you experience over a space of time, then differing amounts of this factor can explain the differences between different countries.

PRELIMINARY STUDIES ON LATE SOMATIC EFFECTS OF RADIOMIMETIC CHEMICALS

A. C. UPTON, J. W. CONKLIN, T. P. McDONALD, AND
K. W. CHRISTENBERRY

Biology Division, Oak Ridge National Laboratory,[†] Oak Ridge, Tennessee, U.S.A.

SUMMARY

The life-span of mice surviving mid-lethal doses of nitrogen mustard (HN2), triethylene melamine (TEM), and X-rays was reduced. The extent of life shortening was greatest in the irradiated mice, intermediate in the TEM-treated animals, and smallest in those given HN2.

The decrease in longevity was not ascribable to any single lesion or group of lesions but was associated with premature onset of diseases of senescence. Moreover, mice without neoplasms died as early as those with neoplasms.

All three agents increased the incidence of certain neoplastic growths, but the oncogenic effects varied from one organ to another, no two agents affecting all organs similarly.

Lens opacities were induced by each agent. However, those in the HN2-injected mice were barely distinguishable from spontaneously occurring senile cataracts.

Although late effects of ionizing radiation on the life-span and on the incidence of neoplastic and other diseases have been recognized for years, there is little information concerning the physico-chemical and biochemical changes responsible for initiating such effects. Attempts to reproduce them with radiomimetic chemicals have led to conflicting conclusions (see Curtis and Gebhard, 1959; Alexander and Connell, 1960; Stevenson and Curtis, 1961). The present study was undertaken, therefore to explore systematically the somatic effects of several radiomimetic chemicals compared to X-rays.

Preliminary results from one phase of this investigation are presented here.

METHODS

Female RF/Up mice were subjected to 500–600 r of whole-body X-radiation, triethylene melamine (TEM), or nitrogen mustard (HN2) at 10 weeks of age (Table I). The factors of irradiation were as follows: 250 kVp; 30 mA; TSD, 93·7 cm; hvl, 0·44 mm Cu; dose-rate, 80–90 r/min with maximum scatter. TEM was administered intraperitoneally, 3·0 to 4·0 mg/kg, as an 0·03 to 0·04% solution in physiological saline. HN2 was administered intravenously in the form of methyl bis-(β-chloroethyl)amine,

[†] Operated by Union Carbide Corporation for the United States Atomic Energy Commission.

0·10 – 0·12 mg/mouse, dissolved in 1 mg/ml physiological saline solution. The doses of each agent were selected, on the basis of earlier pilot experiments, to approximate the $LD_{50}/30$. After treatment, the mice surviving thirty days

TABLE I. *Late somatic effects of HN2, TEM, and X-rays*

Agent	30-day survival Treated total	(%)	Mean age at death† (days)	Incidence of neoplasm (%)† Leukaemia		Tumour	
				Myeloid	Thymic	Lung	Ovary
HN2	160/270	59	561	9	5	59	13
TEM	157/360	44	508	6	16	52	58
X-Rays	259/360	72	396	7	26	8	37
None	130/130	100	638	3	5	20	7

† Based on thirty-day survivors only.

were housed 8 to 10 per cage and observed until death. Some animals in each treatment group were examined periodically with the split-lamp for lens opacities, according to methods described earlier (Upton *et al.*, 1960). After death, every mouse was necropsied, and histological studies were carried out, as necessary, for diagnosis.

RESULTS

All three agents shortened the life-span (Fig. 1), the effect being largest for X-rays and smallest for HN2. The life-shortening effects were equally

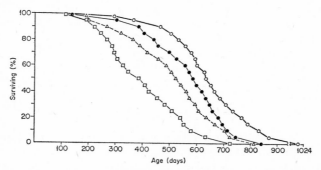

FIG. 1. Longevity of thirty-day survivors, with and without neoplasms, in relation to time after treatment. □, X-rays; △, TEM; ●, HN2; ○, controls.

evident, moreover, in mice free of neoplastic disease (Fig. 2), although the incidence of certain types of neoplasms was greatly increased (Table I). The tumorigenic effects of the agents differed fom one organ to another, no agent appearing oncogenic for all organs (Table I). Effects on the lens tended to parallel effects on the life-span, but the cataractogenic effects of HN2 were barely detectable (Fig. 3).

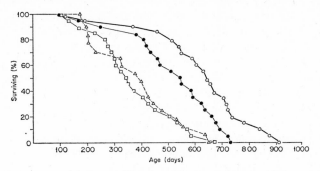

FIG. 2. Longevity of thirty-day survivors without neoplasms, in relation to time after treatment. □ X-rays; △, TEM; ●, HN2; ○, controls.

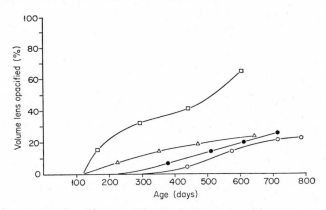

FIG. 3. Average severity of lens opacities in relation to time after treatment. □, X-rays; △, TEM; ●, HN2; ○, controls.

DISCUSSION

These results indicate that the "late effects syndrome" caused by ionizing radiation is not peculiar to this agent alone but may be induced by radiomimetic chemicals. It is evident, however, that neither of the compounds reproduced all the effects of X-rays at the dose levels tested. Instead, the effects of these agents varied on different organs. Whether such variations resulted from differences in the uptake and distribution of the chemicals in different organs or from other factors is not yet known.

From these observations, it may be inferred that oncogenesis constitutes but one of many late somatic effects of radiomimetic chemicals. As important as oncogenesis are effects on the severity and age distribution of non-neoplastic changes, which mimic the effects of radiation. Whether both classes of effects may reflect common mechanisms cannot be decided without further information. It is tempting, however, to interpret them in terms of chromosomal and genetic hypotheses of carcinogenesis and ageing. Viewed in this context, the greater life-shortening effectiveness of TEM, as compared with HN2, may derive from the greater potency of TEM as a mutagen and chromosome-breaking agent (Cattanach, 1957).

Without speculating further about the mechanisms of the effects reported herein, we may envisage new approaches to the study of carcinogenesis and ageing through comparative studies of radiomimetic chemicals and radiation. Systematic application of radiomimetic agents varying in chemical specificity may provide added insight into the molecular basis of effects appearing months or years after exposure. The significance of such comparisons must be interpreted with reservation, however, in view of the possibility that the same functional or morphological lesion may come about through diverse biochemical mechanisms (Koller and Casarini, 1952).

ACKNOWLEDGMENTS

We are grateful to R. C. von Borstel for suggesting the use of TEM and for helpful discussion of the experiment, to M. A. Kastenbaum and J. F. Parham for data processing, and to W. D. Gude and T. Mack for histologic assistance.

REFERENCES

ALEXANDER, P., and CONNELL, D. I. (1960). *Radiation Res.* **12**, 38.
CATTANACH, B. M. (1957). *Nature, Lond.* **180**, 1364.
CURTIS, H. J., and GEBHARD, K. L. (1959). In "Progress in Nuclear Energy Series VI Vol. 2.—Biological Sciences", p. 210. Pergamon Press, London.
KOLLER, P. C., and CASARINI, A. (1952). *Brit. J. Cancer* **6**, 173.
STEVENSON, K. G., and CURTIS, H. J. (1961). *Radiation Res.* **15**, 774.
UPTON, A. C., KIMBALL, A. W., FURTH, J., CHRISTENBERRY, K. W., and BENEDICT, W. H. (1960). *Cancer Res.* **20**, No. 8 (Part 2), pp. 1-62.

DISCUSSION

ALEXANDER: I just wanted to raise the point about these lung adenomas. Don't they occur very late in life? These different incidences you show mean that the radiomimetics and the X-rays are both carcinogenic in so far as they induced lung adenomas but this reduction in the X-ray cases may be because the animals didn't live long enough to get the lung adenomas.

UPTON: This is conceivable I think. We have endeavoured to correct statistically the incidence of lung tumours so as to allow for difference in survival, but this doesn't restore the incidence to normal. I think there is an effect, certainly, of intercurrent mortality, but I am not sure it is the whole explanation for the differences in tumour incidence.

ALEXANDER: But you see; we have found that the latent period for these lung adenomas, particularly with radiomimetic chemicals, was of the order of 550 days. So that cutting off 150 days might lose you all your lung adenomas.

ROTBLAT: I would like to support that, because there are data with X-rays on mice where you find a decrease with dose, but after adjustment for the time of the occurrence there is no change with dose at all.

LEJEUNE: I would like to ask a question about the meaning of life-span. If the mice were as well examined as human beings are, would you say that this life-span reduction is specific such as we could detect in man, or is it some general effect that you cannot define at all?

UPTON: We are making a preliminary attempt to answer Dr. Lejeune's question, but I regret to say that the state of the art isn't very advanced yet; that is, we don't, unfortunately, know very much about our dead mice. We are endeavouring now to do serial sacrifice experiments so that we are not limited to autopsy material. Frequently it's not possible, however, at least in my experience, to say confidently why a given mouse died. Replying to your question about the death rate in relation to age: again, only looking at the mice with lung tumours at the time of death we see that this neoplasm is a late-occurring disease, and the curve is of the same general character as the so called "Gompertz type" curve. Nitrogen mustard displaces the curve to the left, and TEM even more markedly. Now one thing worth noting is that the death rate with lung tumours in the X-ray group is not as greatly displaced as it should be, if effects on the age-distribution of lung tumours parallel the effects on over-all survival, that is, you recall the life-table for all the X-ray mice was displaced much further to the left. So I think in answer to your question, Dr. Alexander, there is in fact less advancement in the age-distribution of lung tumours, in other words, many mice are dying from other causes before they have had a chance to develop lung tumours.

SESSION IV

Chairman: I. BERENBLUM

NON-NEOPLASTIC LATE EFFECTS
By R. BRINKMAN

Concept and Criteria of Radiologic Ageing
By G. W. CASARETT

Initial X-Ray Effects on the Aortic Wall and their Late Consequences
By H. B. LAMBERTS

The Response of Tissues to Continuous Irradiation
By L. F. LAMERTON

The Cholesterol Concentration in Adrenal Glands of Hypophysectomized-grafted Rats (with and without Destroyed Median Eminence) after Whole-body Exposure to a Lethal Dose of X-rays
By P. N. MARTINOVITCH, D. PAVIĆ, D. SLADIĆ-SIMIĆ, AND N. ŽIVKOVIĆ

The Cellular Basis and Aetiology of the Late Effects of Irradiation on Fertility in Female Mice
By W. L. RUSSELL AND E. F. OAKBERG

NON-NEOPLASTIC LATE EFFECTS

R. BRINKMAN

Radiopathologisch Laboratorium der Rijksuniversiteit, Groningen, Holland

SUMMARY

A short comparison of ageing and late radiation effects leads to considerations on "molecular", "cellular" and "nuclear" life-spans. The "vieillissement moleculaire" of haemoglobins is mentioned.

Most attention is paid to ageing and radiation changes of mucopolysaccharide tissue matrix with embedded fibres. This converges on the hypothesis of basal membranes being an important structure from which ageing by irradiation might be understood. Examples are aortic intimal and glomerular basal membranes.

The programme of this symposium now requires us to discuss non-neoplastic late somatic effects, arising from early radiation exposure. From those I will have to select the effects which may lead to damages shortening life-span. So changes in vital organs (arteries, nervous centres, kidneys) or severe breakdown of homeostasis will prevail over late effects in skin, in sensory organs, or in the efficiency of co-ordinated mobility.

Current opinion often considers late somatic effects as not much different from "normal" ageing syndromes, the main effect of radiation being an advancement of all causes of death, but at different rates (Lindop and Rotblat, 1961). Curtis (1961) defines "radiation-induced ageing" as a true" shortening of the natural ageing process".

To my knowledge, gerontology up to the present does not give much help in analysing causes of ageing, apart from considering them partially as non-inevitable. But wherever some possible symptoms of ageing have been ascertained, their similarity to late irradiation effects is great and one is inclined to look for common general causes. It is said, that animals which normally die predominantly of a particular disease such as nephrosclerosis, will also die predominantly of the same disease following irradiation (Curtis, 1961).

Starting from the "dynamic existence" of most biological structures with turnovers, varying from a few hours to possibly years for the "life-span" of their constituting macromolecules, it is conceivable, that this "molecular life-span" might be influenced by age and by radiation. If the turnover of intracellular and extracellular macromolecules increases with cellular metabolism, its retardation (for a certain period) by a hypocaloric diet appears to increase life-span in rats and in pigs (McKay, 1952). If this points to a metabolically increased probability of not completely exact replication, irradiation might have an analogous effect. In small macromolecules like the human haemoglobins it is surprising to see how, during an individual lifetime, its composition will change: foetal Hb is replaced by adult Hb, in at least two possible forms and pathologically many variations may occur, many genetically determined (Muller, 1962). French authors speak of "vieillissement moléculaire de l'hemoglobine".

Roughly the macromolecule here is a well-defined mixture which changes with age. It is not known to me if radiation can influence this process; in my opinion it would be a good subject for research on animals, as the techniques for "recognizing" the various haemoglobins in an animal are excellent. A splendid example of chemically induced haemoglobin alteration is the production of sickle-cell haemoglobin in immature human red cells, caused by incubation with ribonucleoproteins, isolated from a genetically different marrow (Weisberger, 1962).

With regard to individual "cell life", distinction must be made between cells in the various phases of their cycle and between the duration of these phases for various cells.

It may be useful to remember Cowdry's grading of cells with their increasing differentiation: vegetative intermitotics, producing differentiating intermitotics ending up in the definite goal, post-mitotics. Many of these, like liver cells, capillary endothelium, fibroblasts and others still retain mitotic potentiality (reverting post-mitotics) but others, on the highest plane of differentiation have lost the mitotic way of rejuvenation (fixed post-mitotics) and it may also be among those, which cannot be replaced that we will have to look for delayed vital radiation damage (neuronic cells, cardiac muscle cells) (Korenchevsky, 1961).

For a short survey of the theory which considers ageing chiefly as a "nuclear" phenomenon, I can do no better than quote Sinex (1961):

> Adherents of the theory that aging is centered in the nucleus generally believe either that aging is an extension of normal differentiation or that it is due to accidental genetic noise.
>
> The first group points out that, while we as individuals may view aging as a catastrophe, it probably serves a useful evolutionary purpose in insuring succession of generations. Insect physiologists and plant physiologists are particularly likely to hold this view. Many insects, in the adult form, have relatively short life-expectancy and may even be born without mouth parts. In such insects differentiation produces a phenotype with a limited life-expectancy. The death of an annual plant often appears to be the final step in an orderly development. One may therefore argue that aging is a deliberate event, consisting of differen-

tiation to a point where the interdependence of tissue and cells is incompatible with the indefinite life of the total organism. The deaths of individuals could, however, contribute to the survival of the species by insuring a progression of generations and reducing competition for the food supply between young and old. Dobzhansky (1958) has ably presented aging as an adaption of evolution.

A second group hold the view that aging arises from genetic noise or random somatic mutations. Henshaw et al. (1957), Failla (1958), Szilard (1959) and Strehler (1959) have all discussed theories of aging based on somatic mutation. These are reviewed by Glass (1960) in the recent AAAS publication on aging.

The rate constant for somatic mutations viewed as chemical reactions would be very small, possibly of the order of 10^{-13}. If genetic material had the thermal stability of purified DNA (Doty et al., 1960), there would be little probability of the thermal mutation, because of the great stability of the hydrogen-bonded DNA helix. On the other hand, in certain cells, genes may be considerably less stable than purified DNA. The rate constant for thermal mutation of *Escherichia coli* and *Bacillus subtilis* is of the order of 1×10^{-6} at temperatures between 55° and 60°C (Zamenhof and Greer, 1958). Human genes of this order of stability might undergo considerable spontaneous somatic change at 38°C. Such deductions, however, must remain speculative until they can be made to rest on firmer experimental evidence. It will be difficult to demonstrate that random somatic mutations do occur in aging tissue, particularly if such mutations are truly random. However, an effort should be made to ascertain whether clones of cells from aging individuals have altered biochemical properties. The greying of hair might be an example of a somatic mutation in aging melanocytes (Fitzpatrick et al., 1958).

A particular aspect of genetic interest concerns the instructive theory of antibody formation. Is there an impairment in self recognition in aging animals due to alteration in either antigen or antibody?

Somatic theories of aging have appeal to those who feel that ionizing radiation also produces somatic mutations, for such theories would explain the similarity between aging and radiation injury.

For the extracellular processes of ageing, much attention has been paid to the intercellular medium in its composition of fibres (collagenic, reticular and elastic) and amorphous ground substance, soft and firm, giving structure to the extracellular liquids.

As expressed in Ham and Leeson's textbook, it forms the edifice in which the cells are housed. A continuous renewal during life of this structure seems to be beyond the capacity of the organism and it is first the amorphous substance which gradually changes and disappears. Also the fibres deteriorate and for our discussion an important property of collagenic fibres is their metabolic renewal. Neuberger and associates (1951, 1953), with the aid of ^{14}C-labelled glycine showed that even in adult rats collagen is continually, but very slowly rebuilt. The turnover mainly occurs in the intracellular "procollagens"; if this collagen has been fixed in the insoluble fibrous form no rebuilding can be found. It is very probable, that the formation of fibrous collagen requires the presence of matrix mucopolysaccharides. If the production of mucopolysaccharides decreases with age, or if it is influenced by irradiation, the build-up of fibres may be changed also for this reason. Many fibre-properties have been shown to change with age (hydrophilia, heat contraction, chemical contraction). This may correlate with the increased content of hydroxyproline in fibres of older animals, possibly giving rise to increased cross-linking (Korenchevsky, 1961).

Some analogy may be found between the physical properties and the very high dose radiation deterioration of tendon fibrils (Braams, 1961, Braams et al., 1958).

For elastic fibres, so important for the function of large arteries, Lansing et al. (1951) have described their deterioration and calcification with age and after Jellinek's (1958) description of their extraordinary sensitivity to X- and γ-irradiation an interesting analogy appeared to have been found. Unfortunately to my knowledge confirmation of Jellinek's results has not been published and I know of attempts to reproduce them which did not succeed.

It is a general opinion (Korenchevsky, 1961), that for the intercellular medium the quotient amorphous ground substance/collagenic fibres, or the chemoanalytically expressed quotient hexosamines/hydroxyproline decreases with age. In the skin of rats, this process is strongly accelerated by moderate irradiation, not only by disappearance of ground substance but also by absolute increase of the collagen content (Kitagawa et al., 1961).

It has been shown by Glicksman and collaborators (Kitagawa et al., 1961) that this preponderance of collagenic fibres after irradiation may be reversed by treatment with 3-iodothyronine. If a similar application holds for the relation ground substance/fibres in the walls of large arteries, this may be still more interesting.

In the opinion of many investigators (Perez-Tamayo, 1961), the mucopolysaccharide ground substance, in its fibrous framework is not only a two-phase mechanical support for parenchymal and endothelial cells but a controlling and controlled environment. For the extracellular tissue fluids, it again forms a functionally important macromolecular polyanionic reticulum. It is partly firmly bound to the fibres, especially to the elastic and reticular fibres, and this often results in membraneous structures in the intimal fenestrated elastica of arteries, in the numerous subcutaneous membranes (Day, 1959) and, above all, in the basement membranes around most capillaries.

The direct mechanical importance of ground substance for properties like capillary pinocytosis or emigration is apparent. The electrostatic effect on the diffusion of positive ions is demonstrated by the work of Joseph and collaborators (1952). In the polyanionic matrix, fixed by a tight fibrous network, e.g. of the symphysis pubis of guinea-pigs, the mobility of K ions appeared to be halved. Under the influence of sex hormones the relaxation of this tissue increases ionic mobility to normal. It must be important to know, if this extracellular inhibition of cationic mobility may be generalized as a contribution to the slowing-up of ionic capillary exchanges in ageing tissues, or also in tissues where the ground substance fibres quotient had been decreased by irradiation.

The turnover of acid mucopolysaccharides, sulphated or not appears to be a few days or less (Dorfmann and Schiller, 1960), but in older animals the composition of the mixture of acid mucopolysaccharide is changed and the rate of production is slowing down. According to Korenchevsky quoting Sobel and others (Bunting and Bunting, 1953 for human aorta), the decrease in the ratio hexosamines/collagen is characteristic for the ageing of many tissues including aorta in rats and also in man.

Sobel and Marmorston (1956) and Sobel et al. (1960) have already shown that the decrease of the quotient ground substance/fibres is much accelerated in irradiated rats. This held for aorta, lungs, skin and sternum, but not for caval veins. The radiative ground substance deterioration is of vital importance and is mainly expected in the vascular system. We will have the opportunity later of discussing the mucopolysaccharides gel-filtration layer, preventing pressure infiltration of the aortic wall by large molecules. For arterioles, capillaries and venules much evidence can be found in recent reviews; I might quote Casarett (1959): "the only wide-spread histopathological effect of a permanent and progressive type, which could be traced continuously from the time of radiation exposure to the time of death was the effect on small blood vessels, especially arterioles and capillaries". Further, Rhoades (1948) said that vascular changes, following treatment with ionizing radiation, are probably secondary to changes in the connective tissue, surrounding the vessels. And lastly, it must be concluded from the work of Bargmann (1958), Gersh and Catchpole (1961), Illig (1962) and especially of Magno and Palade (1961) that capillary permeability in many regions is chiefly determined, not by the endothelium but by the basement membrane, surrounding the capillary as a formation of the ground substance.

It is in these, still somewhat mysterious, basement membranes that I venture to suspect an important front where ageing and also radiation effects may have serious delayed consequences.

In many organs the basement membranes become markedly thicker during ageing (Gersh, 1952) and this must influence blood/cell exchange. As the membrane is built up as a lamellated dense polymerization of mucopolysaccharide ground substance around a very fine reticular network, a radiation sensitivity might be expected in analogy, with that of cutaneous or aortic matrix. Enzymes like the hyaluronidase complex not only hydrolyze the diffuse ground substance but attack basement membranes as well (Gersh, 1952), so the presence of mucopolysaccharides is to be expected there.

If the basement membrane is a dense network of very elastic (Nagel, 1934) and very fine (40–50 Å) reticular fibres, bedded in mucopolysaccharide ground substance its permeability for water and small molecular solutes will depend on the ground substance/fibre ratio. If the ground substance is disappearing slowly (ageing) or somewhat more quickly (irradiation) elastic retraction of the network might decrease basement membrane permeability. Gersh and Catchpole (1960) explain the apparent difference in thickness of basement membranes as concluded from their visibility either in the light or in the electron microscope by suggesting that this membrane normally occurs in a slightly corrugated state. Also from this one might expect a tightening up if the ground substance disappears.

I must now restrict further consideration of a possible radiation influence on the permeability of basement membranes to one instance, radiation nephrosclerosis. Another instance, not a delayed but a very early effect will be described in the "basement membrane" of the aortic endothelium, by Dr. Lamberts.

Radiation nephrosclerosis has been known for some time and results of recent work appear to agree on the cause: i.e. intercapillary glomerular sclerosis, not distinguishable from the same process in ageing.

Electron microscopy has definitely proved that the chief ultrafiltration barrier in glomeruli is the basement membrane between capillary endothelium and visceral epithelium. The endothelium is highly fenestrated and the epithelium is not a continuous layer. The membrane is not a simple sieve, but a gel-like structure with two fine fibrillar components embedded in an amorphous matrix (Farquhar et al., 1961). It is not a perfect ultrafilter and the few macromolecules which have permeated appear to be collected by pinocytosis in the elaborate foot process arrangement, formed by the visceral epitherlium adjoining the basement membrane. This epithelial activity is greatly enhanced when challenged, e.g. when increased quantities of proteins appear in the basement membrane filtrate such as in the nephrotic syndrome. If, initially, the membrane permeability increases by irradiation a similar process might be expected and this could be at the base of the glomerular intercapillary hypertrophy described by recent authors (Guttman and Kohn, 1960; Nüssel and Schunk, 1961) as the late characteristic change after total-body irradiation.

The literature on radiation nephrosclerosis as a late effect in many animals is increasing; I would like to direct special attention to its study in pigs (Nüssel and Schunk, 1961) and to the demonstration of radiation-produced failure by chronic salt loading (Lamson et al., 1962). Whether the decrease of resistance to cold stress in mice, caused by age as well as by irradiation (Trujillo et al., 1962) belongs to the same group of causes is not known at present.

In suggesting basement membrane damage as a general cause of late lethal effects, I have chiefly mentioned the renal consequences. A second target organ may be the wall of large arteries, where the endothelial basement membrane is shown to be damaged.

REFERENCES

BARGMANN, W. (1958). *Dtsch. Med. Wschr.* **83**, 1704.
BRAAMS, R. (1961). *Int. J. Rad. Biol.* **4**, 27.
BRAAMS, R., HUTCHINSON, F., and RAY, D. (1958). *Nature, Lond.* **182**, 1506.
BUNTING, C. H., and BUNTING, H. (1953). *Arch. Path.* **55**, 257.
CASARETT, G. W. (1959). In "Radiobiology at the Intracellular Level", p. 132. Pergamon Press, London.
CURTIS, H. J. (1960). "Medical Physics", Vol. 3, p. 492. The Yearbook Publishers, Chicago.
CURTIS, H. J. (1961). Proc. 3rd Australasian Conf. on Radio-biology, p. 114. Butterworths, London.
DAY, T. D. (1959). *Quart. J. exp. Physiol.* **44**, 182.
DOBZHANSKY, T. (1958). *Ann. N.Y. Acad. Sci.* **71**, 1243.
DORFMANN, A., and SCHILLER, S. (1960). "Biological Structure and Function", p. 327. Academic Press, New York.
DOTY, P., MARMUR, J., KIGNER, G., and SHILDKRAUT, C. (1960). *Proc. nat. Acad. Sci., Wash.* **46**, 461.
FAILLA, G. (1958). *Ann. N.Y. Acad. Sci.* **71**, 1124.
FARQUHAR, M. G., WISSIG, S. L., and PALADE, G. E. (1961). *J. exp. Med.* **113**, 47.
FITZPATRICK, C. T., BRUNET, P., and KUKITA, A. (1958). In "The Biology of Hair Growth" (W. Montague and R. A. Ellis, eds.). Academic Press, New York.
GERSH, I. (1952). *Harv. Lect.* **45**, 216.
GERSH, I., and CATCHPOLE, H. R. (1960). *Perspect. Biol. Med.* **3**, 282.
GLASS, B. (1960). In "Ageing—Some Social and Biological Aspects", Publ. Amer. Ass. Adv. Sci. No. 65.
GUTTMAN, P. H., and KOHN, H. I. (1960). *Amer. J. Path.* **37**, 293.
HENSHAW, P., STAPLETON, G., and RILEY, E. F. (1947). *Radiology* **49**, 349.
ILLIG, L. (1961). "Die terminale Strombahn". Springer, Berlin.
JELLINEK, S. (1958). *Lancet i*, 1149.
JOSEPH, N. R., ENGEL, M. B., and CATCHPOLE, H. R. (1952). *Biochim. biophys. Acta* **8**, 575.
KITAGAWA, T., GLICKSMAN, A. S., TYREE, E. B., and NICKSON, J. J. (1961). *Radiation Res.* **15**, 766.
KORENCHEVSKY, V. (1961), "Physiology and Pathology of Ageing", p. 50. Karger, Basel, New York.

LAMSON, B. G., LANG, D. A., BILLINGS, M. S., GAMBINO, J. J., and BENNETT, L. R. (1962). *Radiation Res.* **16**, 54.
LANSING, A. I., ROBERTS, E., RAMASARMA, G. B., ROSENTHAL, T. B., and ALEX, M. (1951). *Proc. Soc. exp. Biol., N.Y.* **76**, 714.
LANSING, A. I., ROBERTS, E., RAMASARMA, G. B., ROSENTHAL, T. B., and ALEX, M. (1961). *Circulation* 1283.
LINDOP, P. J., and ROTBLAT, J. (1961). *Proc. roy. Soc.* **B 154**, 332.
MAGNO, G., and PALADE, G. E. (1961). *J. biophys biochem. Cytol.* **11**, 571, 607.
MCKAY, C. M. (1952). In "Problems of Ageing" (Cowdry-Lansing, ed.), p. 139. Williams and Wilkins Comp., Baltimore.
MULLER, C. J. (1962). "Molecular Evolution". van Gorcum, Assen.
NAGEL, A. (1934). *Z. Zellforsch.* **21**, 37.
NEUBERGER, A., and SLACK, H. G. (1953). *Biochem. J.* **53**, 47.
NEUBERGER, A., PERRONE, J. C., and SLACK, H. G., (1951). *Biochem. J.* **49**, 199.
NÜSSEL, M., and SCHUNK, J. (1961). *Strahlentherapie* **116**, 502.
PEREZ-TAMAYO, R. (1961). "Mechanism of Disease", p. 260. W. B. Saunders Comp. Philadelphia, London.
RHOADES, R. P. (1948). In "Histopathology of Irradiation from External and Internal Sources" (W. Bloom, ed.). McGraw-Hill, New York.
SINEX, T. MARROTT (1961). *Science* **134**, 1402.
SOBEL, H., and MARMORSTON, J. J. (1956). *J. Gerontology* **11**, 2.
SOBEL, H., GABAY, S., and BONORRIS, G. (1960). *J. Gerontology* **15**, 253.
STREHLER, B. L. (1959). *Quart. Rev. Biol.* **34**, 117.
SZILARD, L. (1959). *Proc. nat. Acad. Sci., Wash.* **45**, 193.
TRUJILLO, T. T., SPALDING, G. F. and LANGHAM, W. H. (1962). *Radiation Res.* **16**, 144.
WEISBERGER, A. S. (1962). *Proc. nat. Acad. Sci., Wash.*, in press.
ZAMENHOF, S., and GREER, S. (1958). *Nature, Lond.* **182**, 611.

DISCUSSION

UPTON: I should like to enquire about the electron microscope changes in radiation nephrosclerosis. Do you know how soon one might visualize such changes? Do you think they would become detectable first in the basement membrane or perhaps first in the foot processes?

BRINKMAN: Well, the intercapillary hypertrophy of tissues can be shown in pigs about 4 months after irradiation, then it starts and goes on rapidly, especially in young pigs, and it is until now not further analysed as far as I know, but if you look at the beautiful pictures of Palade and his collaborators, there is no choice at all. The basement membrane is the only barrier—there is no barrier in the epithelium, there is no primary barrier in the endothelium at all, so it has to be there. I don't think there are any other references, but if you look at those beautiful pictures—well, I, at least, was convinced.

UPTON: The question is, do you visualize that the changes occur as a result of cellular injury, or as a result of injury to the intracellular material? Does the basement membrane find its origin in the foot processes of the cells? Is it, if you will, an excretion, or is it a deposition from the vascular system? What do you think to be the chain of events or what would you speculate about the chain of events that would ultimately manifest itself as a defect in basement membrane?

BRINKMAN: I think that this basement membrane is a formation of the amorphous ground substance in the space between the endothelium and the epithelium. You can see its formation there and it has to be a formation of the ground substance. It is not formed by the endothelial or the epithelial cells at all. It is a definite formation of the ground substance, and we know that there, the sensitivity to irradiation is large. The rest is speculation.

COTTIER: I should like to emphasize that the question of radiation nephrosclerosis in late stages after mid-lethal doses of radiation is still very open to doubt. First, there is not a 100% incidence of this change, and, second you can observe this change also to a certain extent in non-irradiated animals. As far as rats are concerned, this is another problem, because these rats are apt to become hypertensive while other animals don't. Another point is that the morphology of this particular hypertensive nephrosclerosis is different from the morphology of glomerular changes in mice, at later stages of irradiation. In the former you find an increasing thickness of the strongly PAS-positive basement membranes, while in mice you find hyaline material—in some mice, not all—which is not strongly positive for PAS. I should like to mention the paper of Rosen, which appeared last year. In other strains of rats they did not find the consistent thickening of the basement membrane. Their main finding was cellular changes—giant cells in glomerular loops and things like that. The only thing I would like to stress is that nephrosclerosis in the later stages after whole-body irradiation with these doses in the middle range is not well established. It remains to be shown whether some of these changes are actually complications like the hyaline disease which has been described for mice, and other, possibly infectious complications of the glomerular loops.

BRINKMAN: Of course this was only meant as a speculation. It might be worthwhile to say that at any rate you should look at basement membranes and try to see them by special methods of correlation and confirmation, because in most preparations you cannot see them at all. The preparation should be specially treated in order to be able to see them, and I think in many researches on this topic this has not been tried.

UPTON: A number of workers have reported the effects of radiation on extracellular or intracellular material, and such effects as the thermodenaturation of collagen are well known. The question I think to which we should address ourselves is this. Can one account for effects seen *in vivo* in the ground substance or in the basement membrane on the basis of radiation effects on extracellular materials predominantly, or could it be supposed that such effects are indirect—ascopal in a sense—mediated through cellular injuries?

CONCEPT AND CRITERIA OF RADIOLOGIC AGEING

G. W. CASARETT

School of Medicine and Dentistry, The University of Rochester, New York, U.S.A.

SUMMARY

Critical general comparison of various manifestations of ageing and late radiation effects strongly suggests premature againg as an effect of irradiation, on a generalized or localized basis. Presented is an hypothesis of the process of "radiologic ageing" at the tissue level, based on histopathological studies of the development of manifestations of ageing and of late radiation effects in tissues prior to disease development. This hypothesis maintains that non-specific injury of endothelium of fine vasculature by direct or indirect mechanisms leads to increase in density and amount of collagenous substance interstitially and in subendothelial regions of arterioles. These changes constitute a temporal advancement in the increase of the histohaematic barrier and in the development of arteriolocapillary fibrosis, which are progressive processes in "normal" ageing. Eventually these processes cause progressive reduction in number of dependent parenchymal cells due to relative hypoxia and malnutrition. Secondary to parenchymal loss is a process of replacement fibrosis and reduction of fine vasculature, with consequent further increase in histohaematic barrier and arteriolocapillary fibrosis. Concomitant with parenchymal loss is progressive reduction of functional reserve capacities and a corresponding progressive increase in susceptibility of tissues to trauma, stress, and disease.

Although it is not yet possible to define and compare the essential processes of ageing and of late radiation effects, their late manifestations can be compared.

The manifestations of ageing in adult mammals comprise a progressive deterioration of tissues, with concomitant decline of functional reserves and adaptive powers, which leads eventually to disease and inevitably to death.

Ageing is not uniform with increasing time, but varies in rate among individuals and among organs of an individual. The rate of ageing, the development of disease, and the life-span are the net results of many variable, modifying, conditioning and correlating forces, both environmental and inherent, including genetic constitution. The integration of these forces determines the physical status of the ageing individual.

There are four general types of data which are often used as criteria of alteration of ageing process by irradiation. These are: data pertaining to mortality, pathology and disease incidence, subclinical histopathology, and physiology (including biochemistry). It is fruitful to examine these criteria critically in relation to one another, in a general manner, to determine their relative usefulness with respect to assessment of the ageing process.

According to ideal concepts, an agent is regarded as causing premature ageing if it causes the force of age-dependent mortality to increase earlier in the treated than in the non-treated control population, brings forward in time the age of onset of diseases which affect the controls, without altering the sequence or the incidence of diseases and causes of death, and causes characteristic morphological and physiological manifestations of the ageing process to appear and develop at proportionately earlier chronological ages.

If the agent causes these manifestations to develop prematurely, but does not alter their rate of development thereafter, the effect is simply one of *precocious* ageing. If the agent causes also an increase in rate of development of the manifestations, the effect is not simply precocious but one of *acceleration* of ageing.

MORTALITY DATA

With increasing age in the adult there is generally a progressive increase in the probability of disease and accident and in the probability of death.

Inherited body constitution establishes essentially the baseline in an individual with respect to the ageing process and its rate and the maximal life-span even under optimal conditions.

However, a comparison of mean or median longevities alone is meaningless in terms of the process or rate of ageing, since many age-independent factors are capable of reducing or increasing the median or mean life-span of a population. Comparison of the temporal distribution of death is a little more meaningful but still very limited in usefulness with respect to assessment of the ageing process.

For a group of mammals maintained under excellent environmental conditions the shape of the arithmetic survival or mortality curve tends toward the rectangular (Fig. 1, A), indicating a relatively low incidence of age-independent causes of death. At the other extreme, a group of mammals kept under poor environmental conditions and subject to a very high incidence of age-independent causes of death, with few living to senescence, exhibits an arithmetic survival curve which resembles a logarithmic decay curve (Fig. 1, D). When multiple life-shortening factors independent of age modify an arithmetic rectangular survival curve they tend to reduce it in the direction of a straight line (Fig. 1, B and C) or, if the effect is marked, toward the logarithmic decay curve (Comfort, 1959).

Combinations of the effects of premature ageing and of increases or decreases in incidence of age-independent causes of death at various times in the life-span can result in survival curves of many shapes. Furthermore, when an agent is capable of producing a prophylactic or therapeutic effect in relation to any serious age-independent or age-dependent disease in a population, this can result in change in survival curves and even in increased aftersurvival. On the other hand, an activating effect on pre-existing chronic or latent infectious disease can alter survival curves in other ways. The most that can properly be said of a survival or mortality curve, is that it is compatible or incompatible with a supposed process (Comfort, 1959).

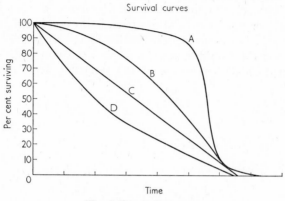

FIG. 1. Survival curves.

It is well-established experimentally in mammals that exposure of the wholebody or parts of the body to ionizing radiation in substantial but sublethal doses can shorten life-span. Numerous mathematical analyses and interpretations of such data have been performed in relation to the ageing problem (Blair, 1956; NAS-NRC, 1961; Brues and Sacher, 1952; Mole, 1957; Furth et al., 1959; Neary et al., 1957; Sacher, 1955). In the case of partial-body exposure, the life-shortening effect is variable in degree, depending on the kind and amount of tissue irradiated as well as on the dose (NAS-NRC, 1961).

The observation of radiation life-shortening, the reduced radiation LD_{50} of irradiated groups as compared with their non-irradiated controls of the same chronologic age, and the residual tissue damage and delayed effects of irradiation, all indicate that some of the radiation injury or its consequent damage is irreparable.

According to Blair's theory of radiological life-shortening (Blair, 1956), the injury of ageing and the irreparable injury by irradiation are additive in producing or contributing to radiation lethality, and the irreparable component of the injury is equivalent to premature ageing (at least in an actuarial sense), in that it ultimately deprives the animal of part of its expected life-span.

Limited experimental observations in this laboratory suggest that irreparable injury is detectable, after an interval of maximal tissue recovery as a reduction in acute lethal dose, and that the LD_{50} dose for irradiation decreases with increasing age (Hursh *et al.*, 1958). Limited experimental experience with measurement of irreparable injury by decrement in radiation LD_{50} at various times after brief total-body irradiation suggests the possibility that the irreparable radiation injury may remain more or less constant, in comparison with non-irradiated animals of the same age (Hursh *et al.*, 1958), at least until the point in time when age-dependent diseases increase considerably in incidence prior to this occurrence in the non-irradiated population (Fig. 2). After this time it seems probable that there may be greater differences in LD_{50} between the two groups when compared at the same chronological ages, but perhaps not when compared at the median death times.

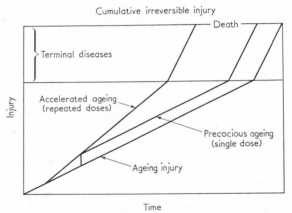

FIG. 2. Cumulative irreversible injury.

If these observations prove to be true, and if it is true also that the irreparable injury of irradiation is equivalent to ageing injury in an actuarial sense, then in the absence of any effect of the irradiation on the incidence and distribution of age-independent causes of death it may be expected that a brief total-body life-shortening radiation exposure would result in a survival curve which is similar in shape or slope to the control curve, but displaced to an earlier time period in a manner compatible with the concept of *precocious ageing*.

This often seems to be the case in single-dose experiments. Sometimes, however, the arithmetic survival curves of briefly irradiated groups become flattened and must be considered compatible with either the concept of accelerated ageing or of increased incidence of age-independent mortality or a combination of precocious ageing and increase in age-independent mortality.

On the other hand, protracted irradiation, e.g. daily or weekly irradiation with substantial doses for a long period of time, may involve the accumulation of irreparable injury in a manner such that there would be increase in the slope of a plot representing irreparable injury with passing time, as compared with the simple persistence of a constant amount of irreparable injury following a brief exposure (Fig. 2). However, under these conditions of exposure, the injury picture is complicated by the constant production or existence of unrepaired injury of reparable types as long as irradiation is continued and is probably different in animals irradiated until their death as compared with animals whose irradiation is stopped far short of death. In either case the unrepaired injury of reparable type may alter the susceptibility of the animals to age-independent or age-dependent causes of death, the direction and extent depending upon the dose-rate, changing tissue sensitivity, and the balance between injury and recovery rates.

If there is accumulation of irreparable injury with long protracted or repeated irradiation it may be expected that such exposure may result in a survival curve for experimental animals which is steeper in slope than the control curve, i.e. a curve compatible with the concept of accelerated ageing or increased incidence of age-independent causes of death or a combination of the two. This is sometimes the case in protracted-exposure experiments. However, in other instances the survival curves have been similar in shape to the control curves, but simply advanced in time, compatible with the concept of simple precocious ageing. This latter finding has led some investigators (Neary et al., 1957) to the interpretation that at comparatively low daily dose rates, shortening of life-span may be determined mainly by the irradiation received during the early part of the life-span.

From these general considerations it can be seen that mortality data alone are greatly limited as criteria of effect of total-body irradiation on the ageing process. Such data are even more limited in their usefulness for this purpose in the case of partial-body irradiation, caused for example by internally administered radioactive isotopes.

PATHOLOGY AND DISEASE INCIDENCE

Survival curves coupled with data on the incidence and temporal distribution of diseases and causes of death are more enlightening. However, in order to bring such data to bear on the ageing process, there must be knowledge of the relative age-dependence of the causes of death, including recognition of the hereditary susceptibilities to causes of death which are more or less age-dependent or age-independent.

A well-kept ageing population of genetically heterogeneous animals tends to die as a result of a relatively wide variety of age-dependent causes of death. Many highly inbred, or genetically more homogeneous, ageing populations of animals tend to show a lesser variety of causes of death and usually a high incidence of certain specific causes of death for which they have a strong genetic susceptibility. In general, the age-dependent diseases or the so-called diseases of ageing are essentially either degenerative or neoplastic in character, and it is well-known that many of the neoplastic diseases develop in relation to previously developing degenerative tissue states either at the site of origin of the neoplasms or in some physiologically related tissue.

As in the case of ageing manifestations, the propensity for development of diseases is not uniform with respect to individuals of a population or organs of an individual. Some diseases are more or less progressively age-dependent. Other age-dependent diseases show considerable age-specific propensity, the incidence rising rapidly to a peak at certain ages and then declining rapidly after those individuals with a genetic tendency for the disease have died.

In order to establish a baseline for interpretation of the ageing process by means of data on disease incidence or cause of death, it is necessary to know the age-dependent diseases in a well-kept ageing population, and to also recognize the possibility that diseases or causes of death which are relatively age-independent, whether they be infectious diseases or degenerative or neoplastic diseases, for which there is very strong genetic propensity, may be extraordinarily increased in incidence or time of onset by relatively little tissue injury, and be relatively independent of change in the ageing process. It is necessary to recognize that increased or decreased incidence of age-independent causes of death may change, in relative fashion, the age-specific incidence of age-dependent diseases expected in an undisturbed control population. Furthermore, for a better understanding of the disease picture in relation to the ageing process, it is necessary to be able to distinguish between temporal advancement of disease and true induction of disease.

If an agent results in the earlier appearance of a disease in a population, with increase in age-specific incidence at earlier ages and without a significant increase in absolute or life-time incidence, as compared with controls in a well-kept ageing population, then the disease may be regarded as having been temporally advanced by the agent. When the agent results in a considerable increase in absolute incidence of a disease as compared with the incidence expected within the maximal life-span of the species or strain, then the excess incidence of the diseases may be regarded as having been induced by the agent.

It is well-known, however, that the observed mean life-spans of experimental animal populations often fall far short of their potential averages because infectious diseases kill, or damage, large numbers of individuals well before the senescent period of life. Some age-dependent diseases of long latency may rarely, or never, develop spontaneously within the observed life-span in many individual experiments. Consequently, it is possible in some instances that some of the diseases regarded as induced by the experimental treatment under these circumstances may have been instead diseases of relatively long latency with their time of onset greatly advanced temporally in individuals with some propensity for them.

The involvement of alteration of the ageing process in the effects of total-body irradiation on life-span is often judged according to ideal standards and assumptions which may not be true. Shortening of life by total-body irradiation is conceded to be an effect of premature ageing only if there is no induction of disease and if there is a proportionately equal temporal advancement of all diseases common to the species or strain. In practice this ideal concept implies, among other assumptions, the assumptions that all of the diseases in question are age-dependent to the same degree, that all of the diseases and causes of death under conditions of maximal longevity are known, that the irradiation is always applied uniformly throughout the tissues, and that the irradiation, if it alters the degree, or rate, of ageing at all, must alter the relative rates of ageing processes in various parts of the body to a degree proportionate to their relative rates of ageing in non-treated animals. The truth of this latter assumption depends greatly on the fundamental nature of the ageing process, which, of course, is not yet known.

In the case of total-body irradiation experiments on rats and mice, some experiments, especially among those with single dose exposures shortening life with a simple temporal displacement of the survival curve, have shown approximately the same diseases in approximately the same incidence in irradiated and control groups (Blair, 1956), i.e. a simple temporal advancement of disease. Other such experiments have shown similar results, about the same diseases, although not necessarily in the same incidence. Still other experiments have shown, in addition to advancement of specific diseases, a considerable induction of certain other diseases. This induction tends to occur more often in inbred strains with a high genetic susceptibility to certain diseases, e.g. leukaemias or ovarian tumours in certain inbred strains of mice.

The observed strain variation in life-shortening in irradiated mice can be partially attributed to genetic differences in sensitivity to the leukaemogenic effects of irradiation (Grahn, 1958). When leukaemia mortality is excluded, life-shortening due to all other causes is found to vary comparatively little between strains.

Usually in the case of total-body irradiation of genetically more heterogeneous animal populations, the temporal advancement of disease is relatively of greater importance in life-shortening than induction of disease; while, in the case of highly localized irradiation, particularly with the use of the more intense, or larger doses that it is possible to administer without acute lethality, the induction of disease at the site of irradiation, or indirectly in a physiologically related site, is relatively of greater importance.

Intensive highly-localized irradiation, as with internal radioisotopes which concentrate in certain tissues, greatly enhances the tendency to diseases related to the part irradiated, with relatively less enhancement of the tendency to disease development in unrelated parts of the body. Whether we regard these radiation-linked diseases as induced or temporally advanced, the incidence depends greatly on the latent periods for these diseases in relation to the temporal proximity of development of other terminal diseases common to the species or strain in other parts of the body. This, in turn, depends on the age of the animal at the time of irradiation.

In general, total-body life-shortening irradiation tends to increase the incidence and/or the severity of age-dependent diseases at given chronological ages (Blair, 1956; NAS-NRC, 1961; Furth et al., 1959; Casarett, 1952, 1956; Upton, 1957; Alexander, 1957). One could perhaps assume that total-body irradiation induces each of the diseases of advanced age separately. However, it is more reasonable to regard such uniformity of response as a temporal advancement of the diseases and as evidence that total-body irradiation causes a nonspecific diffuse, subclinical deterioration of the body tissues that advances the onset of many diseases to a roughly equal degree (Blair, 1956). There are data which suggest that similar effects may occur in man (Warren, 1956).

SUBCLINICAL HISTOPATHOLOGY

In general, the so-called diseases of ageing do not develop suddenly to clinical proportions to be recognized as pathological entities, but are the eventual results of slow, insidious, subclinical, deteriorations in tissues or organs. The various specific terminal diseases which emerge are often only more or less random expressions of more generalized tissue disorders.

Ageing animal populations are seen by gross examination and by microscopic examination to deteriorate slowly but generally in progressive fashion before clinically or pathologically recognizable age-dependent disease entities occur. These changes have been observed to occur prematurely following irradiation (Casarett, 1952, 1956, 1957, 1958, 1960; Russ and Scott, 1939; Henshaw, 1944; Jones, 1956).

A fundamental ageing change in the adult animal may be defined as a change which occurs consistently and progressively with the passage of time in all temporally ageing individuals of the population, in the general phase of life in which the change may be expected, and which is qualitatively independent of variations in clinical history among individuals. Such a change may be detectable initially at different chronological ages among individuals or may vary quantitatively with disease history or variations in genetic constitution or environment.

The fundamental histopathological changes of ageing seem to be degenerative and atrophic or involutional changes, the end-result of which may be described generally as fibro-atrophy. This fibro-atrophy involves a decrease in number of parenchymal cells associated with a decrease in fine vasculature in a process of arteriolocapillary fibrosis and with an increase in density and amount of interstitial connective tissue, constituting an increase in the histohaematic barrier (connective tissue barrier between blood and parenchymal cells).

The hypertrophic cellular changes and hyperplastic or metaplastic tissue changes seen with increasing age are generally not among the fundamental ageing manifestations and do not occur in all senescent individuals. These changes seem to be either normal physiological compensatory responses of less affected cells to degenerative changes in related cells or tissues or, like many degenerative changes observed, are secondary to specific disease processes.

It is not yet clear at the histopathological level which of the three components of the tissue fibro-atrophy of "normal ageing", i.e. the changes of parenchymal cells, of connective tissue, and of fine vasculature, are primary and secondary with respect to one another; or what are the relative contributions of one component to another, since they have mutual or reciprocal influences; or to what extent these relationships or contributions differ among tissues of different kinds.

Loss of parenchymal cells may be followed by a process of replacement fibrosis and reduction in fine vasculature secondary to reduced parenchymal cellularity. Also, non-specific damage to the endothelium of fine vasculature may result in interruption or impedence of circulation, increase in interstitial colloid and in fibrillar density of connective tissue, increase in histohaematic barrier, consequent loss of parenchymal cells through hypoxia and reduced nutrition, then replacement fibrosis and reduction in fine vasculature. The success of regeneration of parenchymal cells depends greatly on the adequacy of the microcirculation and on the permeability or integrity of the histohaematic barrier.

None of these histopathological components of the tissue fibro-atrophy of ageing are due necessarily to inherent changes in the tissue components involved. Even the increasing fibrillar density of interstitial connective tissue may be caused by forces originating elsewhere in the body. All of these changes are non-specific changes which may be brought about by a variety of agents or factors, including adrenal cortical reactions and perhaps any agent eliciting a response of the adrenal cortex as in stress phenomena. The non-specificity of basic histopathological ageing changes is compatible with the concept proposed by Jones (1956) that ageing is a result of the accumulation of non-specific injuries.

Sublethal, life-shortening total-body irradiation and localized irradiation, by means of external or internal sources of radiation, produce permanent changes in the tissues of experimental mammals preceding the appearance of age-dependent disease entities, which, according to the histopathological criteria discussed, seem to this author to be essentially identical with the histopathological manifestations of premature ageing (Casarett, 1952, 1956, 1957, 1958, 1960).

This radiation fibro-atrophy comprises an increase in amount and density of connective tissue, and increase in amounts of interstitial mesenchymal elements, constituting an increase in histohaematic barrier, reduction of fine vasculature in a process of arteriolocapillary fibrosis, and reduction in number of parenchymal cells.

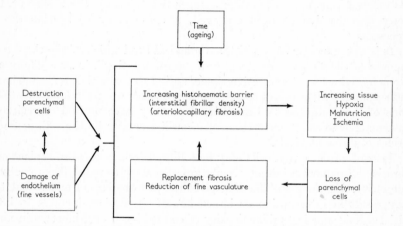

FIG. 3. Histopathological hypothesis for the tissue fibro-atrophy of ageing and "radiologic ageing".

The following is a description of the author's histopathological hypothesis of ageing and premature ageing caused by radiation (Fig. 3). At radiation dosage levels causing temporal advancement of age-dependent diseases and life-span shortening, the nonspecific damage to the endothelium of fine vasculature, by direct or indirect mechanisms, seems to be the change of greatest importance, among the early radiation changes, in the eventual development of the tissue fibro-atrophy of "radiologic ageing". Although some of the damage to the fine vasculature may be due to relatively direct effects of the radiation on endothelial cells, much of it is probably indirect. Some of the indirect damage to fine vasculature in some tissues seems to be due to the initial destructive effect of radiation on parenchymal cells, which varies in degree according to their nature and radiosensitivity from one tissue to another. Secondary to damage to capillary and arteriolar endothelium, there is some degree of pericapillary or interstitial oedema and an increase in the amount of interstitial colloid, followed by an increase in fibrillar density of the interstitial connective tissue. Some increase in subendothelial connective tissue in arterioles also develops. These changes in connective tissues constitute an increase in the histohaematic barrier and a temporal advance in the development of arteriolocapillary fibrosis, which are progressive processes in "normal" ageing. These manifestiations increase progressively with passing time in the irradiated tissues as they do in non-irradiated tissues, but remain temporally advanced in degree in the former as compared with the latter. As these progressive changes gradually reduce the capacity of the vasculo-connective tissues to support the full complement of parenchymal cells, owing to the development of relative hypoxia and malnutrition, the parenchymal cells gradually become decreased in number either as a result of precocious ageing and death or as a result also of decrease in cell reproduction in the case of renewable cells. This loss of parenchymal cells occurs earlier in the irradiated tissues than in the non-irradiated tissues, according to the earlier changes in histohaematic barrier and fine vasculature. With the gradual decrease in number of parenchymal cells there occurs a "replacement fibrosis" with increase in interstitial mesenchymal elements and further increase in the histohaematic barrier and with further reduction of fine vasculature secondary to the loss of parenchyma.

The influence of the early parenchymal damage on the early changes in vasculoconnective tissue and on the late effects or ageing manifestations is best described for different types of tissue in terms of the nature and relative radiosensitivity of the parenchymal cells they contain.

Vegetative inter-mitotic parenchymal cells, which divide regularly but differentiate little or not at all, e.g. basal cells of epidermis, are generally highly sensitive to the destructive action of radiation. *Differentiating inter-mitotic cells*, which divide regularly and differentiate to some extent between divisions, e.g. myelocytes, are somewhat less sensitive but still relatively sensitive cells in general. Certain mesenchymal elements, including endothelial cells of fine vasculature, are intermediate in sensitivity between these highly sensitive cells and the relatively resistant *reverting post-mitotic parenchymal cells* and the extremely resistant *fixed post-mitotic parenchymal cells*. The *reverting post-mitotic parenchymal cells* are variably differentiated cells which do not divide regularly but are capable of dividing upon appropriate stimulus, e.g. hepatic cells when a partial hepatectomy is performed. The *fixed post-mitotic parenchymal cells* are highly differentiated cells which have lost completely their ability to divide under any circumstances, e.g. the neuron. Some of these, like the neurons, are long-lived, age and die without replacement; others, like polymorphonuclear leucocytes, are relatively short-lived, age and die, but are replaced by the activity of vegetative and differentiating inter-mitotic precursor cells.

In tissues containing the radiation sensitive vegetative and differentiating inter-mitotic parenchymal cells, such as epidermis, gastro-intestinal mucosae, and haematopoietic tissues, the early radiation destruction of these cells depends relatively little on the early radiation damage of vasculo-connective tissue, but the degree of damage to fine vasculature and the change in connective tissue seems to be generally greater in such tissues than in tissues with radiation-resistant parenchymal cells receiving similar doses. Following total-body doses in the sublethal life-shortening range, the radiation sensitive parenchyma is regenerated after destructive effects to normal or subnormal levels of cellularity while vasculo-connective tissue-changes are irreversibly "fixed" and progress with passing time to an advanced level as compared with non-irradiated tissue (Fig. 4; Hursh et al., 1958). Eventually there is another phase (phase IV) of gradual loss of parenchymal cells secondary to these progressive changes in supporting tissues and premature as compared with non-irradiated tissues. The intermediate period (phase III) of maintained parenchymal cellularity between the period of regeneration and the beginning of the period of the age involution is shorter the larger the dose, owing to the fact that the vasculo-connective-tissue changes are greater the larger, or more intense, the dose. This intermediate period corresponds to the period of relatively low mortality rate between the period of acute, or subacute, mortality and the late period of age-dependent mortality.

With the larger doses that can be administered to such tissues under conditions of localized irradiation, there is greater damage to vasculo-connective tissue and greater destruction of parenchyma, that is, directly and sometimes also secondary to marked vascular damage. There is also reduced and delayed regenerative activity of remaining parenchymal cells, due to damage of fine vasculature and connective tissue, and a shorter intermediate period between the regenerative phase and the later involutional phase. In fact, with sufficiently high doses, there is no intermediate period due to failure of regeneration and the development of early fibro-atrophy of the tissue.

In the case of tissues containing the radiation-resistant reverting post-mitotic parenchymal cells, such as liver and kidney and many other epithelial glands, and tissues containing the irreplaceable, radiation-resistant, fixed post-mitotic parenchymal cells, such as muscle and brain and spinal cord, early radiation destruction of considerable numbers of parenchymal cells in direct, or indirect, fashion requires relatively large doses. Consequently, early destruction of considerable numbers of these parenchymal cells is not seen with total-body irradiation in the sublethal dose range, but only with more intensive localized, or generalized, irradiation. However, with sublethal irradiation of the whole-body the vasculo-connective tissue changes may be advanced temporally, and the parenchyma, although not appreciably damaged histopathologically in earlier phases, may show precocious loss of parenchyma in the later phase of age-involution secondary to these changes. With increasing doses in localized exposures the temporal advancement of these processes is progressively greater.

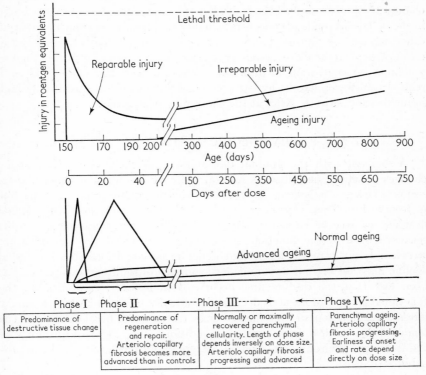

Fig. 4.

Localized ageing changes of this kind have been produced in many of the organs of the body by external sources of radiation (NAS-NRC, 1961), in radio-therapy patients and experimental animals, and by internally-deposited radioactive isotopes (Casarett, 1952, 1956). The tissue changes involved in the development of radiation-induced, or temporally advanced, nephrosclerosis following localized, or generalized, irradiation from external sources (NAS-NRC, 1961; Furth et al., 1959), or internal administration of ^{210}Po (Casarett, 1952, 1956) are histopathologically of the ageing type, since the fundamental histopathological process in ageing of the kidney is essentially a nephrosclerotic process.

Internal radioactive substances distributed more or less diffusely throughout the body, such as ^{32}P, tend to mimic the picture produced by total-body irradiation from external sources.

Intermediate types of exposure from internal emitters which, although diffusely deposited throughout the body, tend to concentrate differentially among a variety of tissues, may result in different degrees of premature ageing change in different parts of the body. For example, ^{210}Po is widely distributed in the body, but tends to concentrate highly in spleen and kidney and certain other organs. Consequently, although the fibro-atrophy may be a widespread effect, it tends to be more advanced generally in tissues and organs of highest polonium concentration or radiation exposure, and the causes of death related to nephrosclerosis, hypertension and renal failure are relatively high in incidence (Casarett, 1952) at life-shortening dose levels.

PHYSIOLOGICAL CHANGES

Associated with the gradual development of the degenerative ageing changes generally described as fibro-atrophy or involutional change, there seems to be a gradual decline in functional abilities, or, more exactly, a decline in reserve functional capacities in related tissues and organs. This gradual decrease of reserve capacity may be detectable early provided suitable sensitive tests of the functions are used and provided the functions are stressed. If not stressed but tested under basal conditions, the functional decline may not become apparent until so much of the reserve capacity has been lost that the decline first becomes manifest as symptoms. As the functional reserve of tissue is decreased gradually to a point where ordinary stresses tax function, or function is deficient even under basal conditions, the tendency to disease from internal conditions and the susceptibility to diseases from environmental factors are gradually increased, as is the probability of death.

A decrease in functional reserve of one vital part by precocious localized ageing far in advance of other vital parts tends to increase relatively, the probability of disease in the affected part or dependent parts with respect to the natural probability of other common diseases of the species.

Once the degenerative changes and decreasing functional reserve capacities of ageing have reached the point at which serious chronic diseases begin to emerge, especially the various chronic progressive disorders of later life, the correlated effects of ageing and disease on physiological processes tend to contribute to the pathogenesis of related disorders, perpetuate themselves by circular reactions, exacerbate other pre-existing difficulties, and cause further changes characteristic of ageing, further decreasing functional reserves. The pathogenesis of nephrosclerosis and the renal-hypertensive syndrome with consequent generalized arteriosclerosis and its consequences following irradiation is a good example (Casarett, 1952).

Unfortunately there has been relatively little study of the long-term effects of irradiation on body functions, especially functional reserve capacities, directed toward the problem of ageing. However, some of the functional studies which have been done suggest that functions which decline with age tend to decline prematurely as a result of life-shortening irradiation, in accordance with the deterioration of tissue as observed histopathologically.

There is also a paucity of biochemical information on effects of irradiation directly related to manifestations of ageing. However, some pertinent data have been obtained which indicate premature ageing manifestations following irradiation. One of the most notable examples is the work of Sobel (1960) showing premature decrease of the hexosamine–collagen ratio following irradiation of skin, indicating a relative decrease of ground substance and a relative increase in collagen. Whether this change is primary or secondary to histopathological effects of irradiation in vasculature or parenchyma, or represents only the increase in collagenous tissue in replacement fibrosis, is not yet clear.

CONCLUSION

The foregoing general considerations of actuarial, pathological, histopathological, and physiological changes following irradiation, in comparison with the manifestations of "normal" ageing, indicate a strong resemblance between the late effects of life-shortening total-body irradiation and the manifestations of premature ageing, especially with respect to the histopathological manifestations preceding age-dependent disease. The histopathological changes preceding disease development at the site of localized irradiation from external sources, or from internal radioisotopic sources, also bears a strong resemblance to the histopathological manifestations of ageing and therefore suggest the concept of premature localized ageing resulting from localized irradiation. Although the processes and manifestations of "normal ageing" and "radiologic ageing" at the histopathological level are similar, their nonspecificity makes it impossible to predict whether or not the more fundamental underlying processes, whatever they may be, will be similar or quite different.

REFERENCES

ALEXANDER, P. (1957). *Gerontologia* **1**, 174.
BLAIR, H. A. (1956). USAEC Document UR–442.
BRUES, A. M., and SACHER, G. A. (1952). In "Symposium on Radiobiology" (J. J. Nickson, ed.), p. 441. Wiley, New York.

CASARETT, G. W. (1952). USAEC Document UR–201.
CASARETT, G. W. (1956). *J. Gerontology* **11**, 436.
CASARETT, G. W. (1957). USAEC Document UR–492.
CASARETT, G. W. (1958). USAEC Document UR–521. Also (1959). "Radiobiology At The Intra-Cellular Level" (T. G. Hennessy *et al.*, eds.), p. 115. Pergamon Press, New York.
CASARETT, G. W. (1960). In "The Biology of Aging", (B. Strehler, ed.), p. 147. A.I.B.S., Washington, D.C.
COMFORT, A. (1959). *Radiation Res.* Suppl. **1**, 216.
FURTH, J., UPTON, A. C., and KIMBALL, A. W. (1959). *Radiation Res.* Suppl. 1, 243.
GRAHN, D. (1958). 2nd UNIC on Peaceful Uses of Atomic Energy, Vol. 22, p. 394.
HENSHAW, P. S. (1944). *J. nat. Cancer Inst.* **4**, 513.
HURSH, J. B., CASARETT, G. W., CARSTEN, A. L., NOONAN, T. R., MICHAELSON, S. M., HOWLAND, J. W., and BLAIR, H. A. (1958). 2nd UNIC on Peaceful Uses of Atomic Energy, Vol. 22, p. 178.
JONES, H. B. (1956). "Advances in Biology and Medical Physics", Vol. 4, p. 281. Academic Press, New York.
MOLE, R. H. (1957). *Nature, Lond.* **180**, 456.
NAS–NRC (1961). Report of Subcommittee on Long-Term Effects of Ionizing Radiations from External Sources, Committee on Pathologic Effects of Atomic Radiation, NAS–NRC Publication 849, Washington, D.C.
NEARY, G. J., MUNSON, R. J., and MOLE, R. H. (1957). "Chronic Radiation Hazards". Pergamon Press, London.
RUSS, S., and SCOTT, G. M. (1939). *Brit. J. Radiol.* **12**, 440.
SACHER, G. A. (1955). *J. nat. Cancer Inst.* **15**, 1125.
SOBEL, H. (1960). In "The Biology of Aging" (B. Strehler, ed.), p. 274. A.I.B.S., Washington, D.C.
UPTON, A. C. (1957). *J. Gerontology* **12**, 306.
WARREN, S. (1956). *J. Amer. Med. Ass.* **162**, 464.

DISCUSSION

CURTIS: I agree with most of the points which Dr. Casarett has raised. There is one thing, however, to which I should like to enter an objection and that is the concept that he proposed in the first part where he indicated that radiation may be used as a tool for measuring the residual injury which is present in a group of animals. I think at several times we've shown, and it has also been shown by a number of other people, that the LD_{50} really changes remarkably little with age. There are certain strains of animals apparently in which it does go down, but there are many strains in which it's almost flat right to the end and then goes down again. I think that perhaps a fallacy in reasoning might be brought out by a much over-simplified concept. You could think for example that the acute 30-day mortality of mice might be primarily due to the bone-marrow effect, whereas perhaps in the long-term effects which we are discussing here it might be due to failure of completely different organ systems—the kidney has been mentioned as one possibility. So that when you test the residual injury in these animals, what you really want to do is test the residual injury in the organ in question, which might be the kidney, whereas when you give an LD_{50} dose what you are doing is testing the residual injury in the haematopoietic system which may have almost completely recovered from the disease.

CASARETT: This is a very good point. I did emphasize however that this test, limited as it is, was essentially best applied after maximal tissue regeneration, in other words when the acute and sub-acute death periods have passed and there has been maximal tissue regeneration. If done periodically, of course, one is testing different degrees of change and also different diseases in terms of resistance to lethality. Of course we are again limited when the age-dependent diseases begin to take their toll. We start then to get a difference in the degree of difference between LD_{50}'s if we measure at the same chronological ages, but if comparison is made at the median death times, which essentially is a more comparable situation after these diseases begin to take their toll, one can find, at least in some cases, still a similar degree of difference in the LD_{50}'s. This has to be taken into account also.

ROTBLAT: For the first point, the use of the LD_{50}'s as a criteria, you are probably right in saying that it is not a very good tool unless you use such a large dose that you almost end the life, in which case the animal died, so you can't use it in any case. On the other hand, the remarkable thing is that there is a correlation between the long-term effects which you express in the life-shortening, and the sensitivity to the acute effects. This is quite remarkable.

INITIAL X-RAY EFFECTS ON THE AORTIC WALL AND THEIR LATE CONSEQUENCES

H. B. LAMBERTS

Physiologisch Laboratorium, der Rijksuniversiteit, Groningen, Holland

SUMMARY

Effects of X-irradiation *in vivo* and *in vitro* on the aortic wall are described, e.g. increased atheromatosis in hypercholesterolaemic rabbits drop of injection-pressure in the aortic wall and increased permeability for several dyes. The importance of the radiosensitivity of the mucopolysaccharide matrix is stressed. The possibility of chemical protection of this matrix by $Na_2S_2O_3$ (sodium thiosulphate) is discussed.

In considering late vascular damage by irradiation we can start with the statement that no doubt is accurate about the end effect: a single aortic irradiation in dogs with 1,500–2,500 r will cause atherosclerosis in 30 to 40 weeks, not to be distinguished from the "natural" process, but strictly localized to the irradiated field (Lindsay *et al.*, 1962). There also appears to be general agreement that the initial damage, both in the ageing and the irradiation process must be traced to the intimal elastic tissue and its mucopolysaccharide matrix. The lipid infiltration is an early, or late, secondary occurrence; in men it is early and very important and the relation to plasmalipids, mostly measured as hypercholesterolaemia is well proved. In rats (Gold, 1961) and in rabbits (our own observations) evidence points to increase of hypercholesterolaemic atheromatosis by irradiation (Figs. 1, 2 and 3) but this may become complicated by the apparent chemoprotection against irradiation by increased cholesterol (Koch and Schenk, 1961; our own observations), so the time-relations between the two atherogenic processes must be important.

As we are chiefly concerned here with the initial damage by irradiation, leading up to late atheromatosis; the changes in the extracellular elastic or reticular fibres, narrowly related to changes in their ground substance must be central. Further we have to supplement the results of measurements on static material like excised "dead" aortic walls with observations of the dynamic, living situation. Observations with uptake and turnover of ^{35}S-sulphate in living animals or surviving intimal aortic tissue have demonstrated the dynamic existence of the polysulphate matrix. In human foetal tissue the polysulphate synthesis is high, in adult tissue it is slower but still indicating a turnover of 1 to 2 weeks for elastic fibres. So formation and destruction of these fibres must be going on continuously and for this an intact mucopolysaccharide matrix is fundamental. With age, turnover decreases sharply; with development of atheromatosis sulphation may again increase markedly (Kirk, 1959).

It is important to know how the sulphation rate behaves after irradiation. In the study of Lindsay et al. the earliest observable radiation lesion is located in the intimal fenestrated membrane and the intimate generic relationship of this membrane to the enveloping mucopolysaccharide matrix is stressed by them. This study was mainly structural and I will now supplement it by some functional observations on the irradiated, non-living aortic intima, indicating the functional importance, and the radiation sensitivity of the subendothelial "gel filtration layer" or "basement membrane", covering the fenestrated elastic membrane on the luminal side.

Firstly, direct evidence of a radiation effect was found in measuring the injection pressure in the aortic wall during irradiation in a way analogous to that used for skin (Brinkman et al., 1961). The instantaneous drop of the injection pressure during the irradiation clearly demonstrates an increased permeability to water (Fig. 4), caused by depolymerization of the mucopolysaccharides in the matrix and possibly in the membraneous structures which have a close relation to the matrix. The possibility that radiochemical changes in the matrix substance will cause changes in the development of its fibrous and lamellar structures, e.g. in the basement membrane, should not be overlooked and in fact this possibility is really a probability (Perez-Tamayo, 1961; Gersh and Catchpole, 1960; Lindsay and Chaikoff, 1955; Weiss, 1962).

Impairment of the function of the aortic basement membrane or "gel filtration layer" by irradiation is shown in the results of the following experiments in which we studied the diffusion rate of several dyes by different techniques. In the simplest experiment in this series we irradiated pieces of excised aorta (cow, pig, sheep) with a dose of 2,000 r (50 kV) after which one drop of a saturated trypan-blue solution was placed on the endothelial surface and was left there, the pieces being kept for 15 hours in a wet chamber at room temperature. In the irradiated pieces the trypan-blue penetration was regularly 2 to 3 times as deep, in the same time, as in the non-irradiated control pieces (Fig. 5).

In the next experiment intimal membranes having a thickness of ± 0.35 mm were prepared by peeling them off freshly excised aortae. They were fixed over 15 mm bore lucite tubes, which were afterwards filled to 1 cm height with a salt solution and immersed in the same solution. The diffusing solute, haematoporphyrine, was added to the salt solution in the tubes. Diffusion of this dye through the intimal membrane could easily be detected and estimated by the observation of red fluorescence in the immersion bath. The probability of photodynamic effects, made it necessary to perform these

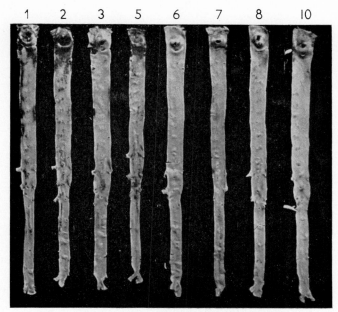

Fig. 1. Aortae of rabbits fed on a high cholesterol diet. The first four aortae are of rabbits irradiated once a week with 100 r over the aortic region (200 kV, 20 mA, 0·5 mm Cu) for 10 weeks. The last four are of non-irradiated control animals. There is an increase in atheromatous lesions (stained with Sudan-red) in the irradiated group.

Fig. 2. Aorta with atheromatous plaque in a rabbit fed on a high cholesterol diet for 3 months.

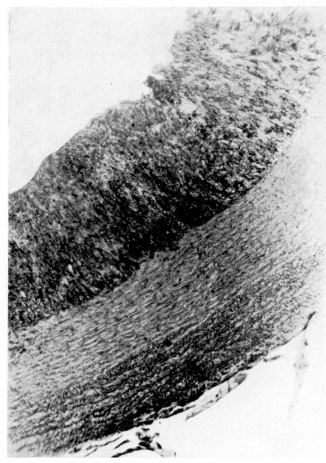

FIG. 3. Aorta with enormous atheromatosis in a rabbit fed on a high cholesterol diet for three months and irradiated with 3,000 r over the aortic region 1 week before sacrifice.

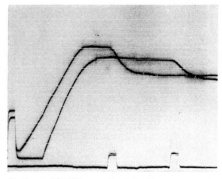

FIG. 4. Effect of irradiation with 500 r on the injection pressure in the medial portion of a cow aorta. On the left is the calibration signal corresponding to a pressure of 100 mm Hg. The lower line gives the irradiation signals for both needles.

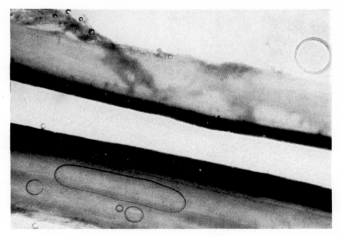

Fig. 5. Difference in trypan blue diffusion between control and irradiated piece of a cow aorta. Both sections are taken from the same region of the same aorta.
Upper control; *lower* irradiated with 2,000 r.

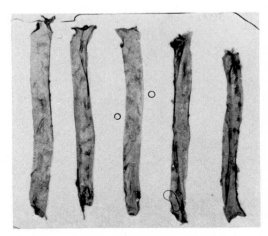

Fig. 7. Rat aortae; first three, non-irradiated controls; last two, total-body irradiation with 3,000 r; all were injected with 0·7 ml 1% Congo-red solution and killed 30 minutes after injection.

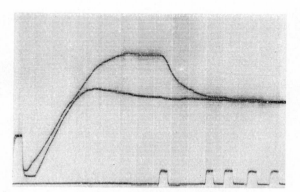

Fig. 9. Protective action of 0·001 M $Na_2S_2O_3$ injected through one of the two needles in the aortic wall. There is only a pressure drop in the needle injecting saline without thiosulphate while repeated irradiations on the other needle point have no result.

experiments in a dark room. The results are shown in Fig. 6. Irradiation of the intimal membrane with 1,000 r at once caused a much more rapid permeation.

This result encouraged us to examine a possibly analogous radiation effect *in vivo*. If a 1% solution of Congo-red in 5% glucose is injected intravenously in a rat directly after a whole-body irradiation with 1,000–3,000 r and the rat is killed 30 minutes later, the aorta is distinctly more stained than

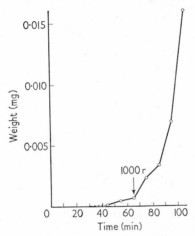

FIG. 6. Rate of haematoporphyrin diffusion through the intimal membrane of a pig's aorta, before and after irradiation with 1,000 r.

the aortae of non-irradiated control animals (as may be seen in Fig. 7). To prevent confusion, caused by possible staining of the deeper aortic layers through diffusion of dye from the vasal vessels, we modified the procedure. After the aorta is excised, it is opened by a longitudinal cut at the dorsal side between the origins of the intercostal branches. Then a thin intimal membrane can be peeled off, which is put on a slide and dried. Immersion of this membrane in glycerol makes it quite transparent and spectrophotometric estimation of dye diffusion is possible. An example of such an estimation is shown in Fig. 8. Here haematoporphyrin was used as the injected dye and the rats had been exposed to 3,000 r.

With regard to a possible chemoprotection against this vascular damage by ionizing radiation I might say that in our work $Na_2S_2O_3$ (sodium thiosulphate) has given some very promising results. In all of the radiation effects on mucopolysaccharide systems we have studied, the vascular effects included, $Na_2S_2O_3$ acts as an efficient protector (Fig. 9). Intracellular radiation effects are not influenced by thiosulphate (the LD_{100} 30 days for C57 black mice does not change), but there might be considerable protection against the extracellular radiation effects on vascular walls. The toxicity of thiosulphate is very low which gives us the possibility of flooding animals, and eventually patients, with the substance. These considerations make it worthwhile to study the influence of $Na_2S_2O_3$ on non-neoplastic long-term radiation effects, in particular on the vascular effects and shortening of life-span.

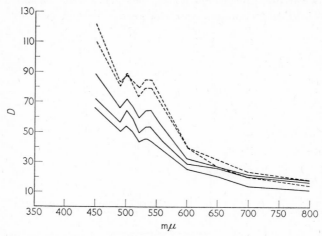

Fig. 8. Spectrophotometric estimation of the amount of haematoporphyrin diffusion in the intimal membrane of the rat aorta.

Broken lines: aortae of rats irradiated with 3,000 r.
Solid lines: aortae of unirradiated, control rats.

REFERENCES

Brinkman, R., Lamberts, H. B., Wadel, J., and Zuideveld, J. (1961). *Int. J. Rad. Biol.* **3**, 205.
Gersh, L., and Catchpole, H. R. (1960). *Perspectives Biol. Med.* **3**, 282.
Gold, H. (1961). *Arch. Path.* **71**, 268.
Kirk, J. E. (1959). "The Arterial Wall" (A. J. Lansing, ed.). Williams & Wilkins, Baltimore.
Koch, R., and Schenk, G. O. (1961). *Strahlentherapie* **116**, 364.
Lindsay, S., and Chaikoff, I. L. (1955). *Arch. Path.* **60**, 29.
Lindsay, S., Kohn, H. J., Dakin, R. L., and Jew, J. (1962). *Circulation Res.* **10**, 51.
Perez-Tamayo, R. (1961). "Mechanisms of Disease." W. B. Saunders, Philadelphia.
Weiss, P. (1962). "The Molecular Control of Cellular Activity" (J. M. Allen, ed.). McGraw–Hill, New York.

DISCUSSION

UPTON: I would like to enquire about the relationship between the observed effects and dose. First, can you get demonstrable effects with doses substantially lower than 1,000 r *in vitro* or *in vivo*? Secondly, do you have any idea how much dose reduction you got with thiosulphate?

LAMBERTS: 1,000 r is the dose with which we have an effect in almost all experiments, but sometimes an effect can be found with a dose of a few hundred r. From experiments with synovial fluid in which we measured the X-ray effect on the viscosity, I might say that the dose reduction factor is about 2 or more.

MOLE: May I ask one or two questions about the technique? This stems from an interest in whether oxygen was present at the place where the irradiation was producing its effect. In the *in vitro* experiments, were the aortae in a bath or were they in air, and how deep was the needle within the aortae?

LAMBERTS: There is an oxygen effect in this system, but it is a negative one. In experiments with synovial fluid and solutions of pure hyaluronic acid, we found a larger drop in viscosity by irradiation under anoxic conditions than under hyperoxic conditions. In Professor Bacq's laboratory the effect of X-irradiation on the injection pressure in the rat's skin was studied under complete anoxia. The effect was at least as strong as in the presence of oxygen. In the *in vitro* experiments the aortae were kept in air and the depth of the needle in the injection experiments varied, but was usually less than 1 mm.

CURTIS: I was going to ask about the injection experiments. The effect that you get is very rapid there, a matter of seconds as I interpreted it. Could not this be a smooth muscle reaction?

LAMBERTS: Certainly not. The needle is inserted in the intima where practically no muscle is present. Furthermore the effect is still present in aortae that had been kept in the 'fridge for several days.

CURTIS: Nevertheless, the fluid has to go somewhere. As the smooth muscle contracts would not that increase the resistance?

LAMBERTS: The amount of injected fluid is a few mm^3. Sometimes a small bubble is formed on the intima. Smooth muscle contraction would cause a rise in injection pressure but the effect after irradiation is a decrease.

COTTIER: My first question concerns the reversibility of the effect. The second question—you have shown these atheromata and in correlation with your findings here, how strong are your feelings that these two are strongly correlated? Atheroma is a focal process and it is still in discussion what part in its pathogenesis is played by local thrombosis. I think that irradiation can also produce local thrombi; either due to endothelial, or some other, damage.

LAMBERTS: In our experiments we used excised dead pieces of aortae, so the drop of injection pressure was not reversible. In the skin of living rats you do find reversal—rebuilding of your injection pressure in about 4–5 hours, and then if you irradiate again you get the same effect. As for your second question, I was afraid somebody would ask that. I myself am not quite certain about the relationship between these effects, but I think the most important thing here is to prove that something is produced in this field by irradiation. If the development of atheromatous lesions has something to do with endothelial damage or with the production of thrombi or the depolymerization of mucopolysaccharides under the endothelium well, they all might play a part in this process, I think.

UPTON: This is to observe that radiation has been noted to cause lipaemia in irradiated animals, and this effect coupled with some injury to the endothelium might conceivably be additive or synergistic in setting up the cholesterol plaques.

THE RESPONSE OF TISSUES TO CONTINUOUS IRRADIATION

L. F. LAMERTON

Physics Department, Institute of Cancer Research, Royal Cancer Hospital, London, England

INTRODUCTION

The topic I am going to discuss very briefly is the response of the rat gut and bone-marrow to continuous irradiation, with particular reference to the extent of maintenance of cell population and function. This topic perhaps does not fall directly within the subject matter of this session, but is certainly very closely related to it.

There are great differences in the response of the various renewal systems of the body to continuous irradiation. From the point of view of maintenance of function the extremes, in the rat, may be represented by the epithelium of the small intestine (and possibly of other parts of the gut) which can maintain function under a dose-rate of at least 400 rads/day until death supervenes from other causes, and the testis, which shows progressive cellular depletion at doses of a few rads per day.

RESPONSE OF THE EPITHELIUM OF THE SMALL INTESTINE

Work on the epithelium of the small intestine of the young rat, 6 to 8 weeks old (Quastler *et al.*, 1959; Lamerton *et al.*, 1961) shows that at a dose-rate of 415 rads/day (from a ^{137}Cs source) the mitotic index in the crypts falls during the first day or so and then rises to a value about 60% of normal. At the same time the cell population in the crypts drops to about one-half of normal at which value it remains steady until death occurs from bone-marrow failure (see Fig. 1). One would expect that this apparent adaptation was brought about by an increase in the mean proliferation rate of the crypt cells, as a homeostatic response to cell death. However, measurements using tritiated thymidine labelling, with the method of percentage-labelled mitoses (Lamerton *et al.*, 1962) indicate no substantial difference either in minimum, or mean, generation time in the crypts of control rats, or of rats which have been exposed for 5 days at 415 rads/day. Another possibility which must be considered is that the crypt cells, for some reason, develop an increased radio-resistance—a suggestion that was made some time ago by Margaret Bloom (1950), when studying the effect of fractionated exposure on the small intestine of the rat. This suggestion receives some support from the observation that the number of cell fragments in the crypts decreases with time under continuous irradiation. If there is, in fact, an increase in radio-resistance in this tissue under continuous irradiation, the mechanism is unknown. An increase in polyploidy has been looked for, but not found (Kember et al., 1962).

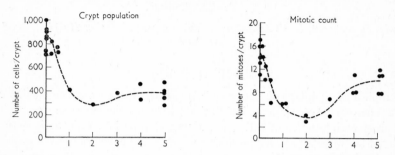

FIG. 1. Response of small intestine of young rat (6 to 8 weeks old) to a continuous dose-rate of 415 rads per day from a ^{137}Cs source. Data obtained from squash preparations of crypts.

It is not known how long the gut will maintain function at a dose-rate of 415 rads/day, but in studies after 100 days irradiation at 50 rads/day, a normal mitotic index has been found with evidence of only a small reduction in crypt population. The generation time of the cells under these conditions is at present being measured.

RESPONSE OF THE BLOOD-FORMING TISSUES

The blood-forming tissues behave very differently from the gut under continuous irradiation. The dose-rate at which they maintain function is much lower than for the gut, and there is also some evidence that they respond by a substantial decrease in mean generation time of the cells. The dose-rates which can be tolerated for considerable periods of time appear to differ for the different types of blood-forming tissues (Lamerton et al., 1960); erythropoietic tissue being the most resistant type. At 415 and 176 rads/day the count of white cells and platelets in the blood falls rapidly without any evidence of stabilization, and studies with ^{59}Fe indicate that the erythropoietic activity also collapses rapidly. At 84 rads/day there is an initial fall in white count and platelets followed by a transient recovery preceding the final fall to 50 or 60 days when the animal dies. However, even terminally, when white cell count and platelet count are very low, ^{59}Fe studies show a net erythropoietic activity little different from normal, though there may be some difference in the relative activities of spleen and bone-marrow. This point will have to be investigated.

At 50 rads/day the white cell count and platelet count, after an initial fall and recovery, are generally maintained at normal levels or at levels only slightly below normal until the animal dies at 150 to 200 days (Fig. 2). Studies with ^{59}Fe (Lamerton *et al.*, 1961) have demonstrated an almost normal erythropoietic activity after 140 days of irradiation.

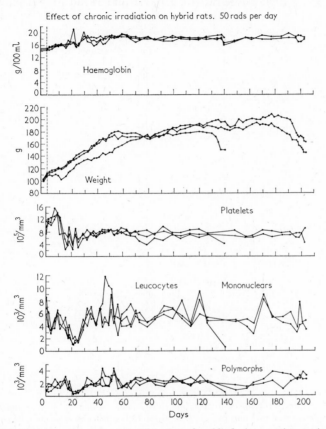

Fig. 2. Blood changes in young rats (6 to 8 weeks old) during continuous irradiation at 50 rads per day.

It is of interest to study the response to a stress of the blood-forming tissues in a continuously irradiated animal. The stress used has been removal of one-third of the blood volume, by cardiac puncture, and Fig. 3 compares the response of normal animals with those of the same age but irradiated at 50 rads/day for 130 days. It can be seen that the rate of haemoglobin recovery is almost the same in normal and irradiated animals. One difference observed in these bleeding experiments between normal and irradiated animals was the lack of a marked leukocytosis in the irradiated animals, suggesting that the leucocyte stores in the body had been considerably reduced.

At the moment it is not possible to say whether the megakaryocyte-platelet and white cell systems differ with respect to the dose-rates that can be tolerated. At the higher dose-rates of 415, 176 and 84 rads/day the low platelet

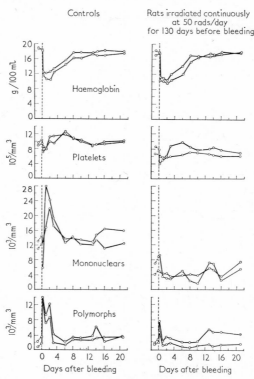

FIG. 3. Response of rat to removal of one-third blood volume by cardiac puncture. Comparison between normal rats and rats irradiated for 130 days at 50 rads per day.

count appears to be the major cause of death (leading to a rapid haemorrhage) but at the same time the monocuclear and polymorphonuclear counts are also very low.

Studies of cell proliferation in the bone-marrow, using tritiated thymidine as a label for dividing cells, have given some indication, at 50 and 84 rads/day, that the mean proliferation rate is increased for both red and white cell precursors (Lord, unpublished). More detailed studies of generation time are at present under way, but one would suspect, from various lines of evidence, that the decrease in generation time is the result of the more slowly dividing cells speeding up, or of latent cell population called into division, and not of any change in minimum generation time.

We have not found any evidence of increased radio-resistance in the blood-forming tissues after exposure at 50 rads/day. The effect of a single dose of acute radiation appears, if anything, to be slightly more severe than in normal animals.

DISCUSSION

It is not yet possible to come to any firm conclusion about the reason for the different response of gut and bone-marrow to continuous irradiation. Possible reasons are:

(i) a different radiosensitivity of the precursor cells with respect to continuous irradiation,

(ii) a difference in the homeostatic forces which influence speed and efficiency of regeneration following the radiation injury,

or (iii) a difference in the significance of general tissue damage produced by radiation.

With regard to (i), the generation time could be a very important factor. With a shorter generation time less radiation dose will be accumulated between divisions, which would lead to a decrease in radiosensitivity if there were a substantial "wiping out" of damage at division. To determine whether this is the case requires much more data than we have at present on the response of cell populations with different mean generation times. The extent of the accumulation of chromosomal or other damage in the stem-cell compartment of tissues is obviously one of the major factors that has to be taken into account, and more work on tissues of low proliferation rate (such as the liver prior to partial hepatectomy) is required.

Generation time, in stem-cell and in multiplicative compartments, will also be one of the factors influencing (ii), the speed of regeneration, but there will be many other factors as well, such as the nature and efficiency of feedback mechanisms. The greater tolerance of the red cell system to continuous irradiation, as compared with the white cell system, could be the result of a more rapid and efficient homeostatic mechanism. For an assessment of the importance of this factor in comparing the responses of gut and blood-forming tissues much more information is required with regard to both systems. However, it could be that the rapid recovery observed in the gut after injury results from a homeostatic mechanism not based on feed-back, but on a normally very high proliferation rate together with a short cell precursor line. In the blood-forming tissues, feed-back mechanisms and a relatively long line of precursors would both tend to delay the onset of cell repopulation.

Also one must recognize the possible importance of generalized tissue dose. In the bone-marrow, capillary changes and haemorrhage might lead to a collapse of cell proliferation at a dose-rate lower than the limiting value for individual cell function, whereas the gut architecture might be of a less susceptible type.

ACKNOWLEDGMENTS

This paper is a short summary of work carried out by a number of investigators in this Department and I would like to express my appreciation particularly to Miss K. Adams, Dr. J. P. M. Bensted, Mr. J. Blackmore, Dr. N. M. Blackett, Mr. B. I. Lord and Mrs. Doris Wimber and to Professor W. V. Mayneord, Director of the Physics Department, Institute of Cancer Research, Royal Cancer Hospital.

REFERENCES

BLOOM, M. A. (1950). *Radiology* **55**, 104.
KEMBER, N. F., QUASTLER, H., and WIMBER, D. R. (1962). *Brit. J. Radiol.* **35**, 290.
LAMERTON, L. F., PONTIFEX, A. H., BLACKETT, N. M., and ADAMS, K. (1960). *Brit. J. Radiol.* **33**, 287.
LAMERTON, L. F., STEEL, G. G., and WIMBER, D. R. (1961). "Fundamental Aspects of Radiosensitivity", p. 158. Brookhaven Symposia in Biology, no. 14.
LAMERTON, L. F., LORD, B. I., and QUASTLER, H. (1962). 5th Int. Symposium on Radioactive Isotopes in Clinical Medicine and Research, Bad Gastein, 1962. *Strahlentherapie* (in press).
QUASTLER, H., BENSTED, J. P. M., LAMERTON, L. F., and SIMPSON, S. M. (1959). *Brit. J. Radiol.* **32**, 501.

DISCUSSION

CURTIS: This is a most interesting and a very puzzling finding—the ability of an animal to withstand a much, much larger dose of radiation if it is given over a period of time than if it is given instantaneously; and the concept which Prof. Lamerton has presented, of a tissue which is dividing very rapidly being able to withstand a lot more continuous radiation because it can, so to speak, wipe the slate clean at the time of cell division and then start all over again. This is an interesting concept and I think that Dr. Sparrow and his group at our laboratory have got a very considerable confirmation of this concept from work with plants in which it is possible, by varying conditions such as temperature and nutrient and things of this kind, to alter the rate of cell division and not to alter much of anything else. In that case there is a very nice substantiation of this concept that a plant can withstand chronic radiation almost in direct proportion to the rate at which cell division takes place.

MULLER: I found this very enlightening also, but also a bit puzzling because, although of course I accept it—it seems to me to run counter to the principle which was put forward something like 60 years ago by Bergonié and Tribondeau, and received a lot of support from different sides, that is, that tissues which divide rapidly are themselves more injured by radiation. This can be understood on genetic grounds, but perhaps there is some way to reconcile it because I think, in addition to that principle, we also have to remember that there are also selection processes taking place so that injured cells tend to be replaced by cells which are not so injured. As a result there is a kind of balance of one of these processes against the other to give the final result. For different tissues that balance might well be different.

LAMERTON: Yes, Dr. Muller, but I believe Bergonié and Tribondeau based their conclusions on the work of Regaud with the testis of the ram. In the testis you have apparently very little feed-back and this, I suggest, is why the testis is so susceptible to continuous radiation—there you can see the results of the injuries and they are not masked by any subsequent rapid regeneration. If the early workers had looked at the epithelium of the small intestine they might have come to another conclusion because there the immediate effects of radiation can so easily be masked by the results of rapid regeneration.

DRASIL: I would like to ask two questions with regard to the adaptation in the intestine to which you were referring. Firstly, could it be some kind of selection of more resistant cells originally present in small numbers? Secondly, have you done more detailed cytological examinations to show increasing chromosome numbers after irradiation?

LAMERTON: With regard to the first point, the problem about selection is that the increased radioresistance, if it be such, seems to come on rather too quickly. However we did consider the possibility of polyploidy developing and this was supported by the fact that the uptake of tritiated thymidine per cell in the continuously irradiated animals doubled or trebled very quickly. Dr. Kember, working with Dr. Quastler at Brookhaven, did a cytophotometric study but found no evidence of increasing ploidy and we have now come to the conclusion on other evidence that the increase in tritiated thymidine uptake is due to an availability difference rather than to any chromosomal difference. It may be due to some sort of selection but if so, then it would appear to happen very quickly.

There is still one other possibility and this is that a tissue which is regenerating at an abnormal rate may not be the same, biochemically speaking, as normal tissue—I mean, for instance, you may have anoxia or some other factor.

To return for a moment to your second question about chromosome studies, Dr. Wimber of Brookhaven, who was working with us for a year, did examine the bone-marrow of animals which had been exposed to continuous irradiation at 50 rads per day for 100 days but he found very little evidence of chromosomal abnormalities in these animals.

BRINKMAN: I should like to ask if the rapid increase in radioresistance in the gut cells could be explained by the liberation of serotonin from these regions.

LAMERTON: It may indeed. At the moment we have no satisfactory explanation and this is certainly a possibility.

DEVIK: Do you think there could be any correlation between the dose given to the cell during one cell cycle? Do you find that there is a critical dose delivered during one cell cycle? It might appear to be so from the curve you showed.

LAMERTON: Well, in order to study this, one wants to look at cells of different generation times. We are not sure about the generation times of the bone-marrow cells, one of the problems being that we have such a complex system. What we are still hoping is that with the gut we shall find cells of different generation times and be able to compare their response though I think it is very unreasonable to believe that sensitivity to continuous radiation will increase indefinitely with the length of the cell generation time.

ILBERY: Perhaps this is a good time to think of chromosome changes in normal tissues after radiation damage. Have you found evidence of increased chromosome numbers in normal tissues after radiation?

LAMERTON: Well, we've done very little work on this. The only work we have done is to look for more obvious changes such as micro-nuclei in bone-marrow cells of rats exposed to continuous irradiation for 100 days.

THE CHOLESTEROL CONCENTRATION IN ADRENAL GLANDS OF HYPOPHYSECTOMIZED-GRAFTED RATS (WITH AND WITHOUT DESTROYED MEDIAN EMINENCE) AFTER WHOLE-BODY EXPOSURE TO A LETHAL DOSE OF X-RAYS

P. N. MARTINOVITCH, D. PAVIĆ, D. SLADIĆ-SIMIĆ AND N. ŽIVKOVIĆ

Institute of Nuclear Sciences, "Boris Kidrič", Beograd, Yugoslavia

SUMMARY

It has been found that the pituitary transplants in hypophysectomized rats grafted in a heterotopic position (anterior eye chamber) will restore to a considerable degree the weight and the function of the testes and the adrenal glands. When these rats were exposed to a whole-body lethal dose of X-rays (850 r) a significant fall in the concentration of cholesterol in the adrenal glands was obtained. When the *median eminence* of hypophysectomized-grafted rats was destroyed by a micro-coagulation technique and 3 months later the experimental animals were exposed to 850 r of X-rays a fall in cholesterol was obtained that was still significant but considerably lower than if the *median eminence* were left intact. Results obtained with testes after the *median eminence* had been destroyed proved too inconsistent to allow interpretation.

The prevailing opinion nowadays is that the anterior pituitary gland grafted into positions distant from the hypothalamus will show very little gonadotropic, adrenotropic and thyrotropic activity (Harris, 1955; Benoit and Assenmacher, 1954). In collaboration with Mrs. D. Pavić (1960) we have found that the anterior pituitaries of infantile rats grafted immediately following hypophysectomy into the anterior eye chamber of 100 g rats will in about 75% of animals prevent the involution of the testes of the host and, if the grafting takes place one or two months following hypophysectomy, restore the weight and the function of at least some of the glands. Similarly, a clear cut restorative effect on the weight of the adrenal and the thyroid glands was observed. The return of the hypophysectomized-grafted rats and of the endocrine glands under the pituitary control into their original hypophysectomized state following the enucleation of the eyes containing grafts produced the necessary evidence for the functional activity of the transplanted pituitaries (Tables I, II).

In a second series of experiments (Bacq et al., 1956a, b, 1957a, b; Martinovitch, 1960; Martinovitch et al., 1961), the fall in cholesterol in the adrenals of the hypophysectomized-grafted rats exposed to a whole-body

TABLE I

Treatment	No. of rats	Body weight (g)	Weight of testes (g)	Weight of testes (g/100 g body weight)	Adrenal weight (mg)	Adrenal weight (mg/100 g body weight)	Thyroid weight (mg)	Weight of thyroid (mg/100 g body weight)
Normal controls	10	339 ± 8	2·50 ± 0·68	0·72 ± 0·22	34·2 ± 1·2	10·1 ± 0·4	14·4 ± 0·6	4·2 ± 0·2
Hypophysectomized controls	43	90 ± 3	0·17 ± 0·01	0·18 ± 0·01	6·3 ± 0·2	6·9 ± 0·2	4·8 ± 0·2	5·5 ± 0·3
Hypophysectomized grafted testes involuted	39	202 ± 6	0·26 ± 0·02	0·13 ± 0·01	15·7 ± 0·6	8·1 ± 0·3	11·6 ± 0·6	5·9 ± 0·3
Hypophysectomized grafted testes good	28	226 ± 9	1·67 ± 0·09	0·75 ± 0·14	13·8 ± 0·8	6·2 ± 0·3	12·0 ± 0·7	5·4 ± 0·3

lethal dose (850 r) of X-rays was studied. The exposure to X-rays produced a highly significant drop in cholesterol in the majority of the glands demons-

TABLE II

Protocol No. of experimental rats	Before removal of grafts Left testis weight (g)	After removal of grafts		
		Right testis weight (g)	Adrenal weight (mg)	Thyroid weight (mg)
B. V–2, 1958	0·80	0·16	7·9	8·6
B. VII–8, 1958	1·07	0·54	5·9	5·3
B. IX–3, 1958	1·21	0·16	8·1	8·3
B. X–10, 1958	0·95	0·18	5·9	7·3
B. IX–16, 1958	1·12	0·76	8·7	11·6
B. X–6, 1958	1·08	0·58	9·4	8·0
B. XII–2, 1959	1·01	0·16	7·3	12·2
B. XIV–1, 1959	1·02	0·16	6·7	8·2
B. XIV–3, 1959	1·17	0·48	6·9	8·0
B. XIV–8, 1959	1·04	0·19	6·1	6·7
Mean values	1·05 ± 0·04	0·34 ± 0·08	7·3 ± 0·4	8·4 ± 0·7
Mean values for hypophysectomized grafted controls			13·7 ± 0·8	11·2 ± 0·6

trating a stimulating effect on the part of the pituitary grafts upon the functional activity of the adrenal glands (Table III).

Recent histological and histochemical studies on the hypothalamus presuppose and endocrine activity by the neurones of the nuclei of the

TABLE III

Treatment	Gland before irradiation of rat		Gland after irradiation of rat	
	No. of glands	Cholesterol (mg/100 mg)	No. of glands	Cholesterol (mg/100 mg)
Hypophysectomized controls	26	3·5 ± 0·2	23	3·6 ± 0·4
Hypophysectomized grafted	10	3·5 ± 0·2	10	1·6 ± 0·1

hypothalamus, the function of which has been linked with the secretory activity of the anterior lobe of the pituitary. The evidence so far obtained has served to place the anterior pituitary in the same dependent position in respect to the hypothalamus as some other glands with internal secretion in their relation to the anterior pituitary.

It is not our purpose to discuss here modern views on the hypothalamus-anterior pituitary relationship. What we wish to do is to point out some facts that have become evident from our work on the hypophysectomized-grafted rats. One of them is the capacity of the grafts to stimulate the growth of the adrenal glands arrested at hypophysectomy. The other fact is the capacity of the adrenals of such rats to react with a highly significant drop in the concentration of cholesterol under stress mediated through a whole-body exposure of the host animal to a lethal dose of X-rays.

The implications of these findings, as it appears to us, are quite obvious. It seems that the anterior pituitaries grafted at a site (anterior eye chamber) removed from the hypothalamus will synthesize sufficient amounts of ACTH to stimulate adrenal growth and elicit the classical cholesterol reaction in the gland when a whole-body lethal dose of X-rays is applied as the stressor.

Under the assumption that the dependence of the anterior pituitary on the hypothalamus for its functional activity is an established fact we thought it worthy of effort to see if this dependence finds its expression in situations where the anterior pituitaries are grafted into positions removed from the hypothalamus (heterotopic grafting). The next logical step, therefore, involved a search for the region in the hypothalamus responsible for the activation of the pituitary gland. The electro-coagulation technique, widely applied in the study of the specific activity of the various areas of the brain mass, we hoped might give an answer to this question, and the *median eminence* was selected as the region to be destroyed. Three months following the destruction of the *median eminence* the rats were exposed to a lethal dose of X-rays. The results obtained are given in Table IV.

Before offering these results for discussion we should like to call attention to the fact that the number of animals upon which a discussion can be based is rather small. Unfortunately, a number of animals succumbed to the *median eminence* destruction operation. In some other rats *in vivo* we failed to find the adrenal gland owing to the presence of a heavy mass of fatty tissue in the region of the kidney.† The number of animals from which one gland had been removed before the destruction of the *median eminence* and the other 3 hours after exposure to X-rays was reduced to 6. Four out of these 6 animals after irradiation showed a distinct drop in cholesterol concentration in the remaining gland; two did not. The adrenal gland of one of the latter animals weighed somewhat less than the gland extirpated months earlier, which is not usual. To draw conclusions from 6 animals on the role played by the *median eminence* region in the processes that control the synthesis and release of ACTH by the anterior pituitary after transplantation into the anterior chamber of the eye would not be a right thing to do. If, however, cholesterol

† For the cholesterol concentration test one adrenal was removed on the day of operation upon the *median eminence*, and the other 3 months later.

TABLE IV

Protocol No.	Body weight (g)		Weight of adrenal gland (mg)		Cholesterol concentration (mg%)		Weight of testes (mg)		Weight of thyroid gland (mg)
	Before destruction of median eminence	When killed	Before destruction of median eminence	When killed	Before destruction of median eminence	When killed	Before destruction of median eminence	When killed	
2	240	—	12.6	11.8	1.60	—	877	506	—
6	215	200	6.8	14.4	2.19	—	1061	294	—
11	200	190	6.8	10.6	4.58	—	1140	745	9.8
17	230	218	8.6	11.5	—	2.73	1220	237	11.8
22	205	140	—	15.8	—	—	1190	463	8.4
24	260	295	—	8.4	—	3.34	1081	282	7.6
10	270	308	8.2	10.0	—	2.14	1430	1423	16.4
12	245	295	5.9	5.8	—	1.66	1350	1378	14.7
18	193	250	12.0	17.7	—	—	1340	1339	13.0
19	185	—	12.0	12.6	—	—	1076	1029	—
4	285	290	5.8	6.8	—	2.54	1100	1392	12.8
13	250	420	6.8	10.4	5.28	3.51	1280	1750	17.0
14	220	285	4.3	6.0	3.57	2.80	1092	1231	13.6
20	155	315	—	11.6	—	1.08	1027	1502	9.7
16	153	185	6.0	5.6	2.60	3.39	960	1103	7.8
1	210	275	7.3	10.0	4.96	3.70	270	244	10.7
5	240	235	5.6	9.4	5.27	1.89	—	85	9.7
9	245	275	8.6	11.2	—	1.89	—	132	14.6
25	165	—	12.0	14.4	3.26	—	138	111	29.6
23	205	307	6.4	11.5	4.50	4.35	—	99	14.5
Mean values			8.0 ± 0.64	10.8 ± 0.73	3.78 ± 0.46	2.69 ± 0.26			
				$P < 0.001$		$P < 0.02$			

concentration in all tested adrenals, i.e. those that run, and those that do not run, in pairs, removed before and after exposure of rats to X-rays, are correlated, a difference of 30% will be observed, which, we found, is statistically significant. It should be noted that this fall is considerably less than in hypophysectomized-grafted rats with undisturbed *median eminence*, where the drop fell to 55%, and was even greater than in normal rats. These findings, because of the incomplete factual background, however clear the implications seem to be, we do not consider sufficiently convincing to draw the conclusion in favour of a hypothalamic control of the anterior pituitary activity; or, in other words, we think that the experiments must be repeated.

Bacq *et al.* (1960) destroyed the *median eminence* in normal rats by the electro-coagulation technique. Two days later such rats were exposed to a lethal dose (whole-body irradiation) of X-rays. Three hours later no drop in cholesterol concentration in the adrenals of these rats was observed. It would be interesting to know what would have happened if the rats had been irradiated after the healing process on the *median eminence* had taken place.

For the inconsistent reactions on the part of the functional testes in hypophysectomized-grafted rats (Table IV) following the destruction of the *median eminence* more than one explanation can be offered. One of them is that the injury inflicted on the *median eminance* did not in every individual case cover the same area or was not of the same severity. Another, less likely but not to be excluded, is that following hypophysectomy small areas of the anterior pituitary tissue had remained adherent to the hypophyseal stalk and were destroyed by micro-coagulation. The involution of the remaining testis following the destruction of the *median eminence* that took place in 3 rats containing pituitary grafts is of special interest for us, for if the phenomenon turns out to be reproducible it may prove to be of real importance. The end point of the involutionary process in these cases is quite comparable with the value obtained following the removal of the pituitary graft. It should be noted here that an involution of the testes following their recovery in animals with healthy pituitary grafts was never observed. We regret to say that a histological study of the stalk region has not yet been made.

REFERENCES

BACQ, Z. M., MARTINOVITCH, P., FISCHER, P., PAVLOVITCH, M., and SLADIĆ-SIMIĆ, D. (1956a). *Arch. Int. Physiol. Biochem.* **64**, 278.

BACQ, Z. M., MARTINOVITCH, P., FISCHER, P., SLADIĆ-SIMIĆ, D., PAVLOVITCH, M., and RADIVOJEVITCH, D. (1956b). *Bull. Acad. R. Med. Belg.*, VI° série, **21**, 328.

BACQ, Z. M., MARTINOVITCH, P., FISCHER, P., PAVLOVITCH, M., and SLADIĆ-SIMIĆ, D. (1957a). In "Advances in Radiobiology", p. 237. Oliver and Boyd. London.

BACQ, Z. M., MARTINOVITCH, P., FISCHER, P., PAVLOVITCH, M., SLADIĆ-SIMIĆ, D. and RADIVOJEVITCH, D. (1957b), *Radiation Res.* **7**, 373.

Bacq, Z. M., Smelik, D. S., Goutier-Pirotte, M., and Renson, J. (1960). *Brit. J. Radiol.* **33**, 618.
Benoit, J., and Assenmacher, I. (1954). *J. Physiol.* **47**, 424.
Harris, G. W. (1955). *Monograph Physiol. Soc.* no. 3.
Martinovitch, P. N., and Pavić, D., (1960). *Nature, Lond.* **185**, 155.
Martinovitch, P. N., Bacq, Z. M., Pavić, D., and Sladić-Simić, D. (1961). *Arch. Int. Physiol. Biochem.* **69**, 9.

DISCUSSION

BACQ: The nero-endocrine factors are important both for acute reactions and for long-term effects. One must not forget that the whole biochemistry of connective tissue is controlled by the pituitary gland and the adrenal cortex. The extent to which the pituitary is controlled by the CNS is an important factor also. It seems that the pituitary has a certain amount of independent capacity to secrete its hormones. The fact that you have taken out the hypophysis and grafted a foreign gland into the eye does not necessarily mean that the graft is absolutely deprived of control from the CNS because we do know that the control of the hypothalamus over the pituitary is mainly achieved through a kind of neurohormone. It may well be that, also in your case, the neurohormone goes into the blood and affects the graft. It is known that the sensitivity of denervated tissue to its chemical stimulus is very much increased, sometimes ten-fold or more. Dr. Martinovitch's results seem to suggest that the hypothalamus is capable of secreting a substance which goes into the blood and may affect heterotopic grafts.

THE CELLULAR BASIS AND AETIOLOGY OF THE LATE EFFECTS OF IRRADIATION ON FERTILITY IN FEMALE MICE

W. L. RUSSELL AND E. F. OAKBERG

Biology Division, Oak Ridge National Laboratory,† Oak Ridge, Tennessee, U.S.A.

SUMMARY

The degree of irradiation-induced shortening of the reproductive life-span in female mice varies with dose-rate, age of female, and other factors. The basic pattern of this response is accounted for by the amounts of killing in the pool of oocytes in immature follicle stages. Thus, an effect on fertility, which, at low doses or low dose-rates, is long-delayed in appearance, traces back to cell deaths occurring within 24 hours after irradiation.

One of the late somatic effects of irradiation for which it may be instructive to consider the cellular basis is the effect on reproductive life-span in female mammals. No attempt will be made here to review all the publications on the late effects of irradiation on female fertility. The purpose of this paper is to emphasize that what, until recently, seemed to be a complex phenomenon, with conflicting results from different investigators, now appears to have a relatively simple aetiology, at least in its broad aspects. For the demonstration of this point, it will be necessary to cite only cogent samples of the experimental findings. Discussion will be restricted to the mouse and rat, because the details have been worked out more thoroughly in these species. The general nature of the aetiology may turn out to be similar in other mammals, but, since marked differences have been noted in the stage at which oocyte development is arrested in the adult (Oakberg and Clark, 1961a), the quantitative nature of the response may vary considerably in different mammals.

Sterility induced in female mice by acute irradiation was clearly shown by Brambell and Parkes (1927), and by Murray (1931), to be due to destruction of oocytes. The same results were obtained by Ingram (1958) for the rat. When later workers turned to the study of dose fractionation and lowered dose-rate, however, some of the results appeared to be in disagreement with one another. Thus, some investigators reported that fractionation of dose lessens the sterilizing action of radiation; others reported no effect of lowered dose-rate; and still others reported an increased damage when dose-rate is reduced. It has been demonstrated, however, that these apparent

† Operated by Union Carbide Corporation for the United States Atomic Energy Commission.

conflicts, and the implied complexity, evaporate when it is realized that the predominant effect of radiation on female fertility is a shortening of the breeding period (Russell et al., 1959a). There are, of course, other effects, for example, some reduction in litter size as a result of dominant lethal mutation induction in the oocytes, but the most striking is the shortening of the reproductive life-span.

When the results of dose fractionation and of various intensities of continuous irradiation were investigated, the effects on the reproductive life-span formed a consistent picture of lowered damage compared with that from a single dose of acute irradiation (Russell et al., 1959a). After a moderate dose of acute irradiation, productivity may be cut down to only one or two litters; whereas, with moderate doses of chronic irradiation, there may be merely a slight curtailment of the normal reproductive life. An example of this effect is given in Table I which contains hitherto unpublished data from

TABLE I. *Number of litters produced by control and irradiated female 101 × C3H F_1 mice†*

	Age of female at mating (days)	No. of females	Mean No. of litters per female
Control	60–79	37	14·08
	80–99	132	13·80
	100–119	154	12·80
	120–139	82	12·05
Irradiated	60–79	32	9·16
	80–99	100	8·63
	100–119	206	7·68
	120–139	76	6·59

† γ-Rays from ^{137}Cs source. Total dose 258 r. Dose-rate 0·009 r/min. Matings were made at the completion of irradiation to multiple recessive test stock males used in specific locus mutation experiments. Females that died before 620 days of age are excluded.

one of our mutation experiments. Here, the mean number of litters per female is about 40% less in the irradiated group. In this experiment, as in others referred to, litter size and number of females having litters, stayed normal, or near normal, until the beginning of a sharp decline that ended in sterility. Thus, little or no effect is observed until the onset of this decline, which, as is apparent for the data in Table I, can be long delayed.

What is the aetiology of this late effect? The known killing of oocytes by acute irradiation the fact that there is no replacement of oocytes in the adult mouse (Brambell and Parkes, 1927; Schugt, 1928; Murray, 1931; Oakberg, 1960) and rat (Ingram, 1958), and the observation that fertility is high until just before the onset of sterility, all suggested that the results might be due simply to reduction in the number of oocytes in immature follicle stages. This has turned out to be the case in both rat (Ingram, 1958) and mouse. Investigations of the adult female mouse (Oakberg and Clark, 1961b) have shown that the variations in effect on reproductive life-span with different doses, dose fractionations, and dose-rates, are closely paralleled by the amounts of cell killing in the early oocytes. This parallelism is also seen in the highly sensitive infant mouse, Table II. The marked effect on fertility of irradiation at this age (Russell et al., 1959b) is closely correlated with an unusually high frequency of oocyte killing (Oakberg, 1962).

TABLE II. *Effect of dose-rate on survival of primary oocytes of the mouse* (condensed from Oakberg and Clark, 1961b)†

Age at exposure (days)	Dose (r)	Radiation	r/min	Cell survival $\frac{\text{Experimental}}{\text{Control}}$
10	25	^{60}Co γ	2·85	0·030
10–12	26	^{137}Cs γ	0·009	0·132
59	50	X-rays	87	0·0013
133–136	44	^{137}Cs γ	0·009	0·312
125–131	87	^{137}Cs γ	0·009	0·279
119–132	164	^{137}Cs γ	0·009	0·074

† All mice killed 72 hours after exposure. Ratios were computed on the basis of normal cells counted in serial sections.

The exact mechanism of the oocyte killing is not understood. What is known can be summarized briefly. It has been amply demonstrated that the cells die shortly after irradiation (Oakberg, 1960). Since the oocytes are in a modified resting stage, it is also clear that cell division and DNA synthesis, which are frequently invoked as playing a role in cellular death from irradiation, can be ruled out of the oocyte response. The marked reduction in cell killing that is observed when the dose-rate is lowered, indicates that the initial radiation damage can be partially repaired before it leads to cell death.

Later work may reveal some minor complexities in the effect of radiation on reproductive life-span. For example, degeneration of oocytes is always going on in normal, unirradiated females, and it is possible that irradiation-induced depletion in the population of oocytes may have some effect on the rate of this process. Thus, it might be expected that the rate of spontaneous degeneration would decrease and so compensate, to some degree, for the radiation-induced loss. This possibility is being tested in a current experiment. However, even if modifying factors working in the other direction are revealed, it seems clear that the main cause for the shortening of the reproductive lifespan is partial destruction of the pool of oocytes.

Here, then, is a late somatic effect of irradiation that has been traced back to the immediate damage, which is simply early cell killing. This suggests that serious consideration should be given to the possibility that the cellular basis for some of the other apparently complex late effects of irradiation may also trace back to nothing more mysterious than cell death immediately following irradiation.

REFERENCES

BRAMBELL, F. W. R., and PARKES, A. S. (1927). *Proc. roy. Soc. B*, **101**, 316.
INGRAM, D. L. (1958). *J. Endocrinol.* **17**, 81.
MURRAY, J. M. (1931). *Amer. J. Roentgenol.* **25**, 1.
OAKBERG, E. F. (1960). Suppl. to *J. Dairy Sci.* **43**, 54.
OAKBERG, E. F. (1962). *Proc. Soc. exp. Biol., N.Y.* **109**, 763.
OAKBERG, E. F., and CLARK, E. (1961a), *Amer. Zoologist* **1**, 465.
OAKBERG, E. F., and CLARK, E. (1961b). *J. cell. comp. Physiol.* Suppl. 1, **58**, 173.
RUSSELL, L. B., STELZNER, K. F., and RUSSELL, W. L. (1959a). *Proc. Soc. exp. Biol., N.Y.* **102**, 471.
RUSSELL, W. L., RUSSELL, L. B., STEELE, M. H., and PHIPPS, E. L. (1959b). *Science* **129**, 1288.
SCHUGT, P. (1928). *Strahlentherapie* **27**, 603.

SESSION V

Chairman: W. L. RUSSELL

MECHANISMS OF LIFE-SPAN SHORTENING
By H. J. MULLER

Detection of Segmentary Heterochromia in Foetuses Irradiated *in utero*
By J. LEJEUNE AND M.-O. RETHORE

Chromosome Aberrations in Liver Cells in Relation to Ageing
By H. J. CURTIS AND CATHRYN CROWLEY

The Failure of the Potent Mutagenic Chemical, Ethyl Methane Sulphonase, to Shorten the Life-span of Mice
By PETER ALEXANDER AND MISS D. I. CONNELL

Life-span Shortening from Various Tissue Insults
By H. J. CURTIS AND CATHRYN CROWLEY

Does Radiation Age or Produce Non-specific Life-shortening?
By R. H. MOLE

Differences between Radiation-induced Life-span Shortening in Mice and Normal Ageing as Revealed by Serial Killing
By PETER ALEXANDER AND MISS D. I. CONNELL

Age-Specific Death Rates of Mice Exposed to Ionizing Radiation and Radiomimetic Agents
By A. C. UPTON, M. A. KASTENBAUM, AND J. W. CONKLIN

MECHANISMS OF LIFE-SPAN SHORTENING

H. J. MULLER

Zoology Department, Indiana University, Bloomington, Indiana, U.S.A.

SUMMARY

Reasons are presented for concluding that spontaneous ageing is a part of normal development caused, like most other developmental changes, by other factors than permanent genetic alterations such as point-mutation, deficiency, chromosome loss or inactivation, or segregation, even though it does involve the point-wise death of many individual somatic cells. These reasons comprise the partial reversibility of natural ageing, and its independence of ploidy and of other features of chromosome structure.

Judged by the same criteria, radiation-induced shortening of the life-span is an expression of point-wise losses of individual cells that are caused by actual genetic changes. That the changes are for the most part recessive, depending on either point-mutations, deficiencies, or whole-chromosome losses, is shown by results in *Drosophila*, *Habrobracon*, and plant material, when effects on individuals of different ploidy are compared. Tests of diverse kinds carried out with *Drosophila* having chromosomes of different structural constitution show clearly that the mechanism here at work is that of chromosome loss, caused by radiation-induced chromosome breaks. It is believed that the same basic mechanism accounts also for most of the acute damage that is produced by radiation.

First let us ask: what causes natural ageing? Although our ignorance on this question is profound, it is no more profound than our ignorance of what casuses adolescence, or what causes invagination, or any other major stage or feature of development. Development is a continuous process, of which senility forms the last stage, but one that can be delimited only arbitrarily. The average length of life allotted for any given species is a resultant of the past action of natural selection in having fixed it at the possible amount most conducive to the survival and multiplication of that species as a whole, just as is true of every other normal stage and feature of every type of organism. For death is an event necessary to allow genetic evolution, and some types of organism can profit more than others by postponing death for a relatively long time. These latter types tend to be the ones which, as individuals, can, by living longer, achieve relatively greater *cumulative* successes on behalf of species survival. Moreover, the fact that the timing of senility is a very labile trait, genetically, is shown by the considerable differences in regard to it in different species, including even some that are rather closely related.

Not for many years has there been much serious advocacy of the view that development changes in general reflect permanent changes, of the nature of directed mutations or segregations, in the genetic material of the somatic cells. Too many cases were found of reversible developmental changes of diverse kinds: for example, dedifferentiations both *in vitro* and *in vivo*; the regeneration of tissues and organs from germ layers other than those normally producing them; cases of somatic cells or their nuclei proving their virtual totipotence by serving, in place of germ cells or their nuclei, for the whole of normal development. To be sure, such work still fails to tell us what does cause development, yet it is very important in telling us something that does *not* cause it: namely, permanent change in the genetic constitution.

The genetic-change interpretation has persisted more obdurately in its application to senility than to other stages and features of development. There are at least two reasons for this. The first is that mutations and genetic changes, in general, must to some extent accumulate with time, even though they do not do so steadily (Muller, 1946a,b). The second reason is that ionizing radiation is known to cause not only genetic changes but also a shortening of the life-span that in considerable measure resembles what would be expected of an acceleration of natural ageing (Henshaw, 1944; Boche, 1946, 1948; Lorenz, 1946). To this fact may now be added Oster's finding (1960, 1961a) in *Drosophila* and Alexander and Connell's (1960) in mice that chemical mutagens likewise cause a life-span shortening, resembling that induced by radiation.

In view of these pertinent-seeming considerations, what is there now to be said for maintaining the view that natural ageing, at any rate, is to be classed with other developmental changes as not being a reflection of true genetic changes? One cogent line of evidence consists in the demonstration, in a wide variety of organisms which undergo clonal ageing when carried through a succession of cycles of asexual reproduction, that these lines of descent are subject to rejuvenation. This can be brought about, as the case may be, either by sexual reproduction or by autogamy, or by special types of asexual reproduction, such as the derivation of the offspring in the next cycle from younger individuals (Jennings and Lynch, 1928; Lansing, 1947) or from more apically placed parts (Sonneborn, 1930). Important in this connection is the fact that these revivals cannot be explained by any selective mechanism acting against the more drastically affected cells or groups of cells, since the frequencies of failures are far from enough to allow such an interpretation. Hence the observed ageing here would not have been based in any permanent genetic alterations. The organisms here in question include ciliates, flatworms, rotifers, fungi, and higher plants. Moreover, it is well-known that given types of vertebrate cells, such as heart muscle cells or fibroblasts, have been maintained in successive tissue cultures without signs of senescence for periods many times the life-spans of the organisms from which they were derived.

More direct evidence of the non-genetic nature of the natural ageing process which occurs *in vivo*, in the soma of any given individual rather than of a clone, is provided by observations made by Clark and Rubin (1961) on a species of the parasitic wasp *Habrobracon*. This particular species readily produces both haploid and diploid males, as well as females, which are all diploid. It was observed that although the median life-span of the diploid males was only two-thirds that of the females, when both were fed alike on honey, nevertheless the median life-span of the diploid males was just the same as that of the haploid males. In other words there was a difference correlated with sex, as is found for so many developmental characters, but none correlated with ploidy *per se*. Yet on any theory of genetic change as a basis of ageing that could apply to these organisms, the haploids should age far faster than the diploids. Moreover, a verification of this expectation was provided when the wasps were irradiated at various stages of their life-cycle, for in this case (if we rule out the immediate deaths induced by irradiation of very early embryos) the life-span was reduced by a much greater fraction in the haploid than in the diploid males, but by just the same fraction in the females as in the diploid males.

Some 5 years ago Failla (1957, Failla and McClement, 1957) put forward the view that natural ageing, as well as its radiation-induced imperfect simulacrum, was an effect of the accumulation of point-mutations. He admitted, however, that quantitatively, the two processes failed to agree by a factor of 24 (the natural ageing being that much too fast), on the basis of the Russells' data (1952, 1954 *et seq.*) on spontaneous and induced mutation rates in mouse germ cells. Failla then proposed (1960) that the data might be reconciled by the *ad hoc* assumption of drastically different mutation rates in early stages from those that had been observed, which pertained to the period of life after the age of 2 months. It could also be assumed that the ratio of spontaneous to induced mutation rates may be very different in somatic cells than in the germ track. However, taking into account our knowledge of the high dominance of the great majority of normal genes it can readily be calculated that neither natural, nor induced, point mutations can be occurring at anything like the rates that would be necessary to cause natural senescence or induced life-span shortening, respectively.

Szilard's proposal (1959) that whole chromosomes or regions of them become inactivated *in situ*, even in non-dividing cells, by "hits" some of which are caused by radiation and others in other ways, assumes a process for which no evidence has ever been adduced. One would expect such evidence to have made itself known if the process were as frequent as would have to be postulated. Although such a process seems unlikely, in the light of present knowledge concerning the genetic material, it is true that it would give results for radiation-induced life-span shortening that resembled in some significant ways those of known chromosome breaks (to be considered presently). However, inasmuch as the proposed inactivations took place in non-dividing as well as in dividing groups of cells, this interpretation would be contradicted by the massive evidence showing the far greater radiation damage done to proliferating than to non-proliferating tissues. One striking example in point here is the enormously higher dose required to produce a given shortening of the life-span when irradiation is applied to adult insects which have virtually no cell divisions than when applied to their earlier stages. As for spontaneous ageing, the diverse arguments already given against this process being based in genetic changes apply equally to views which regard these changes as as point-mutational and as inactivations or losses of whole chromosomes or chromosome regions.

In an analysis of the process of structural change in *Drosophila* the present author (1940) advanced the view that, in general, except for the production of cancers (which are probably of point-mutational nature) the damaging effects of radiation noted in somatic tissues are the results of chromosome breaks. Moreover, he extended this interpretation to include radiation-induced life-span shortening (1950) shortly after the discovery of this phenomenon had been made known. There were, however, two possible variants of this interpretation since genetic studies of the events in germ cells and in the divisions following fertilization in *Drosphila* had shown that chromosome breaks, when not followed by exact restitution, can act in two different ways to cause the death of daughter-cells or of zygotes derived from them.

In the first case, found to be more usual for most cells, the broken chromosome or chromosomes fail to become incorporated in the nucleus of a daughter-cell because the pieces have joined together so as to given rise on the one hand to acentric pieces, those lacking any centromere to mediate their transportation to the poles, and/or, on the other hand, to dicentric pieces, having two centromeres that pull in opposite directions. This can happen even in consequence of single chromosome breaks, which are produced at a rate proportional to the dose (Muller, 1940; Pontecorvo, 1941), for in such cases the two chromatid fragments derived by replication (either before or after the break) from the acentric chromosome fragment may unite at their broken ends to form an acentric isochromosome, as it is called, while the two chromatid fragments derived from the chromosome fragment having the centromere, on uniting similarly, form a dicentric isochromosome that fails to reach either pole. If, now, the missing chromosome material were essential to life, as in the case of the single X-chromosome of a male cell or any chromosome of a haploid cell, the resulting hypoploid cell or zygote would die. This method of killing would obviously be less damaging to cells of higher ploidy.

In the case of the other mode of death, the pieces join up in the same way as already described. However, the dicentric structure, instead of being left out of both daughter-nuclei, becomes so stretched out by the opposing pulls that each of its opposite centromeric portions succeeds in entering the daughter-nucleus nearer it, and the thread between them forms a chromatin bridge that hampers the effectiveness and upsets the orderliness of chromosome distribution in subsequent mitoses. This causes the death of the descendant cells even if the affected chromosome material were not really needed, as when a normal homologous chromosome was present that would have sufficed. The bridge, then, may prove lethal even to a diploid or polyploid cell. Thus with a given dose, cells having higher ploidy, being possessed of more chromosomes to form bridges, are killed by this method at a correspondingly higher rate than cells with lower ploidy. Work by Pontecorvo and me on *Drosophila* (1941; Muller and Pontecorvo, 1942; Pontecorvo, 1942), and by Whiting (1945 *et seq.*) as well as by Heidenthal (1945) on *Habrobracon*, showed that deaths by bridge formation in early zygote stages are the usual cause of the dominant lethals found amoung early zygotes after irradiation of spermatozoa, perhaps because the daughter-nuclei in early zygotes do not usually move so far apart as to render a bridge ineffective.

In experiments reported by Rowena Lamy and myself (1941) on the comparatively high mortality (some 50%) caused by irradiation of early embryos of *Drosophila* with rather low doses (500 r), we could find no connection between ploidy, or sex, and survival—a result recently supported in our laboratory by results obtained by Helen Meyer and me (unpublished), using a different genetic set-up for testing the matter. Lamy and I had concluded that this damage was probably cytoplasmic or at least not by chromosome breakage. It has recently been found by the Valencias (1962a, b) that there is about this same death-rate if eggs are irradiated shortly before they are to be fertilized (by unirradiated sperm), or shortly after their fertilization (in which case the sperm nucleus also is irradiated), or in the stage of embryos containing some 4 to 8 nuclei. This result, as well as the comparatively small number of recessive lethal mutations which they found by genetic testing to have been induced by the same irradiation, helps to confirm the cytoplasmic interpretation as the main cause of death here. But this effect cannot be termed a kind of senescence since the survivors are not permanently weakened (see also Clark and Mitchell, 1952, for results with *Habrobracon* embryos). From species to species this cytoplasmic effect in eggs is very variable, since somewhat more than 15,000 r are needed to cause the egg cytoplasm of *Habrobracon* to give rise to much embryonic mortality if the effective nucleus (the paternal one) of the merogonically developing embryo were not irradiated, while in the silkworm Astaurov and Ostriakova-Varshaver (1957) find that the egg cytoplasm can well tolerate 540,000 r.

That such cytoplasmic effects play a negligible role, if any, in the seeming senescence induced by the irradiation of stages beyond those of the early embryo, has been shown for *Drosophila* in experiments initiated by Oster (Oster and Cicak, 1958; Oster, 1959a, b, 1960, 1961b) at Indiana University. They have been carried further by Ostertag (Ostertag and Muller, 1959; Muller, 1960, 1961; Ostertag 1961, Ph.D. thesis and in press), and more recently by Meyer, in conjunction with the present writer. Oster and Cicak's finding that irradiation of larvae increased the pupal mortality of the males more than that of the females, was followed by the discovery that this was not basically a sex difference since females with attached X-chromosomes also have a high mortality, as expected on the chromosome-loss interpretation. Equally striking was the exceptionally high induced mortality of males with ring-shaped X-chromosomes. For this result is in harmony with the fact that even when rings are restituted the frequent cases where the chromosome thread is subject to an axial twist of one or more turns before union of its broken ends occurs would result in interlocked chromatids, forming an effectively dicentric chromatin structure which failed to reach the daughter-nuclei.

In mice the X-chromosome forms a much smaller portion of the total chromatin than in *Drosophila*. Correspondingly, Lindop and Rotblat (1961) have found that the difference in X-ray-induced life-shortening between the sexes is not statistically significant, but that there is an indication of the male's life-span being reduced more.

Ostertag showed that the increased mortality caused by irradiation of *Drosophila* larvae affected not only the pupae but continued unabated throughout life (that is, over some 3 months), thus constituting a type of premature ageing. Moreover, if one member of any major pair of homologous chromosomes (X, 2nd or 3rd) contains a small deficiency, the mortality is raised as though only the normal member had been present to begin with. Thus, females with a deficiency in one of their X-chromosomes are as much damaged as males, and females with a deficiency in a large autosome are much more damaged, since the autosome, being longer, is correspondingly more subject to breakage and loss. Moreover, as expected, females with one ring X-chromosome and a deficiency in their other, non-ring X-chromosome, are affected as much as males with a ring.

Among other things, this finding shows that in this material (unlike the zygote or the individual as a whole) one member of any pair of chromosomes is enough for cell viability. Secondly, the result shows that it is not the chromatin bridge but the chromatin loss that kills these cells. The qualification should be made, however, that special types of *Drosophila* chromosomes —namely, those with what are called strong centromeres—and the chromosomes in special types of cells—namely, those in early embryonic stages—do give evidence of forming lethal bridges, as has also been found for the early embryos of *Habrobracon* by Clark and Mitchell (1952). Thirdly, and most important of all, it shows that it is not the induction of point mutations or even of deficiencies of chromosome parts that causes the somatic damage, but the loss of the entire chromosome, referable to its breakage. For the chromosome having a small deficiency would in these cases have afforded protection against all point-mutations and small deficiencies in the other member of the pair except those located in the very same small region as it itself occupied. On the other hand, it would have been of no avail at all for giving protection against loss of the entire homologous chromosome and so the individual would react in this case like one which had been haploid for that pair, just as the data showed it did react.

Still further evidence was advanced by Ostertag, and confirmed by Meyer, that anoxia caused by a nitrogen atmosphere, when given as a post-treatment to the irradiation over a period in which (if given by itself) it resulted in little mortality, considerably accentuates the radiation-induced life-span shortening. For this result is in accord with the known effect of anoxia in hindering the restitution of broken chromosomes (Wolff and Luippold, 1955, 1956; Abrahamson, 1959). A similar pre-treatment with anoxia, on the other hand, was found by Meyer to be ineffective.

These conclusions may be applied to the interpretation of the differences in X-ray-induced mortality found by those working with *Habrobracon* of different ploidy (Whiting and Bostian, 1931; Clark and Kelley, 1950; Clark and Mitchell, 1951; Clark, 1957, 1960; Atwood et al., 1956), with *Bombyx* of different ploidy (Tultseva and Astaurov, 1958), and with diverse plants (e.g. Müntzing, 1941; Fröier et al., 1941; Latarjet and Ephrussi, 1949; Tobias, 1952; Mortimer, 1952, 1959). These authors had recognized their results as evidence that the mortality, in effect a life-span shortening, was genetic but it could not be concluded that the mechanism worked by way of the loss of entire chromosomes. Similarly, in the work of Tsunewaki and Heyne (1959a,b) on allohexaploid wheats, the greater vulnerability to radiation of the types with one chromosome missing to begin with is seen to be caused by a recessively acting genetic damage of some kind, which could however be point-mutational, or regional deficiency, or whole chromosome loss.

On the other hand, when we turn to adult insects, with their non-dividing cells and correspondingly far higher radiation resistance, how are we to interpret Clark's recent (1960) result that, at all doses tried on adults, the haploid still proves to be more vulnerable than the diploid? Possibly this is a matter of point-mutations rather than chromosome loss, or the midgut cells may still be dividing. But the enormously higher radiation resistance in this case emphasizes the exceptional nature of the mechanism at work here. In adult as well as young mammals, with their abundance of proliferating tissues necessary for life, it must surely be the chromosome breaks that lie at the basis of the radiation-induced life-shortening.

Finally, it is evident why the symptoms of natural senescence and of radiation-induced life-span shortening are so similar. For both involve the scattered, point-wise loss of cells (and sometimes, no doubt, their defunctionalizing), even though the cell deaths and damages are not mainly based in genetic changes in natural senescence and *are* so based in radiation-induced life shortening. However, the relative incidence of cell losses and impairments in different tissues would doubtless be different for the natural and induced processes. This is especially true because chromosome breaks prove damaging only after cell division, although of course non-dividing cells can be injured indirectly, by the damage of dividing cells that service them (such as the nurse cells of late oocytes). Because of such differences in site of damage the incidence of death from different causes should not have just the same time distribution in natural and radiation senescence—an expectation in agreement with the recent findings of Lindop and Rotblat (1961).

ACKNOWLEDGEMENTS

Work herein reported which was carried out at Indiana University was supported by grants from the Atomic Energy Commission (Grant 2–III) and from the National Institutes of Health of the U.S. Public Health Service.

REFERENCES

ABRAHAMSON, S. (1959). *Genetics* **44**, 173.
ALEXANDER, P., and CONNELL, D. I. (1960). *Radiation Res.* **12**, 38.
ASTAUROV, B. L., and OSTRIAKOVA-VARSHAVER, V. P. (1957). *J. Emb. and exp. Morph.* **5**, 449.
ATWOOD, K. C., VON BORSTEL, R. C., and WHITING, A. R. (1956). *Genetics* **61**, 804.
BOCHE, R. D. (1946). *Manhat. Dist. Declass. Com.* 204.
BOCHE, R. D. (1948). Address (unpub.) at Symposium on Low Level Irradiation at Argonne Nat. Lab., Chicago, 19 Oct.
CLARK, A. M. (1957). *Amer. Nat.* **41**, 111.
CLARK, A. M. (1960). *Biol. Bull.* **119**, 292.
CLARK, A. M. and KELLY, E. M. (1950). *Cancer Res.* **10**, 348.
CLARK, A. M., and MITCHELL, C. J. (1951). *J. exp. Zool.* **117**, 489.
CLARK, A. M., and MITCHELL, C. J. (1952). *Biol. Bull.* **103**, 170.
CLARK, A. M., and RUBIN, M. A. (1961). *Radiation Res.* **15**, 244.
FAILLA, G. (1957). *Radiology* **69**, 23.
FAILLA, G. (1960). In "The Biology of Aging", A Symposium, (L. Strehler, ed.), p. 70.
FAILLA, G., and MCCLEMENT, P., (1957). *Amer. J. Roentgen.* **78**, 946.
FRÖIER, K., GELIN, O., and GUSTAFSSON, A. (1941). *Bot. Notiser* **2**, 199.
HEIDENTHAL, G. (1945). *Genetics* **30**, 197.
HENSHAW, P. S. (1944). *J. nat. Cancer Inst.* **4**, 513.
JENNINGS, H. S., and LYNCH, R. S. (1928). *J. exp. Zool.* **50**, 345.

LAMY, R., and MULLER, H. J. (1941). Proc. 7th Int. Congress Genetics, Edinburgh, 1939, J. Genetics Suppl. Vol., p. 180.
LATARJET, R., and EPHRUSSI, B. (1949). *C.R. Acad. Sci., Paris* **229**, 306.
LANSING, A. I. (1947). *J. Gerontol.* **2**, 228.
LINDOP, P. L., and ROTBLAT, J. (1961). *Proc. roy. Soc.* **B154**, 332.
LORENZ, E. (1946). "CH"-3698.
MORTIMER, R. K. (1952). *Univ. Cal. Rad. Lab.* **1922**, 66.
MORTIMER, R. K. (1959). *Radiation Res.* Suppl. **1**, 394.
MÜNTZING, A. (1941). *D. Fysiograph. Sallskap, (Lund) Forhandl.* **2**, No. 6, 1.
MULLER, H. J. (1940). *J. Genet.* **40**, 1.
MULLER, H. J. (1946a). "Yearbook Amer. Phil. Soc. for 1945", p. 150.
MULLER, H. J. (1946b). *Genetics* **31**, 225.
MULLER, H. J. (1950). *Amer. Scient.* **38**, 399.
MULLER, H. J. (1960). Proc. Conf. on Immediate and Low Level Effects of Ionizing Radiations, Venice, 1959, *Int. J. Rad. Res.* **1**, 321.
MULLER, H. J. (1961). In "Genetic Aspects of Radiation Injury and Processes Leading to Normal Senescence" by Gowen, J. W., *Fed. Proc.* **20** (Suppl. 8), p. 43–44.
MULLER, H. J., and PONTECORVO, G. (1942). *Genetics* **27**, 157.
OSTER, I. I. (1959a). *Science* **129**, 1286.
OSTER, I. I. (1959b). "Proc. 2nd Australas. Conf. On Rad. Biol." p. 268. Academic Press, New York.
OSTER, I. I. (1960). *Science* **132**, 1497.
OSTER, I. I. (1961a). Excerpta Med., Int. Cong. Ser. 32, 2nd Int. Conf. Human Genetics, paper No. 18.
OSTER, I. I. (1961b). Conf. on Res. on the Radiotherapy of Cancer (Amer. Cancer Soc.), p. 45.
OSTER, I. I., and CICAK, A. (1958). *Dros. Info. Serv.* **32**, 143.
OSTERTAG, W. (1961). Ph.D. Thesis, Indiana Univ.
OSTERTAG, W., and MULLER, H. J. (1959). *Science* **130**, 1422.
PONTECORVO, G. (1941). *J. Genet.* **41**, 195.
PONTECORVO, G. (1942). *J. Genet.* **43**, 295.
PONTECORVO, G., and MULLER, H. J. (1941). *Genetics* **26**, 165.
RUSSELL, W. L. (1952). *Cold Spring Harbor Symp. on Quant. Biol.* **16**, 327.
RUSSELL, W. L. (1954). In "Radiation Biology", p. 825, Vol. I, pt. 2 (A. Hollaender, ed.), McGraw-Hill, New York.
SONNEBORN, T. M. (1930). *J. exp. Zool.* **57**, 57.
SZILARD, L. (1959). *Proc. nat. Acad. Sci., Wash.* **45**, 30.
TOBIAS, C. A. (1952). In "Symposium on Radiobiology", (J. J. Nickson, ed.), p. 357 Wiley, New York.
TSUNEWAKI, K., and HEYNE, E. G. (1959a). *Genetics* **44**, 933.
TSUNEWAKI, K., and HEYNE, E. G. (1959b). *Genetics* **44**, 947.
TULTSEVA, N. M., and ASTAUROV, B. L. (1958). *Izv. Akad. Nauk. SSSR* **3** (Biol.), 197.
VALENCIA, R. M. (1962a). *Dros. Info. Serv.* **36**, 126.
VALENCIA, R. M., and VALENCIA, J. I. (1962b). *Dros. Info. Serv.* **36**, 125.
WHITING, A. R. (1945). *Biol. Bull.* **89**, 61.
WHITING, A. R., and BOSTIAN, C. H. (1931). *Genetics* **16**, 659.
WOLFF, S., and LUIPPOLD, H. E. (1955). *Science* **122**, 231.
WOLFF, S., and LUIPPOLD, H. E. (1956). In "Progress in Radiobiology" (J. S. Mitchell *et al.*, eds.), Oliver and Boyd, Edinburgh.

DISCUSSION

MOLE: I would like to raise one or two questions about what I thought Prof. Muller accepted as demonstrated facts. For example, the longevity of cells in tissue culture. This may be true of some kinds of cells but recent work in transplantation from mammal to mammal seems to suggest that there is a limit to the life of some cells. He also said that life-span was fixed by natural selection. I wonder whether this is a wholly unambiguous statement. The life-span observed in nature is not necessarily the same thing as the potential life-span which an organism will show living in some kind of fixed condition. This seems to me to make the life-span of any kind of organism under laboratory conditions not necessarily the same thing as the life-span in—let's call it—nature. Can that be true of man who has only got himself free from some of the selective forces in the last few tens of generations?

MULLER: Well, with regard to the first question I am quite prepared to admit that there may eventually be an ageing of the somatic cells in culture, taking place long after the ageing of the individual as a whole, which was conditioned by other tissues. That would not prove, however, that the ageing in culture was of genetic origin. It might be like the clonal ageing which you referred to.

In regard to the other point there are several considerations. For one thing it must be remembered that man and most higher organisms—organisms in general—have had to face quite varying conditions and become adapted to them. Thus they are quite versatile. And while man under primitive conditions doesn't *usually* live as long as he does among ourselves there are circumstances under which he does. It's very curious, for instance, that we should find the statement in the Bible written so many years ago that threescore years and ten is the natural age of man. We haven't gone very much beyond that yet, really, for most of us. I think studies among the most primitive people show, too, that although the great majority die off early there are some that do live that long and these people generally have a function. I think Haldane and some others too made a considerable mistake in assuming that selection stops after reproduction stops because the reproductive function—in its more general aspect of furthering the survival of the species—goes far beyond the more direct reproductive activities of having offspring. For the grandparents have a great deal to do especially amongst primitive peoples, in helping to ensure survival of the grandchildren, and so I think there has been a use in their old age.

Moreover, one must remember that even though a certain contingency isn't too frequent a species will nevertheless adapt to it—and here I give Haldane credit, for he once pointed out that species might be found adapted to some conditions, such as extreme cold, which came perhaps only once in ten generations. Even that is enough to give a natural selection pressure which will cause that species to become adapted to that degree of cold. We don't generally recognize that as being part of the normal life of the organism but it comes often enough for the organism to have become adapted to it. In the case of life-span there have obviously been considerable genetic changes in our rather recent evolutionary history, as shown by the considerably longer life-span of man than of the apes, the longer life-span of the apes, in turn, than that of the monkeys, and so on. So it is very flexible, genetically, and I think has been fixed by natural selection like anything else.

ALEXANDER: I wonder whether the loss of a chromosome which as you said, is going to be so serious in the case of germ cells and eggs is going to be so disastrous for a somatic cell? For example, tumour cells which seem to be capable of looking after themselves very well often show serious chromosome abnormalities.

MULLER: Well, in our experiments with *Drosophila* the evidence showed that the cells can stand having a chromosome missing provided the homologous member is there. If they could stand having it missing they could surely stand having an extra one. We know that polyploids can exist too, and as soon as you have a polyploid you have much more lability because then the loss or addition of one or two chromosomes isn't proportionately nearly as much of a change as it is to a diploid. Therefore I think there's a lability in this respect but that what the cells can't stand is the loss of any portion of both members of a pair of chromosomes, or of one chromosome that's only present in haploid.

BACQ: In that case one would suspect that polyploids would increase regularly with resistance.

MULLER: Oh, yes.

BACQ: And apparently I had stated that in the book written with Peter Alexander, it seems not to be true in yeast.

MULLER: I think that can be understood in the light of the two different mechanisms which I spoke of by which chromosome breakage can kill the daughter or descendant cells. That is, where you have polyploidy giving you greater radiation sensitivity the deaths are presumably caused by the chromosome bridges. Now in *Drosophila* we have some evidence that occasionally even in the later stages when certain special chromosomes are present their bridges kill the cells, since the presence of one of these chromosomes will cause radiation sensitivity even in heterozygotes.

We have therefore the following seriation: the haploid, which is ordinarily expected to be the most sensitive; the diploid, which is much less sensitive, if the species is one in which the chromosomes usually don't form lethal bridges but are lost completely. But if you have a certain proportion—let us say 10% of the breakages—followed by lethal bridges, that wouldn't be enough to play a big role in the case of the haploid, nor even in the case of the diploid. But when you go from the diploid to the triploid or the tetraploid then you have loss of both or all members of the same type of chromosome—hypoploidy —playing very little role, and now it's the bridges that become more and more damaging as the ploidy increases. And I think your cases correspond with this situation.

ROTBLAT: You say that there is a fundamental difference in the process of radiation-induced life-shortening and natural senescence and natural mutation. Is the loss of cells a sufficient explanation?

MULLER: I should think it would be. I think the matter should certainly be explored with this question in view. In our work with *Drosophila*, anyway (and that's the only thing I have a first-hand experience with in this connection), the radiation-induced life-span shortening so closely follows what must be the loss of cells that very little of it can in that case be due to impairment of cells short of their loss. Now in the case of *natural* life-span shortening I wouldn't know if this held but I would think it very unlikely that in natural senescence the loss of cells resulted from *genetic* changes occurring in them. There might be some other reason for it—some physiological ageing of a totally different kind. And yet the symptoms might be very similar. The cells are lost here and there in a scattered fashion—more and more of them.

DETECTION OF SEGMENTARY HETEROCHROMIA IN FOETUSES IRRADIATED *IN UTERO*

J. LEJEUNE AND M.-O. RETHORE

Faculté de Médecine, Institut de Progénèse, Paris, France

As a marker of possible somatic effects of *in utero* irradiation, segmentary heterochromia of the iris has been studied by us (Lejeune et al., 1960).

A second enquiry complementary to the first is now completed and the convergence of the results seems to warrant this preliminary presentation.

Technical data have been dealt with previously and only the broad characteristics of the enquiry will be repeated here.

SELECTION OF THE DATA

Index cases

Observations of mothers, irradiated on the pelvis or the abdomen during pregnancy were systematically searched for in the files of a large maternity hospital. For each case, the mode of irradiation and the date of application, were recorded. The period from 1946 to 1954 was covered. The irradiated children are now between 8 and 18 years old, a fact which simplifies all the investigations concerning pigmentation.

Control cases

For each indexed case the closest file of a non-irradiated mother of the same class-age and same parity was selected as a control. (Some of these were subsequently found to have been irradiated at another hospital or for another pregnancy.)

Examination at home was performed for the indexed cases and the control cases, as well as for their parents and siblings. The actual total of the two enquiries, subjected to a small increase after definite completion, was 1,101 for children irradiated *in utero*, out of a grand total of 8,193 persons examined.

The absence of statistical differences between control cases, their sibs, the sibs of the index cases, and even the parents of both categories, with regard to some pigment characteristics, allows us to use these subsamples as pooled data.

ANALYSIS OF DATA

For the actual analysis three characteristics have been selected, out of the ninety recorded for each individual.

1. *Segmentary heterochromia of the iris*

A segment, delineated by the pupilla, two radii and the corresponding peripheral arc of the iris, is found to be uniformly in tone colour different from the rest of the iris. This unilateral trait looking like a "slice of cake" can be detected very safely and was known by the parents in most instances. All cases reported by our visiting students, have been controlled secondarily by an ophthalmologist. Many combinations of colours have been found, dark on blue, or white on dark.

2. *Dark lock in fair-haired children*

In three instances (all three irradiated) a lock, the size of 1 to 2 cm on the skin, was dark in contrast with the rest of the blond hairs of the neck. The skin itself did not show any abnormal pigmentation at the site of implantation of the dark lock.

3. *White forelock*

This very characteristic dominant trait was used as a control for the internal consistency of our data. Being related to one gene mutation, and being already well spread throughout the population, such a characteristic should not be influenced by X-raying the foetus, at least with the very low doses received.

TABLE I. *Total number of persons examined in the two enquiries*

	Total	Heterochromia of iris	Dark lock	White forelock
Irradiated *in utero*	1,101	15	3	3
Controls (control cases, sibs, and parents)	7,092	11	0	24
Total	8,193	26	3	27

Table I shows:

(i) No relationship between irradiation *in utero* and frequency of the white forelock. This expected negative finding is in favour of no selection in the irradiated group.

(ii) The dark lock is much more common in the irradiated than among the unirradiated.

(iii) The frequency of heterochromia of the iris is much greater among irradiated than among non-irradiated. χ^2 is 36 for one degree of freedom (after Yates' correction).

4. *Time of irradiation*

The time distribution of irradiation among the heterochromic children is strikingly different from that of the general irradiated population.

Table II shows a highly significant cluster of heterochromia for an irradiation age of 4 to 6·9 months *in utero*. This does suggest the existence of a sensitive period, a fact largely established in embryology.

TABLE II. *The age of the foetus at the time of the irradiation* in utero. *The distribution among children of the second enquiry is very comparable to that of the first*

Month of irradiation *in utero*	First enquiry			Second enquiry		
	Heterochromia of iris	Dark lock	Total of irradiated children	Heterochromia of iris	Dark lock	Total of irradiated children
0–0·9						
1–1·9						
2–2·9			9			
3–3·9			8			
4–4·9	2		14	1	1	not yet
5–5·9	2		32	4	1	tabulated
6–6·9	3	1	81	2		
7–7·9	1		123			
8–8·9			243			
ante-partum			41			
Unknown exactly			16			
	8	1	567	7	2	534

5. *Dose of irradiation*

Due to the mode of ascertainment, all radiological procedures are known precisely, at least with the accuracy of the transcription. The calculation of a mean dose is thus possible, with a general restriction concerning its precision.

From the actual data we can conclude that the mean probable dose received by the whole sample is of the order of 2 to 3 r. The mean dose received by the children showing heterochromia is also of the order of 2 to 3 r.

The main interest of these findings is, in our opinion, to show that even small doses of radiation have a somatic effect, which can be detected if induced at a sensitive period of embryonic development.

The study of the frequency of cancers and leukaemias is in progress, as well as the statistical screening of the 87 other physical particularities recorded.

REFERENCE

LEJEUNE, J., TURPIN, R., RETHORE, M.-O., and MAYER, M. (1960). *Rev. franç. Etudes Clin. Biol.* **5**, 982.

DISCUSSION

MOLE: I would just like to ask a few questions about details. I understand that the children were all irradiated *in utero*, you have the records because they were irradiated in hospital, and from the hospital records you were able to deduce the dose. There are two questions here. What about other kinds of radiation exposure? I'm told that in France, pregnant women are fluoroscoped under the National Health legislation, as presumably happens with all the mothers; I don't know at what period in pregnancy. Secondly, you have lumped all the irradiated people together under one dose. Is it possible to get any kind of a dose reponse?

LEJEUNE: Well, I can answer the two questions. Every child has a personal file and every examination is recorded there as exactly as possible. The chest fluoroscopy of the mother is effectively systematic but is probably not relevant in this respect. First the dose to the foetus is likely to be small and randomly spread among the whole sample of mothers.

ROTBLAT: You said the average dose is 2 r.

LEJEUNE: Yes, of this order.

ROTBLAT: Obviously there must be some spread. Do you know what the actual doses were in the cases which you mentioned?

LEJEUNE: Yes, in three cases, one had two pictures, the other had five and the other had a urography—which means six. So one has received around two roentgens, and two of them around five. The differences are not much greater than that. We do not have in such a sample a relationship between the number of roentgens and the somatic effects.

DRASIL: Have you calculated how many cells must be changed or damaged in order to obtain this changed segment of the iris?

LEJEUNE: I do not have any idea because of the possibility of a different selection value for the mutant cells, but it is possible that the number is very small because a primary cell can have a big progeny.

RUSSELL: I think that in this case calculation could be made in the same way as was made in the case of white segments of *Drosophila* eyes that had been irradiated as young or embryos. I'm thinking of course of just what you said, what portion this segment forms of the total and suppose it forms in the average one-tenth of the total it would mean that there were on the average 10 cells there.

ZELENY: Do you find any differences in the size of the segment and are they related to the time of exposure?

LEJEUNE: Yes, we did, and it was one of the best hopes we had that the earlier the irradiation the bigger should be the segment. And the only thing I can tell you is that the earliest foetus—around 3 months old when it was irradiated *in utero* had quite a large segment of one eye and that the smallest one was irradiated at 7 months, but in-between the relation is linear. I am sorry, but there are variations. But then of course this was expected.

CHROMOSOME ABERRATIONS IN LIVER CELLS IN RELATION TO AGEING†

H. J. CURTIS AND CATHRYN CROWLEY

Biology Department, Brookhaven National Laboratory, Upton, New York, U.S.A.

SUMMARY

The percentage of chromosomal aberrations present in regenerating liver cells of mice has been scored as a function of (1) age in normal animals, (2) time following neutron irradiation, and (3) age of animals subjected to chronic γ-irradiation at 7·5 r/day. It is found that aberrations are very high in neutron-treated animals and even increase to nearly 90% in succeeding months. With chronic γ-irradiation the frequency increases somewhat faster than that of the controls, but not nearly fast enough to be accounted for on the basis of a single hit phenomenon. It is concluded that there is chromosome healing following X- or γ-irradiation.

The somatic mutation theory of ageing was proposed a number of years ago, but because of the fact that direct evidence concerning this theory has been lacking, it has not been widely discussed. This theory postulates that the various somatic cells of the body gradually undergo spontaneous mutation. Most of these mutations are not lethal, so as cell division takes place in these cells the mutations are gradually multiplied until a large percentage of the body cells contain mutations. Since most mutations are deleterious to the cell, and therefore to the organism, by this process the various organs of the body gradually become less well able to perform their functions. This slowly leads to senescence and finally to death.

It is well-known that radiation causes a change in the animal which is either identical to or closely resembles the ageing process. It is also well-known that radiation is a very effective mutagenic agent for all cells, so these facts argue very strongly in favour of the mutation theory of ageing.

Since methods for observing mutations in somatic cells in mammals are lacking, an indirect method has been developed for this purpose (Stevenson and Curtis, 1961). It consists in scoring the numbers of chromosome aberrations visible under the light microscope in regenerating liver cells of the mouse. From work with plants (Caldecott, 1961) it was found that the numbers of aberrations present in the somatic cells of a plant were in all cases directly proportional to the mutations therein as measured by the mutation frequency in subsequent generations. If the situation is the same in mammals as in plants, and there is every reason to believe that such is the case, the

† Research carried out at Brookhaven National Laboratory under the auspices of the U.S. Atomic Energy Commission.

above method should provide a valid estimation of changes in somatic mutation rates, but should give no indication as to absolute numbers of mutations.

Previous work (Stevenson and Curtis, 1961) has shown that the aberrations in liver cells increase approximately linearly with age up to 12 months. Also, mice given a single large dose of X-rays show aberration percentages as high as 70%, which decrease very slowly with approximately half of them present 7 months later. The disappearance of the aberrations might be explained by cell division which occurs slowly in the liver.

It is well-known that chronically administered X- or γ-radiation is much less effective in causing shortening of the life-span than a single massive dose. For neutron irradiation, the shortening of the life-span is independent of the dose-rate. The present work was undertaken in an attempt to reconcile and explain these facts on the basis of the somatic mutation theory of ageing.

METHODS

The mice were of the strain "Charles River CDI" and were either bred in the laboratory or obtained from Charles River Farms. The experiments were begun when the mice were 8 weeks old.

The technique used for preparing liver cells for observation is essentially as previously described (Stevenson and Curtis, 1961). The mice were given subcutaneous injections of 0·005 ml/g of CCl_4 and 72 hours later, at the peak of mitotic activity in the liver, the animals were sacrificed and squash preparations were stained with acetic orcein. The slides were scanned and 150 anaphases and telophases were counted for each mouse. Abnormal mitotic figures were scored as either bridges or fragments or both.

The X-ray exposures were performed with 250 kVp; 30 mA; $\frac{1}{4}$ mm Cu and 1 mm Al filtration. γ-Rays were from a small ^{60}Co source and the 7·5 rads was given in 7·5 hours. The neutron irradiations were performed at the Brookhaven reactor (Curtis et al., 1956) in which a converter plate was used to give fast neutrons with a fission spectrum.

RESULTS

The experiments described here are time-consuming and are incomplete. However, there are several important facts revealed which make it worth while to issue a progress report.

The appearance of the abnormalities is essentially as previously described (Stevenson and Curtis, 1961). The first experiment involved a determination of the increase in aberrations with age for mice subjected to chronic γ-irradiation at a rate of 7·5 r/day starting at the age of 2 months. Control animals, kept in the same room but shielded from the radiation are also shown. It is apparent that radiation given at this rate increases the aberration frequency relatively little above that of the control (Fig. 1).

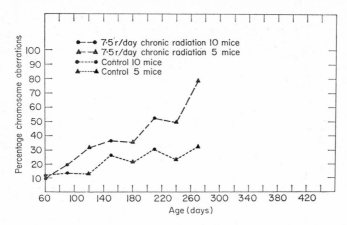

Fig. 1. Increase in percentage of chromosome aberrations with age for control and chronically irradiated mice. The accumulated dose is shown for various times.

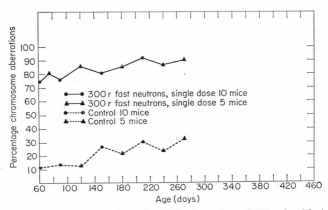

Fig. 2. Percentage of chromosome aberrations following a dose of 300 rads of fission neutrons.

The next experiment, shown in Fig. 2, followed the chromosome aberrations as a function of time following a single dose of 300 rads of fast neutrons (an LD_{10} dose). The curve shown is actually the average of two experiments both of which showed a very significant increase of aberrations with time. In the neutron-treated mice the aberrations were very severe, with many of the cells scored having several chromosomal injuries.

254   H. J. CURTIS AND CATHRYN CROWLEY

The survival curves to date are shown in Fig. 3. It will be noted that so far, except for some acute mortality in the neutron-irradiated animals, there has been little mortality. The weight curves show the experimental animals to have essentially the same weight as the controls.

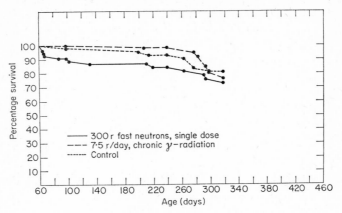

Fig. 3. Survival curves for the animals shown in Figs. 1 and 2.

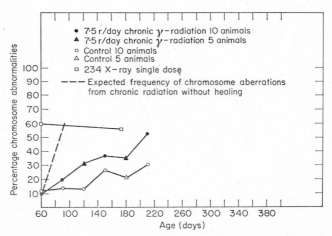

Fig 4. Percentage of chromosome aberrations plotted as a function of age of the mice for (1) control animals, (2) animals chronically irradiated at 7·5 r/day starting at 2 months of age, and (3) animals which received a single dose of 234 r of X-rays at age of 2 months. The time at which the chronically irradiated mice had accumulated 234 r is indicated, and the broken line is the slope one would have expected for the chronically irradiated group if there were no recovery process.

The upper curve in Fig. 4 shows the percentage of chromosome aberrations in mice given a single dose of 234 r of X-rays at 2 months of age, plotted as a function of age. Although the curve represents only 2 points, previous work substantiates the tendency of aberrations induced by X-rays to decrease slowly with time. On the same graph is plotted the data of Fig. 1 in order that a comparison can be made between the effects of chronic and acute irradiation. At 7·5 r/day it takes 31 days to accumulate 234 r; if chromosome aberrations form and are retained for small doses as they are for large doses (follow a single hit curve), the chronically irradiated mice should have accumulated 58% aberrations at 31 days. In other words, under this assumption the chronically irradiated animals should have followed a curve shown by the dotted line in Fig. 4. Since the experimental curve is very far from the same slope as this dotted curve, it follows that the assumption was not correct. It then seems justified to conclude that the chromosome aberrations follow a multihit curve as a function of dose. In other words, chromosome healing must take place.

DISCUSSION

The fact that the aberrations do not decrease and even increase for a time following neutron irradiation is consistent with the fact that there is little or no recovery from neutron irradiation as far as life-shortening is concerned (Curtis, 1961). The chromosomal injury was very severe and most cells scored as abnormal had multiple bridges and fragments. If a large fraction of these cells were so badly injured that they could not be forced into cell division for some time, then the initial scorings would show an abnormally low aberration frequency because of the abnormally large percentage of normal cells undergoing mitosis. In time, the injured cells apparently recover the power to undergo mitosis and thus the percentage abnormality increases.

The finding that chromosomal damage by X- or γ-rays is a multihit phenomenon, implying a chromosomal recovery mechanism, is consistent with several other well-known radiobiological phenomena. First, it has been shown (Steffenson and Arnason, 1954) that for X-irradiation of *Tradescantia* and other plant material, there is a similar effect. Secondly, it has been shown (Russell and Russell, 1959) there is an effect of dose-rate in mutation induction by X-rays in mice. Thirdly, for X- or γ-rays very small doses are proportionately only about one-quarter as effective in decreasing the life-expectancy in mice as are large doses (Curtis, 1961). The present data do not allow an accurate estimate of the proportionately lower effectiveness of small doses, but it must be close to a factor of 4.

These facts support the tentative hypothesis that the acceleration of ageing by radiation is due entirely to the induction of mutations in somatic cells. This follows from the fact that both with X- and γ-rays and with neutrons, chromosome aberrations correspond to the degree of life-shortening produced in each case, and also from the fact that the radiomimetic compound nitrogen mustard produces very few aberrations and no shortening of the life-span.

These facts also support the concept that mutations are an important part of the normal ageing process, but there must be other factors also involved in ageing. This follows from the fact that young irradiated animals may have an aberration frequency of nearly 100%, but quite a long life-expectancy; while old normal mice may have 30% aberrations and a short life-expectancy. These experiments give no information as to the other ageing factors.

In these experiments liver cells are taken as typical somatic cells, so there is every reason to believe that, following a large dose of radiation, almost all somatic cells in the body contain disrupted chromosomes. Yet under these conditions animals are functionally almost normal, and remain so for many months.

REFERENCES

CALDECOTT, R. S. (1961). In "Effect of Ionizing Radiation on Seeds". International Atomic Energy Agency, Vienna.
CURTIS, H. J. (1961). In "Radiobiology", p. 114. Third Australasian Conference. Butterworths, London.
CURTIS, H. J., PERSON, S. R., OLESON, F. B., HENKEL, J. F., and DELIHAS, N. (1956). Nucleonics 14, 26.
RUSSELL, W. L., and RUSSELL, L. B. (1959). Proc. 2nd International Conference on the Peaceful Uses of Atomic Energy 22, 360, 1959.
STEFFENSON, D., and ARNASON, T. J. (1954). Genetics 39, 220.
STEVENSON, K. G., and CURTIS, H. J. (1961). Radiation Res. 15, 774.

DISCUSSION

GRAY: This is an extremely interesting correlation that you've shown but I wonder whether it is in fact quite fair to compare liver with bone-marrow, because may it not be that the nitrogen mustard which undoubtedly produces the chromosome damage does so by interference with the actual turnover, the duplication, of the chromosomes. When it acts on cells which are proliferating it produces more damage. In the case of the radiation, of course it is obviously introducing a defect into the genome which expresses itself as a chromosome abnormality when the cell divides. The X-ray damage may only be repaired to a small extent. But nitrogen mustard which acts only in producing chromosomal instantaneous breaks at the time of turnover wouldn't have its effect on the liver but it could on proliferating tissue, so that this perhaps wouldn't mean that you couldn't produce with nitrogen mustard chromosome defects which would persist if they'd been initiated in cells which were turning over at the time of treatment.

CURTIS: Yes, I quite agree. I feel this work shows clearly that nitrogen mustard does not break chromosomes in non-dividing cells whereas it has been shown many times that it does so in dividing cells such as those in bone-marrow or testes. But rapidly dividing cells constitute a very small fraction of the total number of cells of the body, and this is probably why we see no life-shortening from nitrogen mustard. For mice susceptible to leukaemia induction, nitrogen mustard might be expected to cause some shortening of the life-span, and Dr. Upton has observed this. The real point of the nitrogen mustard experiments reported here is that they removed the objection to the somatic mutation theory of ageing created by the demonstration that nitrogen mustard does not cause life-shortening in this strain.

UPTON: I was very much fascinated by the nice demonstration in the animal that the neutron is a more effective chromosome breaker. This differentiation between the induction of point mutations as opposed to gross chromosomal derangements lead us to try the triethylene melamine and from our work it would seem in fact that this agent, for the same degree of lethality in mice judged on a 30-day survival, did exert a far greater life-shortening effect and this tempts me to correlate it with the findings which you have presented. We think it might be instructive to compare wider varieties of chemical mutagens or alkylating agents, to see whether or not there was a general correlation between chromosome-breaking effectiveness and life-shortening effectiveness.

CURTIS: I quite agree with you. We must know more about the basic mechanism of action of the chemical mutagens before we can make meaningful comparisons.

ALEXANDER: I think I can make a suggestion as to why you found that HN2—the oldest nitrogen mustard—did not shorten life-span, while we found another mustard, and myleran, effective. The reason we chose not to work with HN2 was that, although this is a radiomimetic compound one cannot, in fact, kill animals with acute single doses by a radiomimetic pathway because it also acts centrally. Therefore from such radiomimetic behaviour it isn't at all possible to give a comparable dose of HN2 because if one works at well below the lethal dose then it's radiomimetic but when one comes up to somewhere near the true lethal dose one can't achieve true bone-marrow killing because it kills centrally first. I would like to hazard a guess that the reason why HN2 doesn't show this effect is that at the true radiomimetic activity level you in fact use only a small fraction of the dose which the rest of us using other compounds were able to give.

CURTIS: You are probably correct, but I should point out that our mice were given almost lethal doses of mustard three times a week for more than half their total life-span and if there had been any appreciable chromosomal effect on liver cells it should have been evident.

MULLER: I wanted to make a point following Dr. Curtis's mention of the two-hit curve for chromosome aberrations. This, of course, is true at higher dose-rates because many chromosome aberrations that lead to losses of chromosomes and chromosome bridges are caused by the union of pieces derived from two different breaks. That's the predominant method at relatively high dose-rates but it's theoretically to be expected that when you go down to exceedingly low dose-rates such as you get with ordinary chronic treatment that you hardly ever have two breaks near enough in both space and time ever to get the broken ends connected with one another. Then the predominant chromosome loss which comes from single breaks of the kind that I was discussing (and for simplicity' ssake I left the other ones out of my discussion so as to concentrate on the simpler case) would mean that when you use very massive doses—acute doses—that you had considerably more shortening of life for the given total dose than from the same dose used chronically. This would also mean that this same evidence showed that what you were dealing with was not point-mutation—mutations in genes—because they're not affected by dose-rate.

MOLE: Can I ask one embarrassing question of Dr. Curtis? As far as life-shortening is concerned, isn't there only one way to interpret your experiments, and that is that mitotic abnormality in the regenerating liver has nothing to do with it!

CURTIS: Well you can certainly say that there is a correlation between the numbers of abnormalities that are produced and the age of the animal. Here is a phenomenon which is age-dependent and we know that—shall we say—that ageing process is age-dependent. So that there is that correlation I think, but I certainly agree with you that it's indicated very strongly that there must be other important factors coming in which may be of dominant importance. As a matter of fact I hope to give some further evidence later this afternoon as to what some of these other factors are.

MOLE: If you got 50% or 70% mitotic abnormalities as a result of X-rays, and just 10% more in the controls after your chemical and the life-shortening wasn't very different between those two then surely your evidence has no bearing on life-shortening.

THE FAILURE OF THE POTENT MUTAGENIC CHEMICAL, ETHYL METHANE SULPHONATE, TO SHORTEN THE LIFE-SPAN OF MICE

PETER ALEXANDER AND MISS D. I. CONNELL

Chester Beatty Research Institute, Institute of Cancer Research, London, England

SUMMARY

Mice of the CBA strain were injected with large doses of the very powerful mutagenic agent, ethyl methane sulphonate (EMS). The treatment had no effect on the average life-span although it did elicit a high incidence of tumours late in life. This observation suggests that the pronounced life-span shortening seen in comparable experiments after treatment with X-rays and the bifunctional alkylating agents, myleran and chlorambucil cannot be attributed to the induction of somatic mutations since, by analogy with data from *Drosophila*, the dose of EMS used will induce many more mutations than the three treatments which shortened the life-span of mice. We conclude that a non-genetic cause must be sought for the reduction of life expectancy which occurs in CBA mice after large doses of X-rays and which under the irradiation regime used is not due to a carcinogenic or leukaemogenic action. The fact that all the life-span shortening agents known are cytotoxic to dividing cells suggests that late disorders may be due to incomplete repair of the massive cell death which occurs at the time of treatment in organs where mitosis is high. Delayed cell death in organs with low mitotic rate may be another important factor. The deduction that somatic mutations have no part in radiation life-span shortening refers only to the particular situation studied.

The hypothesis has frequently been put forward that the life-span shortening produced by ionizing radiations is a consequence of the induction of somatic mutations which impair some normal physiological functions of cells and, in particular, of differentiated cells. Our earlier observation (Alexander and Connell, 1960) that two chemicals which are mutagenic, myleran (CH_3SO_2 . O . $(CH_2)_4O$. SO_2CH_3) and the nitrogen mustard, chlorambucil (COOH . $(CH_2)_3$. C_6H_4 . $N(CH_2CH_2Cl)_2$) brought about a shortening of life-span might be considered as support for a mutation mechanism. These chemicals are, however, bifunctional alkylating agents and as such are truly radiomimetic (Haddow, 1955). In particular, these two substances resemble radiations in killing rapidly dividing cells within a few days. They also produce in cells, not undergoing division at the time of treatment, delayed lesions which cause death when the cells enter division weeks or months later. This is seen, for example, in the kidney and liver. Consequently, the finding that myleran and chlorambucil shorten life-span does not eliminate mechanisms other than somatic mutations, which have been put forward for life-span shortening by radiation, such as imperfect repair of radiosensitive organs.

A more decisive test for the somatic mutation hypothesis would be whether the monofunctional alkylating agent derived from myleran (colloquially "half-myleran") ethyl methane sulphonate (EMS) $CH_3 . SO_2 . O . CH_2 . CH_3$ brings about life-span shortening. This substance has a very low toxicity yet it has been shown to be a very powerful mutagen in all the systems in which it has been tested (*Drosophila*, Fahmy and Fahmy, 1956; *Neurospora*, Westergaard, 1957; barley, Heslot and Ferrary, 1958; bacteria, Loveless and Howarth, 1959; bacteriophage, Loveless, 1959). With EMS effects due to cell killing and mutagenicity can be clearly separated since in bacteria even a 1% solution does not kill (Loveless and Howarth, 1959). We find that in mice the LD_{50} dose of EMS is eight times greater than that for myleran. The ratio of the actual cell killing efficiency of myleran and EMS is, however, even greater than 8 : 1 since mice given a lethal dose of EMS do not die from bone-marrow depletion following destruction of dividing cells as is the case for myleran and X-rays. With EMS death is due to acute kidney damage which is probably brought about by the large amount of strong acid released when EMS hydrolyzes in the body. Mustards and myleran also hydrolyze with production of strong acid but at the dose levels used the amount of acid released is physiologically neutralized. The mouse cannot, however, cope with acid from 400 mg/kg of EMS (the LD_{50} dose) and kidney necrosis results. In *Drosophila* also, myleran is much more toxic than EMS (Fahmy, private communication).

EXPERIMENTAL

Males from an inbred CBA strain were bred by brother-sister mating and maintained on rat cake nuts. EMS obtained from Eastman Kodak was suspended in arachis oil and injected intraperitoneally. The LD_{50} dose was found to lie between 400 and 450 mg/kg and a dose corresponding to half the LD_{50} (i.e. 200 mg/kg) was given three times at 3-weekly intervals starting with mice 11 to 14 weeks old. This regime produced no acute deaths. The procedure used with myleran and chlorambucil was exactly comparable and has been described (Alexander and Connell, 1960). The X-ray irradiation was carried out with 250 kV X-rays at a dose-rate of 55 r/min and given in four doses of 300 r, 300 r, 300 r and 200 r at 3-week intervals, a procedure which killed none of the animals.

RESULTS AND DISCUSSION

Table I shows that EMS has no significant effect on the average life-span when defined as the time for half the animals in a given experiment to be dead.

The difference between 693 days for the EMS treated and 762 for the control is not statistically significant. Furthermore, the true value for the life-span in the EMS group would be some days longer since many of the mice in this group were killed when in a moribund state because we wished to examine them histologically. We found that the EMS-treated group developed a high incidence of lung and kidney tumours. The latent period for this carcinogenic action of EMS would appear to be so long that it had no appreciable influence on life-span.

TABLE I

Treatment†	Average life-span‡	Post-mortem studies			
		% Without tumours other than hepatoma	% Kidney tumours	% Lung tumours	% Other tumours
	days				
None	762	77	3	20	5
3 × 300 r + 200 r; total 1,100 r of X-rays	404	62	5	19	24
3 × 25 mg/kg myleran; total 75 mg/kg	593	76	5	18	10
3 × 20 mg/kg chlorambucil; total 60 mg/kg	491	64	0	30	18
3 × 200 mg/kg ethyl methane sulphonate; total 600 mg/kg	693 §	6	33	89	11

† The radiation and the chemicals were administered to 11–14-week-old animals at three weekly intervals.
‡ Time when 50% of the mice were dead.
§ Most of the animals in this series were killed when they reached a moribund state to assist pathological study. The true life-span of the animals that had received ethyl methane sulphonate is therefore longer and approximates more closely to that of the controls.

While we have no direct data showing the relative number of somatic mutations produced in the mouse by the treatments tested in Table I, figures are available (see Table II) for the production of sex-linked recessive lethal mutations in *Drosophila*. If, for purposes of comparison, we equate mice with *Drosophila* then it can be seen from Table II that the EMS treatment used in the mice (i.e. as shown in Table I) provokes very many more mutations than do the treatments with X-rays, myleran and chlorambucil. Though on a weight basis these chemicals do not differ very greatly in mutagenicity, in mice the EMS treatment is likely to be much more mutagenic than the others because this substance could be administered at much higher doses.

While the extrapolation from germ mutations in the fruit-fly to somatic mutations in the mouse involves many uncertainties, the difference in the observed mutation yield in *Drosophila* between EMS and the other treatments is so great that one can feel confident that the same relative order is maintained. A possible objection is, that in the body of the mouse, EMS is distributed differently from myleran and chlorambucil and that the cells of certain organs are not exposed to the mutagenic action of EMS while the other chemicals—and of course X-rays—reach them. Investigations (Roberts and Warwick, 1958) concerning the metabolic fate of myleran and EMS make

TABLE II. *Mutagenicity of different treatments in* Drosophila

Treatment	Concentration used	% Sex-linked recessive lethals
Ethyl methane sulphonate	0.3×10^{-3} ml per fly of 10^{-2} M †	6.1
	600 mg/kg ‡	10.1
Myleran	0.3×10^{-3} ml per fly of 10^{-2} M †	3.8
	75 mg/kg ‡	0.4
Chlorambucil	0.3×10^{-3} ml per fly of 10^{-2} †	10.0
	60 mg/kg ‡	0.7
X-Rays	1,100 r †	3.1

† Values taken from data by Fahmy and Fahmy (1961).
‡ Values computed on the assumption of a linear dose-response curve and assuming one fly weighs 1 mg.

such differences seem unlikely and more recent studies by Drs. J. J. Roberts and G. P. Warwick (unpublished) with ^{14}C-labelled EMS, myleran and nitrogen mustard have shown that in the rat these substances are quickly dispersed all over the body and that by 1 hour after administration there is no evidence of any localization. At later periods the radioactivity is highest in the excretory organs because of the elimination of metabolites, but even in this respect all the compounds behave similarly.

We conclude from these experiments that the induction of somatic mutations is not responsible for the life-span shortening in mice induced by X-rays and the bifunctional radiomimetic alkylating agents used by us, since a treatment with EMS which probably produces many more mutations does not shorten life-span. The available data is consistent with the hypothesis (see Alexander and Connell, 1960) that the observed effect on life-span by X-rays and the bifunctional alkylating agents is a result of their cell-killing action—immediate or delayed—since EMS has virtually no effect on dividing cells.

This deduction cannot, however, be extended to other systems and it may well be that the effect of ionizing radiations on life-span in, for example, insects, may have a genetic origin, especially since the effect only sets in at much higher doses than for mice. In insects cell division only occurs in the brain and testes and massive cell death following irradiation is unlikely. Life-span shortening in insects may therefore only be seen when the dose is sufficiently high for an effect of radiation other than cell death to come into play.

ACKNOWLEDGMENTS

We wish to thank our colleagues, Drs. Fahmy, Roberts and Warwick for valuable discussions and for making their unpublished results available to us.

This investigation was supported by grants to the Chester Beatty Research Institute (Institute of Cancer Research: Royal Cancer Hospital) from the Medical Research Council, the British Empire Cancer Campaign, the Anna Fuller Fund, and the National Cancer Institute of the National Institutes of Health, U.S. Public Health Service.

REFERENCES

ALEXANDER, P., and CONNELL, D. I. (1960). *Radiation Res.* **12**, 38.
FAHMY, O. G., and FAHMY, M. J. (1956). *J. Genet.* **54**, 146.
FAHMY, O. G., and FAHMY, M. J. (1961). *Genetics* **46**, 1111.
HADDOW, A. (1955). Proc. 1st International Conference on Peaceful Uses of Atomic Energy, Geneva **11**, 213.
HESLOT, H., and FERRARY, V. A. (1958). *Ann. inst. natl. recherche agron.* **44**, 133.
LOVELESS, A. (1959). *Proc. roy Soc.* **B150**, 497.
LOVELESS, A., and HOWARTH, S. (1959). *Nature, Lond.* **184**, 1780.
ROBERTS, J. J., and WARWICK, G. P. (1958). *Biochem. Pharmacol.* **1**, 60.
WESTERGAARD, M. (1957). *Experientia* **13**, 224.

DISCUSSION

CURTIS: Is there any information as to whether ethyl methane sulphonate produces chromosome breaks?

ALEXANDER: This substance is not a cytoxic agent, but it does give rise to chromosome damage though the ratio of translocation to recessive lethals is smaller than with the bifunctional compounds.

CURTIS: I think it's perhaps a little premature to say that it does or does not support the somatic mutation hypothesis if you don't know whether ethyl methane sulphonate produces mutations in the animal with which you're dealing.

ALEXANDER: I agree; but you will appreciate the great difficulty of establishing mutagenic action in mammals under these acute insults with large doses of radiation and alkylating agents. One of the important factors in life-span shortening is massive cell death soon after the exposure followed by incomplete repair of the organ as a whole. Ethyl methane sulphonate differs from the other agents in not producing this massive destruction but is a powerful mutagen in all systems.

MAISIN: It would be interesting to know the effect of your fractionated irradiation on lung tumour induction.

ALEXANDER: Under our irradiation conditions few tumours were induced in the CBA mice.

MAISIN: But much more than Dr. Upton reported yesterday?

ALEXANDER: No. We had 23% animals dying with malignant tumours in the controls and 38% in the irradiated. This difference is unlikely to be significant. Our data are in agreement with Dr. Mole's observations reported yesterday. Using the same strain of mice he found that, after large doses of X-rays, no leukaemia was produced while smaller doses gave rise to a large incidence.

ROTBLAT: I think you have more than one tumour in some of the mice.

ALEXANDER: Yes.

MAISIN: If you irradiate one lung only instead of the whole animal you may get much more lung cancer in your mice than with total-body irradiation. Because irradiating the lung only in rats gave us 70% of lung tumours, possibly because the animal's life is not shortened and it survives long enough to have these tumours.

BERENBLUM: In the series with ethyl methane sulphonate the life of the animals was not shortened and yet this compound was very highly carcinogenic. Does that mean that most of the tumours appeared very late in life?

ALEXANDER: We started killing them off at about 700 days because we realized that there was this high tumour incidence and we wanted to make sure of the histology of the tumours. We don't know whether these lung tumours grow very slowly or whether they only appear late in life. You see with no life-span shortening we had virtually no dead mice until about 600 or so days.

BERENBLUM: But you must have started killing them off in large numbers at the end because your figures don't refer to the age at the end of the experiment but to the age of the 50% of the survivors.

ALEXANDER: Yes. As soon as we felt confident that it hadn't produced any life-span shortening we thought we would use the same animals to give us information about the carcinogenicity of ethyl methane sulphonate. We were so puzzled by the fact that myleran and chlorambucil which are carcinogenic under other situations were not carcinogenic when given in three large doses by mouth. It has been suggested that this absence of carcinogenicity arose because it was given by mouth. But this now seems unlikely because the monofunctional agent also given by mouth did produce tumours.

MULLER: Were those medians or averages of life-span?

ALEXANDER: Median, in time at which 50% of the animals were dead.

LAMERTON: Dr. Alexander said it was very carcinogenic but it looked as though the whole of the excess was in the lung adenomas and kidney tumours.

ALEXANDER: They weren't all lung adenomas. About half were carcinomas. Kidney tumours might well have been metastases from the lung.

LAMERTON: Does this correspond to the normal spontaneous distribution of tumours?

ALEXANDER: Yes; in control animals about 25% show lung tumours at death (see my other paper, p. 281).

LAMERTON: I wonder if the fact that you had your excess in kidney and lung only meant that the agent didn't actually get around.

ALEXANDER: I don't know whether I believe this a valid argument—carcinogenic compounds have odd predilections for site of action and these are usually unrelated to distribution. I see no reason why such a small compound shouldn't get around and tracer work by Roberts and Warwick of this Institute has revealed no organ selectivity at all. From the cancer chemotherapy point of view one would of course be delighted if compounds didn't get everywhere and remained in a few places but none have been found to do this to any marked extent.

MULLER: I just want to say we might be talking at cross-purposes if we simply say that this or that observation is for or against the somatic mutation hypothesis because I think you were using the somatic mutation hypothesis to mean somatic point-mutation. In fact Curtis was using it in a more general sense too, including chromosome aberrations.

ALEXANDER: I was stressing point-mutation because, for these, ethyl methane sulphonate is so very effective.

LIFE-SPAN SHORTENING FROM VARIOUS TISSUE INSULTS†

H. J. CURTIS AND CATHRYN CROWLEY

Biology Department, Brookhaven National Laboratory, Upton, New York, U.S.A.

SUMMARY

From experiments in mice, it is shown that a specific organ stress, mercury poisoning of the kidney, can cause irreparable damage which persists indefinitely. The animal is able to compensate for this damage when it is young, but as it ages it fails to compensate and death follows. Tissue scarring is apparently one of the important factors in the ageing process.

One of the proposed hypotheses of the ageing process is that the various organs of the body can recover only partially from a tissue insult, and death finally results from the accumulation of such insults. The life-span shortening caused by radiation would thus be due to its action as a non-specific stress. To test this theory a rather extensive series of experiments was undertaken (Curtis and Healey, 1956; Curtis and Gebhard, 1959) employing various non-specific chemical stresses. Single, large, just sublethal doses of typhoid toxoid and nitrogen mustard failed to produce any shortening of the life-span, whereas a comparable dose of radiation caused marked shortening. In the belief that perhaps there was not enough of a tissue insult to cause appreciable life-span shortening, these and other toxic agents were administered repeatedly over a large fraction of the life-span of the mice. These included intraperitoneal injections of turpentine every 14 days, intraperitoneal injections of nitrogen mustard three times a week, subcutaneous injections of tetanus toxin and also tetanus toxoid every 14 days, intraperitoneal injection of typhoid vaccine twice weekly, and others. All these demonstrated that these non-specific stresses cause no appreciable shortening of the life-span. If the animals were able to withstand the stress and recover from it, their life-expectancy was not shortened even though they were kept under severe stress for most of their lives.

These experiments do not give information on the action of an organ-specific toxin. It is reasonable to suppose that if one vital organ is severely damaged, it might form a "weak link", causing death predominantly by ultimate failure of that organ. The organ chosen for the present study was the kidney.

† Research carried out at Brookhaven National Laboratory under the auspices of the U.S. Atomic Energy Commission.

METHODS

Mice of the Charles River CD 1, specifically pathogen-free, strain were used for the experiments which were begun when they were 2 months old. Mercuric chloride was administered by stomach tube in the first experimental group at the rate of 40 mg/kg of body weight, and at the rate of 20 mg/kg in the second. These doses were administered once a week for four consecutive weeks. Thereafter the mice were set aside for a period of 4 months. At that time the animals were subjected periodically to 24-hour water deprivation, and immediately thereafter pooled urine samples were collected from each of the experimental and control groups. These were analysed for protein by means of the Biuret reaction (Gornall et al., 1949). Also after the 24-hour water fast, urine samples were expressed from 5 individual mice in each group and the specific gravity was measured in a linear density column (Linderstrom Lang and Lantz, 1938).

RESULTS

The results of the protein Biuret analyses are shown in Fig. 1. Three to four months after the administration of mercuric chloride, the protein output began to climb in the experimental mice reaching maximum output in 5–6 months.

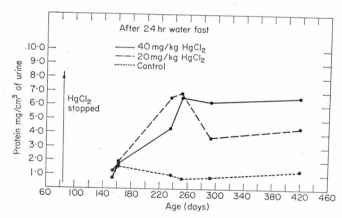

FIG. 1. Variation of protein concentration with age for normal and mercury-poisoned mice.

In Fig. 2 the results of the specific gravity tests are shown. As indicated by the control curve, the ability of the normal kidney to concentrate urine decreases with age. For the points tested up to about 9 months of age the experimental group are less able than the controls to concentrate the urine.

Thereafter, as the controls age, kidney function declines sharply until it approaches experimental values.

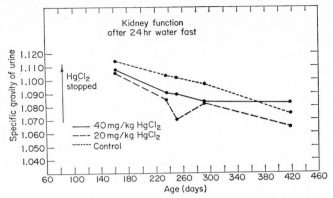

FIG. 2. Variation of urine specific gravity with age for normal and mercury-poisoned mice.

Figure 3 shows survival curves to date and experimental groups show mortality rates advanced over the controls.

Figure 4 shows the weights of experimental animals and controls.

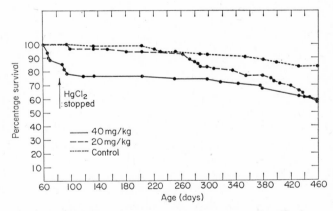

FIG. 3. Survival curves for normal and mercury-poisoned mice.

DISCUSSION

It is thus apparent that if damage is created in an organ which has only limited regenerative power, this organ can act as a "weak link" and eventually lead to organ failure and death. Thus tissue insults can be a factor in the ageing process. However, the fact that function was virtually normal for some time following the insult, and took a large fraction of the total life-span before it became manifest, clearly demonstrates that this factor is not a dominant one in the ageing process.

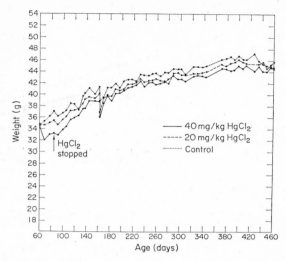

FIG. 4. Weight curves for normal and mercury-poisoned mice.

The nature of at least one of the other factors involved in the ageing process is suggested by other work. It has been shown (Stevenson and Curtis, 1961; Curtis and Crowley, 1962) that chromosomal aberrations in regenerating liver cells increase steadily with age and undoubtedly form part, but by no means all, of the ageing process. In the present experiment, the kidney was damaged when the animals were young and their cells were quite free of aberrations. The damage caused scarring which could be compensated for by an over-activity of normal cells. As these normal cells accumulated spontaneous mutations with age, they were less well able to perform this function, and there were no reserve cells to take over. Functional failure was followed by organ failure and death. This is an attractive hypothesis, and while the present experiment does not prove it, it does give it positive support.

REFERENCES

CURTIS, H. J., and CROWLEY, CATHRYN (1962). This volume p. 251.
CURTIS, H. J., and GEBHARD, K. L. (1959). *Proc. 2nd U.N. Conference on Peaceful Uses of Atomic Energy* **22**, 53.
CURTIS, H. J., and HEALEY, RUTH (1956). In "Radiobiology", p. 261. Oliver and Boyd, Edinburgh.

GORNALL, A. G., BARDAWILL, C. J., and DAVID, M. M. (1949). *J. biol. Chem.* **177**, 751.
LINDERSTRÖM-LANG, K., and LANTZ, H., Jr. (1938). *Mikrochem. Acta* **3**, 210.
STEVENSON, K. G., and CURTIS, H. J. (1961). *Radiation Res.* **15**, 774.

DISCUSSION

POCHIN: One of the problems in this work, I imagine, must be to see what quantitative effect you are getting at a different level of your generalized stresses. I was wondering if you happened to get any information of this sort to give radiation and other generalized stressing agents at an LD_{50} and study or compare the 50% survival and the length of survival of those surviving the LD_{50} irradiation as compared with the life-times of those surviving an LD_{50} from other stresses.

CURTIS: What we did in every case was to have one group of animals in which we determined the lethal dose of the agent in question, and then in the experimental group we gave about half the lethal dose. Thus we were sure in each case that the animals were receiving an almost lethal dose of the agent.

POCHIN: But the survivors from the LD_{50} of the other stress agents survived normally?

CURTIS: Yes.

ALEXANDER: I just wanted to stress that perhaps your experiments had in fact shown that general stress did shorten life-span perhaps not quite so acutely as tetanus toxoid, but maybe just living in America, because your animals had a medium life-span of 400 days whereas I don't think many of their genetically identical English counterparts, would live for quite such a short time, so the hectic American life seems to have an effect on mice as well!

CURTIS: We selected this strain of mouse becaue it is quite a short-lived strain so we do not have to wait three years to obtain results. I see no reason to believe that the conclusions would not be the same for all strains, but would certainly agree that eventually they should be verified on other strains.

BRINKMAN: I understand that if one kidney of an animal is removed the other enlarges, unless it is irradiated. So then the capacity for enlargement of the normal kidney is lost by a small amount of irradiation.

CURTIS: This might be so, but you run into trouble because the hyperplasia of the kidney can be by both cell enlargement and cell division, and the proportion of the enlargement due to each, changes with age. Radiation acts much more strongly on cell division than on cell elongation, so the interpretation of results may not be so easy.

DOES RADIATION AGE OR PRODUCE NON-SPECIFIC LIFE-SHORTENING?

R. H. MOLE

M.R.C. Radiobiological Research Unit, England

SUMMARY

The older mice are at the time of irradiation the less the life-shortening effects of single exposures to radiation and of daily irradiation for the duration of life. Since natural ageing and radiation-induced life-shortening were not additive, it seems unlikely that radiation causes non-specific life-shortening in the sense that it ages the organism in some general over-all way.

The phenomenon of ageing is very general indeed so that even one exception to the popular generalization that radiation causes non-specific ageing may throw real doubt on the validity of the generalization.

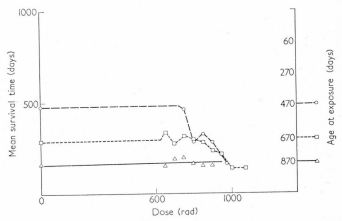

FIG. 1. Mean survival time of female CBA mice, survivors of single X-ray exposures in the acutely lethal range given at varying ages as shown.

In an as yet, unfinished series of experiments by Mole and Thomas (1961) the survival time of female CBA mice aged 70, 270, 470, 670 and 870 days has been determined after single doses of X-rays in the LD_{50} range (Fig. 1) or during continuous γ-irradiation at 10 r nightly (Table I). Seventy- and 470-day-old mice have also been compared at daily doses of 15, 30 and 50 r γ-rays and 2 rads of fast neutrons.

Figure 1 shows that mice aged 870 days which survived the acute effects of 650 to 850 rads of X-rays did not have their lives shortened at all as compared with specific controls. Similarly 670-day and 470-day-old mice did not have their lives shortened by similar doses of $\sim$ 750 rads, although at slightly higher doses in the upper range of acutely lethal exposures there was quite definite life-shortening. Life-shortening was clearly not a linear function of dose and the results do not suggest that radiation and chronological ageing are additive in their effects.

TABLE I. *Survival of female CBA mice of different ages exposed to* 10 r *γ-radiation nightly continuously till death*
(Mole and Thomas, 1961)

Age of start of irradiation (days)	Natural mortality at at start (%)	Expectation of life † (months)	Mean survival time (months)	Life-shortening Absolute (months)	% of expectation
70	0	$(28\frac{1}{2})$	17	$11\frac{1}{2}$	41
270	0	(22)	$16\frac{1}{2}$	$5\frac{1}{2}$	25
470	1	(15)	$(14\frac{1}{2})$	1	7
670	14	10	8	$1\frac{1}{2}$	17
870	31	6	$4\frac{1}{2}$	$1\frac{1}{2}$	23

† Taken to equal the observed mean survival time of the specific controls for each age group. The experiment is not quite complete: the results in brackets are provisional.

These results are easier to understand if it is accepted that radiation exposure initiates ultimately lethal processes which take time to develop to the bulk or intensity which actually kills (cf. tumour development time, Mole, 1962). If so, the 150 days which 870-day-old mice have still to live is presumably too short for the full development of a killing lesion. However this line of argument leads to the conclusion that 280 days and 470 days are also too short a period for the full development of killing lesions in 670- and 470-day-old mice respectively. Alternatively the rate at which killing lesions develop may be thought to depend on the magnitude of the dose. In either case plain survival time or reduction in after-expectancy of life cannot be expected to be any simple function of radiation dose.

With continuous irradiation till death at 10 r nightly (Table I) both the absolute amount of life-shortening and the proportion by which the life expectancy was reduced were smaller the older the animals at the start of their exposure over the range from 70 to 470 days. At still older ages life-shortening was just detectable but was very small, in spite of the magnitude of the accumulated doses. Again there was no additive or synergistic effect of γ-radiation and natural ageing.

Unpublished results show that there was similarly no additive or synergistic effect of whole-body irradiation by fast neutrons and natural ageing.

Any study of ageing processes which uses time of death as the measure of the amount of ageing must face the problem of how to allow for the "accidental" aspects of death. Thus unirradiated reciprocal hybrid A × CBA and CBA × A female mice die at very different ages: their life-spans are in the ratio 2 : 3. The two kinds of mice are genetically identical except for some parts of their sex chromosomes and this should give pause to those who attribute differences in life-span to genetic differences and the life-shortening effects of radiation to induced genetic change.† Taking this suggestion seriously (and ignoring additional information gained by examining individual organisms for possible causes of death) entails the conclusion that sex chromosomes are far more important than autosomes in determining life-span, a conclusion which is at variance with the basic idea that gross loss of genetic information is the determining factor in cellular and organismal ageing.

Actually the difference in life-span of the two hybrids is due to a specific lesion, the murine mammary tumour, and the reason why normally one hybrid and not the other suffers from the tumours, although both are equally susceptible, is that the A × CBA hybrid derives a transmissible cell-free agent (virus) from its mother's milk whereas the CBA mother of the other hybrid does not carry the virus. Thus in this case the life-span depends not on chromosomal or genetic constitution but on the "accident" of infection.

REFERENCES

MOLE, R. H. (1962). In "Some Aspects of Internal Irradiation" (T. F. Dougherty ed.). Pergamon Press, Oxford.

MOLE, R. H., and THOMAS, A. M. (1961). *Radiation Res.* **14**, 487.

DISCUSSION

ROTBLAT: First of all, I would like to say that the results of Dr. Mole don't disagree with our results; slight differences may be explained by the fact that we used a different strain of mice. The interesting thing is that if you take a look not at the absolute shortening of life-span but express it as a proportion of the potential life-span, then it turns out that the variations with age are much less, and it follows almost exactly in relation to age. I would like someone to check Dr. Mole's results with different radiations as well.

MOLE: First of all I did show, in one slide anyway, that the proportional life-shortening, also got less the older the animal, from 40 to 10%, quite a substantial change. Secondly, I don't think that these results necessarily disprove a non-specific life-shortening but what they do seem to show is that radiation and natural ageing don't add. This I think is very difficult to swallow if you think they have the same mechanism.

ROTBLAT: Well, we may have been lucky there, but I don't think people can say they add but they act in a similar fashion, this is what people can say but we are sure that they cannot be identical because they have produced different results. They may act in a similar fashion but it doesn't necessarily follow that they must be additive.

† See H. J. Muller, "Mechanisms of Life-span Shortening", this volume.

MOLE: This is partly a question of the meaning of words. "Similar" must mean to some extent additive mustn't it? Dissimilar, anyway, means non-additive.

CASARETT: I think Dr. Mole has pointed out very well many of the vagaries and the difficulties of using LD_{50} and life-shortening in judging ageing processes, points which I emphasized myself. In regard to the differences in life-shortening when one irradiates animals at different ages, I would like to take up the suggestion of long development time. I think that when you irradiate older animals there is not enough time to get the full development of the life-shortening potential of the dose. This is something to consider besides those points you brought up. I would also like to ask whether Dr. Mole thinks that all of the causes of death of the irradiated animals are entirely induced by the radiation and all of the causes of death in the non-irradiated animals are entirely brought about by ageing processes, if he says there is nothing additive. If I may clarify this, one can take the point of view that in irradiated animals each of the lesions which kill are induced separately by the radiation given, but it seems much more sensible to regard the uniformities of the response, taking full account of the non-uniformities that exist, as evidence of some generalized additive non-specific process that affects the diseases and causes of death in both populations.

MOLE: I don't see how one can possibly say that everything in the irradiated animal is due to the irradiation, everything in control animals to something different. Also, is death ever due to a non-specific cause? If pathology and post-mortems are taken seriously a lot of lesions will be found. Now a really competent human pathologist can put his finger on the cause of death in between one-quarter and one-third of a non-specific collection of human post-mortems. In the remainder there will be many lesions but it would be hard to say which is the cause of death. (It's perhaps much easier to say that post-mortem examinations show you what it's possible to live with.) I think that to equate ageing just with death-time is the basic mistake. We know that ageing of anything is a progressive process and death can only happen once. It often happens in radiobiology, since survival time can be scored so easily, that a technician can do it for you. Then you can sit at your desk and do some calculations, but I don't think that really has any bearing on the subject. You've got to try and do the best you can to decide what did kill the animals. Now suppose you confine your attention to those things which quite clearly do kill, like leukaemias, or tumours which lead to haemorrhage or to gross infection and things like this. When you find regularly a higher incidence of such kinds of lesion in irradiated animals than in control animals, I don't see how you can avoid concluding that irradiation has specifically produced particular forms of disease. The fact that the same disease occurs in control animals is exactly what you would expect because the cells of any one organ of the body can react only in particular kinds of way. It doesn't matter what the agent is which starts a process off, the reaction must depend on the nature of the cells and the organs.

CASARETT: I think you are putting emphasis on life-terminating diseases, on this one event rather than on the pathogenesis of the disease.

COTTIER: I would stress Dr. Mole's view that this life-shortening is not necessarily identical with ageing. If you compare the histopathological picture of spontaneously dying mice or mice killed at several intervals after irradiation throughout life-span there is a discrepancy insofar as a specific change lies in the vascular and other changes which progress very nicely during life, like calcification of rib cartilage or calcification of tendons or of intravertebral discs. They follow the same course in irradiated as in non-irradiated animals. This is just one example of a discrepancy between these two and I think one really has to consider carefully what the causes of death are in these irradiated animals.

DIFFERENCES BETWEEN RADIATION-INDUCED LIFE-SPAN SHORTENING IN MICE AND NORMAL AGEING AS REVEALED BY SERIAL KILLING

PETER ALEXANDER AND MISS D. I. CONNELL

Chester Beatty Research Institute, Institute of Cancer Research, London, England

SUMMARY

Mice that had received in several fractions a total dose of 1,100 r of X-rays were killed at monthly intervals and the histopathology of selected organs examined. The results of the first two years of this investigation are reported. The onset of some diseases and their incidence is quite unaffected by irradiation; for other diseases the latent period is the same, but the incidence is higher in the irradiated group. In yet others radiation advanced the time of appearance without altering the incidence. In addition whole-body irradiation produces diseases that are not found in normal mice. These studies do not support the view that radiation-induced shortening of life-span can be described as an acceleration of normal ageing since this would require that the latent period of all diseases was advanced, but their incidence remains unaffected. It is concluded that other experiments, in which the time-course of the various diseases was determined from post-mortem examinations of mice that had been allowed to die, can be misleading and that the method of serial killing is more reliable.

The process of senescence is usually equated with the progressive impairment of many, and diverse, physiological functions. A consequence of this physiological deterioration is that the probability of an animal dying of certain specific diseases increases. The end effect of senescence is to impose an upper limit on expectation of life and in this sense death and ageing are related. We do not think that it is useful to consider ageing as a direct cause of death. The force of mortality, though related, is not a measure of senescence. A number of authors have referred to any process which shortens life-expectancy (i.e. one which shortens the time axis but not the over-all shape of a typical survival curve of an accident-free population) as an acceleration in ageing. Such an approach leads to obvious absurdities. For example, any carcinogenic stimulus with a long latent period will produce this effect, yet in reality the treatment produces a new disease and need in no way be an acceleration of a normal process. The effect of heavy cigarette smoking—started in adolescence—causes a shortening, by some 15%, of the time needed for 50% of the population to die—the average life-span—but apart from this contraction of the time axis the shape of the survival curve of smokers and non-smokers is very similar (R. A. M. Case, private communication). In man, smoking is not said to accelerate senescence because the change in the mortality curve is unmistakably due to the induction of two diseases, lung cancer and chronic bronchitis, which are almost wholly absent in non-smokers. Yet radiation has been claimed to accelerate the natural processes of senescence in small mammals largely because it increases mortality rate for which, in general, there is no obvious pathological explanation such as a great increase in malignant disease (cf. Blair, 1956).

The most compelling method for examining the hypothesis that radiation ages is to compare the time-course of a number of physiological and biochemical functions that vary with age in an irradiated and unirradiated population. While there are a number of well-defined, quantitative tests for assessing ageing in man (e.g. muscular power, sensory perception, skin elasticity, vital capacity, memory) these cannot be satisfactorily applied to small mammals. At present, only changes in the physico-chemical properties of collagen (see Verzar, 1958) have been quantitatively related to age in rats and mice. The force of contraction which collagen fibres develop on heating increases regularly throughout life. Alexander and Connell (1960) showed that the tail tendons of mice that had received 1,100 r of X-rays were indistinguishable throughout life from those of unirradiated mice of the same age in spite of the fact that the average life-span had been reduced by 40% by the irradiation.

In the virtual absence of physiological tests comparative pathology of irradiated animals can be useful. The hypothesis that radiation ages would be supported if the onset of all diseases was advanced to the same extent and by a factor comparable to that of the life-span shortening. Connell and Alexander (1959) found that neither the time of appearance, nor incidence, of benign hepatomas in CBA mice was influenced by 1,100 r. Yet in this strain the incidence of this condition seems to be a function of age.

Most other publications have been confined to the investigation of the causes of death. If, under the conditions of the experiment, radiation is powerfully leukaemogenic or carcinogenic, then an ageing effect will be difficult to detect by post-mortem examination of animals that have died. However, a number of workers, in particular Lindop and Rotblat (1961), Furth et al. (1959), and Upton et al. (1960) have analysed the causes of death in strains of mice where the carcinogenic action of radiation was not predominant. They found that the general pattern of the causes of death were not altered by radiation, but that it was advanced in time and concluded that radiation accelerated normal ageing processes. All these investigations suffer from the serious drawback that post-mortem examination of animals that have died of old age can provide reliable information for the cause of death in only a fraction of the animals unless tumours or leukaemia were involved.

We have attempted to follow the onset and incidence of disease by killing groups of ten mice at monthly intervals and subjecting them to a histopathological examination of selected organs. We have confined ourselves to an irradiation of 1,100 r of X-rays given in four doses (3 × 300 r + 1 × 200 r) at 3-weekly intervals since the magnitude of the life-span shortening of this treatment (see Fig. 1) was well-known to us (Alexander and Connell, 1960). More than 90% of the animals were alive 30 days after the last irradiation. Under these irradiation conditions malignant disease constitutes only a small fraction of the deaths. The post-mortem results showed

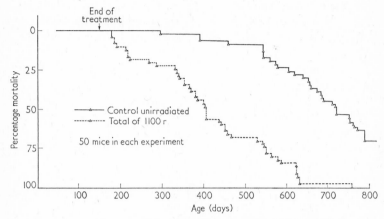

Fig. 1. Mortality curves for CBA male mice after 1,100 r of X-rays.

that only 30% of the animals had tumours, or leukaemia, other than the benign hepatomas (see Table I on p. 261 of our other paper in this symposium). This experiment therefore represented a typical example of life-span shortening which is so often called *radiation-accelerated ageing*.

EXPERIMENTAL

Male CBA mice aged between 11 to 14 months at the time of the first irradiation were used. The animals were exposed at a dose-rate of 55 r/min to 300 r + 300 r + 300 r + 200 r of 250 kV X-rays at 3-weekly intervals. In all respects the experimental conditions were exactly the same as those reported earlier (Alexander and Connell, 1960). After the last irradiation the animals were divided into groups and designated to be killed at a specified date. Batches were killed at 3, 6 and 9 months and thereafter at monthly intervals. It was intended that there should be ten mice alive in each batch at the time killing was due. Consequently, allowance had to be made for intercurrent deaths. From our earlier data on the effect of radiation on the survival curve the number of animals that had to be allocated to each particular batch could be calculated.

RESULTS AND DISCUSSION

At the time of writing, this experiment has been in progress for just under 2 years. Not all of the slides from the mice that have been killed have so far been examined and classified. The present report is therefore limited to the rate of appearance of the following conditions only: (1) benign hepatomas, (2) lung tumours, (3) cataract, (4) papillonephritis.

Figure 2 shows that the time of appearance of the benign hepatomas (*cf.* Connell and Alexander, 1959) is exactly the same in the irradiated as in the control population. The serial killing technique fully confirms our earlier

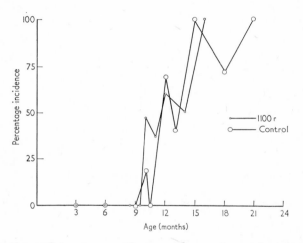

FIG. 2. Incidence of hepatomas with increasing age in serially killed control and X-irradiated CBA male mice.

conclusion, based on examination of animals which had been allowed to die, that radiation has no influence on the time-course or incidence of this condition. Thus if the appearance of hepatomas is used as a criterion for ageing then radiation does not age.

Lung tumours would suggest the opposite. Figure 3 shows that radiation has no significant influence on the incidence of these tumours, but they appear earlier. It could be argued that radiation had speeded up the "internal clock" of the animal about twice in agreement with the observed life-span shortening; in other words radiation accelerates ageing.

The appearance of cataracts (see Fig. 4) illustrates another type of response. Radiation reduces the latent period, but also increases the total incidence. The situation is even more complex since the cataracts in the irradiated group are morphologically quite different from senile cataracts (Upton *et al.*, 1960). It might therefore be more correct to look upon radiation as bringing about a new late somatic change which is quite absent in the unirradiated population.

Another characteristic pathological change occurring with advancing age in several strains of mice, including ours, is to be found in the kidney. The condition is described under several names in the literature, but for the CBA

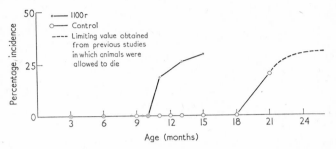

Fig. 3. Incidence of lung tumours with increasing age in serially killed control and X-irradiated CBA male mice.

animals, the term papillonephritis, as defined by Dunn (1944), is thought to be most applicable. The effect of the radiation treatment on the series of developmental changes leading to papillonephritis, as seen in the control mice, is different, yet again, from the other examples quoted above. From 6 months of age, when the first pathological changes become visible, the steady

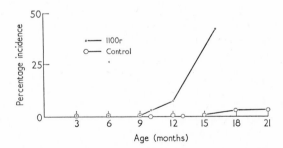

Fig. 4. Incidence of cataracts with increasing age in serially killed control and X-irradiated CBA male mice.

progression of the various stages of papillonephritis follows the same pattern in both control and irradiated animals. Lesions, severe enough to be included in the probable causes of death, begin to appear at 18 months of age. As the serial killing has not gone beyond twenty months, the total number of cases in this experiment is still very small. From previous experiments, 40% of deaths in both control and irradiated animals are probably due to renal failure in the last stages of papillonephritis. As, however, the life-expectancy of the treated mice is much less than for the unirradiated, the fraction of animals developing this kidney lesion from 18 months of age rises much more rapidly amongst the irradiated than amongst the controls. There is, however, no indication that the latent period of this disease is altered by radiation.

Though this study using serial killing is as yet incomplete, the results that have been obtained so far are not consistent with the hypothesis that radiation advances all diseases. The effect of radiation on late somatic changes is extremely complex and most conceivable variants were seen in this investigation.

(1) The time-course and incidence are completely unaffected by 1,100 r in spite of the fact that the disease seems to be related to age (i.e. like the ageing changes in collagen).

(2) The latent period is unaffected but the incidence is increased.

(3) The latent period is shortened but the total incidence is increased.

(4) The latent period is shortened but the total incidence remains unaltered.

If radiation were to induce a process akin to normal ageing then most of the major pathologies should fall into the last category. There is yet another possible variant, namely that radiation delays or prevents certain diseases. An example has not been observed by us but has been seen in strains of mice having a high leukaemia incidence.

None the less an understanding of the biochemical mechanisms that lead to radiation life-span shortening may materially advance knowledge about some of the processes responsible for ageing. It seems unlikely that there is one basis effect at either the chemical or cellular level which is responsible for the physiological changes which *in toto* are changes known as senescence. The relative contributions of the different processes may be different in different animals. Radiation may imitate one or more of the biochemical reactions that lead to ageing, but it does not reproduce the whole complex. Before the aetiology of ageing can be understood at the sub-cellular level it must be broken down into a number of different biochemical pathways each of which causes an upper limit to appear in the life-span of an animal. Radiobiology may pin-point some of these life-limiting factors.

This investigation was supported by grants to the Chester Beatty Research Institute (Institute of Cancer Research: Royal Cancer Hospital) from the Medical Research Council, the British Empire Cancer Campaign, the Anna Fuller Fund, and the National Cancer Institute of the National Institutes of Health, U.S. Public Health Service.

REFERENCES

ALEXANDER, P., and CONNELL, D. I. (1960). *Radiation Res.* **12**, 38.
BLAIR, H. A. (1956). University of Rochester UR-442. Unclassified.
CONNELL, D. I., and ALEXANDER, P. (1959). *Gerontologia* **3**, 153.
DUNN, T. B. (1944). *J. nat. Cancer Inst.* **5**, 17.
FURTH, J., UPTON, A. C., and KIMBALL, A. W. (1959). *Radiation Res.* Supplement 1, 243.
LINDOP, P., and ROTBLAT, J. (1961). *Proc. roy Soc.* **B154**, 350.
UPTON, A. C., KIMBALL, A. W., FURTH, J., CHRISTENBERRY, K. W., and BENEDICT, W. H., (1960). *Cancer Res.* **20**, 1.
VERZAR, F. (1958). *Helv. phys. pharm. acta* **14**, Fusc. 2, 207.

DISCUSSION

CURTIS: I would just like to point out that in some strains of mice with a high spontaneous leukaemia rate irradiation reduces the incidence of this disease.

BERENBLUM: The small amount of radiation dose, which in C57BL mice raises the incidence of leukaemia from about 2% to 40–50%, reduces the leukaemia incidence in AKR mice, from 70–80% in the controls to 40–50%.

CURTIS: I think all we are saying really is that radiation leukaemogenesis is not a simple process, it is a combination of processes and I don't think that we should try to generalize it and put it all in one box.

ALEXANDER: One thing seems clear, radiation does not accelerate ageing in the sense of speeding up the "internal clock" of the animals so that cell changes, physiological and pathological, occur in the same order but are moved forward in time.

AGE-SPECIFIC DEATH RATES OF MICE EXPOSED TO IONIZING RADIATION AND RADIOMIMETIC AGENTS

A. C. UPTON, M. A. KASTENBAUM, AND J. W. CONKLIN

Biology Division, Oak Ridge National Laboratory,† Oak Ridge, Tennessee, U.S.A.

SUMMARY

The life-shortening action of ionizing radiation in the mouse is well documented (Mole, 1957; Neary, 1960; Upton, 1960). In this animal, moreover, irradiation early in life causes a displacement of the Gompertz curve, which some observers say may indicate induction of precocious senescence. As yet, however, there is relatively little detailed information on the age-specific incidence of various diseases in the mouse, their relation to senescence, and how their development is affected by radiation. This report presents preliminary results from a study of age-specific death rates of mice irradiated in earlier experiments.

METHODS

The data reported herein were obtained from LAF_1 mice exposed when 6 to 12 weeks old to γ-rays from an experimental nuclear detonation (Upton et al., 1960); RF male mice exposed at 5 to 10 weeks of age to whole-body X-rays (Upton et al., 1954; Upton, 1959; and unpublished data); and RF female mice subjected at 10 weeks of age to mid-lethal doses of whole-body X-rays (500–600 r), nitrogen mustard (HN2), 0·025–0·030 mg per kg, or triethylene melamine (TEM), 3·0–4·0 mg/kg, (Upton et al., 1962). Unless otherwise specified, the death rates for RF male mice were computed by the classical actuarial method, and those for LAF_1 mice by Seal's method, as described earlier (Upton et al., 1960; Kimball, 1960). The data for RF female mice were computed by a simpler actuarial procedure; i.e. the death rate was expressed as the ratio between the number dying within an interval and the number alive at the start of the interval, the length of the interval being determined by percentage of the total population dying (10 or 20%) rather than by elapsed time. In all cases, the data are based on animals dying naturally or killed in extremis.

RESULTS

In both LAF_1 and RF mouse strains, the rate of over-all mortality tended to increase exponentially with time beyond a certain age and was displaced

† Operated by Union Carbide Corporation for the United States Atomic Energy Commission.

upward by irradiation (Figs. 1–4) and by radiomimetic chemicals (Fig. 4). The displacement by radiation was, furthermore, dose-dependent and evident

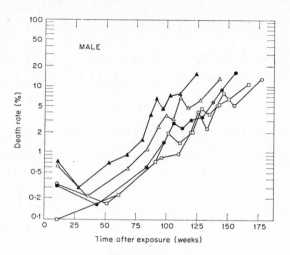

FIG. 1. Mortality rate of LAF_1 male mice in relation to time after γ-irradiation. (Mice were irradiated at 6 to 12 weeks of age.) (Modified from Upton et al., 1960.)

○ Non-irradiated controls (306 mice); □ 223 rads (210 mice); ● 368 rads (316 mice); △ 578 rads (306 mice); ▲ 679 rads (270 mice).

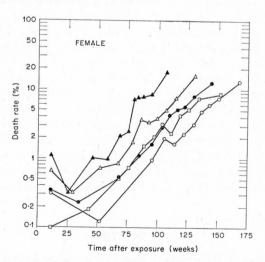

FIG. 2. Mortality rate of LAF_1 female mice in relation to time after γ-irradiation. (Mice were irradiated at 6 to 12 weeks of age.) (Modified from Upton et al., 1960.)

○ Non-irradiated controls (308 mice); □ 223 rads (212 mice); ● 368 rads (290 mice); △ 578 rads (298 mice); ▲ 697 rads (266 mice).

whether mice with, or without, neoplastic disease were excluded from consideration (Figs. 5–8). Inflections in the curves suggested heterogeneity in the various animal populations, which became more apparent on inspection of

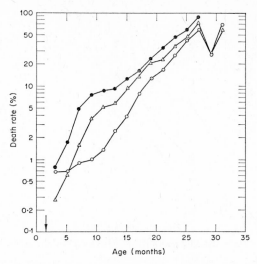

FIG. 3. Age-specific mortality rate of RF male mice, as influenced by X-irradiation early in life (arrow denoted age at exposure).

○ Non-irradiated control (882 mice); △ 100–200 rads (897 mice); ● 300 rads (1905 mice).

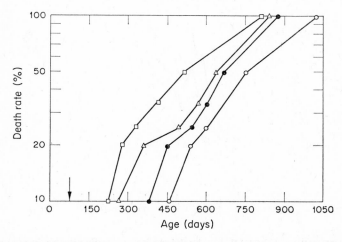

FIG. 4. Age-specific mortality rate of RF female mice as influenced by X-rays, TEM, or HN2 administered early in life (arrow denotes age at treatment). See Methods for doses.

○ Untreated control (130 mice); ● HN2 (160 mice); △ TEM (157 mice); □ X-ray (259 mice).

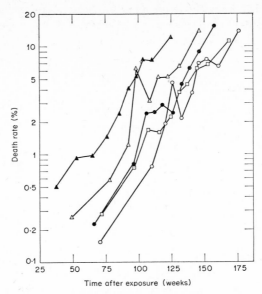

FIG. 5. Mortality rates of LAF_1 males dying with neoplastic diseases in relation to time after γ-irradiation.

○ Non-irradiated control (230 mice); □ 223 rads (178 mice); ● 368 rads (259 mice); △ 578 rads (228 mice); ▲ 697 rads (148 mice).

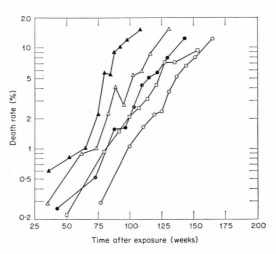

FIG. 6. Mortality rate of LAF_1 females dying with neoplastic diseases, in relation to time after γ-irradiation.

○ Non-irradiated control (208 mice); □ 223 rads (203 mice); ● 368 rads (276 mice); △ 578 rads (268 mice); ▲ 697 rads (200 mice).

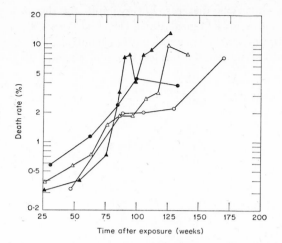

FIG. 7. Mortality rate of LAF$_1$ male mice dying without neoplastic diseases, in relation to time after γ-irradiation.

○ Non-irradiated control (76 mice); □ 223 rads (32 mice—not plotted because of small sample size); ● 368 rads (57 mice); △ 578 rads (78 mice); ▲ 697 rads (122 mice).

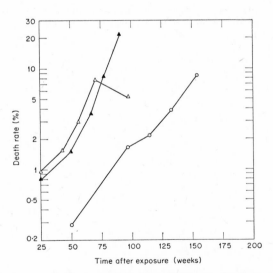

FIG. 8. Mortality rate of LAF$_1$ female mice dying without neoplastic diseases, in relation to time after γ-irradiation.

○ Non-irradiated control (62 mice); □ 223 rads (9 mice—not plotted because of small sample size); ● 368 rads (14 mice—not plotted because of small sample size); △ 578 rads (30 mice); ▲ 679 rads (66 mice).

the mortality of specific diseases. For example, for thymic lymphoma (Fig. 9), mortality rate reached a peak during the first year of life in the irradiated mice, contrasting with that in non-irradiated controls. For myeloid leukaemia (Fig. 10), the trend was similar but less pronounced and a bimodal distribution

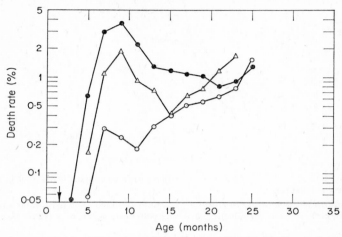

FIG. 9. Age-specific mortality rate of RF male mice dying with thymic lymphoma, as influenced by X-irradiation early in life (arrow denotes age at exposure).

○ Non-irradiated control (51/882 mice); △ 100–200 rads (104/897 mice); ● 300 rads (378/1905 mice).

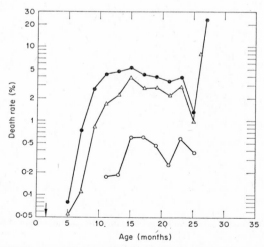

FIG. 10. Age-specific mortality rate of RF male mice dying with myeloid leukaemia, as influenced by X-irradiation early in life (arrow denotes age at exposure).

○ Non-irradiated control (34/882 mice); △ 100–200 rads (184/897 mice); ● 300 rads (533/1905 mice).

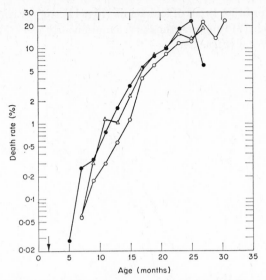

Fig. 11. Mortality rate of RF male mice dying with leukaemia other than thymic or myeloid type, as influenced by X-irradiation early in life (arrow denotes age at exposure).

○ Non-irradiated control (341/882 mice); △ 100–200 rads (268/897 mice); ● 300 rads (371/1905 mice).

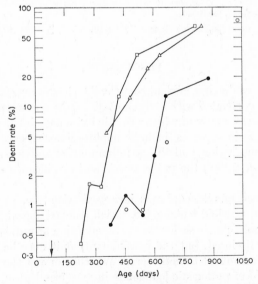

Fig. 12. Age-specific mortality rate of RF female mice dying with ovarian tumour, as influenced by X-rays, TEM, or HN2 administered early in life (arrow denotes age at treatment). See Methods for doses.

○ Untreated control (9/130 mice); ● HN2 (21/160 mice); △ TEM (91/157 mice); □ X-ray (96/259 mice).

was evident. For other types of leukaemia (Fig. 11), and ovarian (Fig. 12) and pulmonary (Fig. 13) tumours, the curves more nearly resembled those of the over-all population (Figs. 1–4).

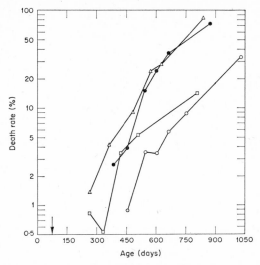

FIG. 13. Age-specific mortality rate of RF female mice dying with pulmonary tumour, as influenced by X-rays, TEM, or HN2 administered early in life (arrow denotes age at treatment). See Methods for doses.

○ Untreated control (26/130 mice); ● HN2 (94/160 mice); △ TEM (82/157 mice); □ X-ray (12/259 mice).

DISCUSSION

From these results, it is evident that the life-shortening effect of irradiation in the mouse is correlated with early mortality from non-neoplastic, as well as neoplastic diseases, many of which are otherwise associated with senescence. The non-neoplastic diseases include nephrosclerosis, cardiac thrombosis, enteritis, intussusception, and diverse inflammatory and degenerative changes (see Upton *et al.*, 1960), the pathogenesis and significance of which require further study.

The incidence of all neoplasms increased with age, except for thymic lymphomas and myeloid leukaemias, which reached a peak at early ages in irradiated animals. The basis for the early induction of radiogenic leukaemias remains to be disclosed, but the possibility that it may be related to viral aetiological factors must be considered. Whether or not the relatively short induction period of radiogenic leukaemia in man (see Upton, 1961) reflects a situation analogous to that in the mouse cannot be decided without further information.

It appears noteworthy that the increase in age-specific mortality associated with pulmonary tumours was less marked in irradiated mice than in those treated with radiomimetic chemicals, despite the greater over-all life-shortening in the former. This may be correlated with the failure of radiation to induce lung tumours.

These results serve to illustrate the complexity of the effects of radiation on the life-span. The variation of the age-specific death rate of different diseases and the effects of radiation on the age distribution of such diseases cannot be characterized by a single simple formula. Nor is it possible to comprehend changes in the incidence of any one disease without taking into account other intercurrent diseases affecting survival, as stressed earlier (Kimball, 1958). The generalization that irradiation advances the onset of senescent diseases is consistent with the effects of radiation on the over-all age-specific death rate but is an oversimplification in the light of variations among specific lesions.

CONCLUSIONS

1. The age-specific death rate increases irregularly with age in mice, tending to rise exponentially with time beyond a certain age.
2. Radiation and radiomimetic chemicals administered early in life increase the over-all age-specific death rate in a manner suggestive of ageing.
3. When the age-specific death rates for individual diseases are analysed, the curves are found to vary in shape and slope, and the effects of radiation appear quantitatively, if not qualitatively, variable.
4. Age-specific death rates from radiogenic leukaemias (thymic lymphoma and myeloid leukaemia) appear unusual in that mortality from these diseases becomes maximal relatively early.
5. Analysis of the age-specific death rates for various other diseases, including non-neoplastic lesions, indicates that the life-shortening effects of radiation are not confined to induction of neoplasia or any other specific type of pathological change. Instead, they appear to be correlated with a general advancement in the age distribution of diseases otherwise associated with senescence.
6. The results emphasize the complexity of the relation between the incidence of a disease and the radiation dose, owing to effects of radiation on intercurrent diseases affecting the survival of the population at risk.

REFERENCES

KIMBALL, A. W. (1958). *Bull. Int. Statist. Inst.* **36**, 193.
KIMBALL, A. W. (1960). *Biometrics* **16**, 505.
MOLE, R. H. (1957). *Nature, Lond.* **180**, 456.

NEARY, G. J. (1960). *Nature, Lond.* **187**, 10.
UPTON, A. C. (1959). In "Carcinogenesis: Mechanisms of Action". Ciba Foundation Symposium, p. 249. (Wolstenholme, G. E. W., and O'Connor, M., eds.). J. & A. Churchill Ltd., London.
UPTON, A. C. (1960). *Gerontologia* **4**, 162.
UPTON, A. C. (1961). *Cancer Res.* **21**, 717.
UPTON, A. C., FURTH, J., and CHRISTENBERRY, K. W. (1954). *Cancer Res.* **14**, 682.
UPTON, A. C., CONKLIN, J. W., MCDONALD, T. P., and CHRISTENBERRY, K. W. (1962). This volume, p. 171.
UPTON, A. C., KIMBALL, A. W., FURTH, J., CHRISTENBERRY, K. W., and BENEDICT, W. H. (1960). *Cancer Res.* **20**, 1.

DISCUSSION

ROTBLAT: Dr. Upton's results show the complexity of this problem. We have had a number of results given to us in papers which seemingly disagree with other findings, but we must remember that they don't involve the same conditions. If one finds an agent which brings forward the time of death naturally one says that it is similar to ageing, it produces the same results as ageing, but all of us who have been working on these problems know that the different causes of death may be due to the fact that the different treatments were made at different ages and therefore, by calculating at the age at which the exposure was done and on the dose-rate you get different effects. I don't think that these results necessarily contradict each other, it simply indicates that this is a very complex problem.

UPTON: I should like to emphasize the same thing. The fact that not all age-dependent changes are affected in the same way or to the same extent need not contradict the argument that the effects of radiation and the effects of time have certain basic factors in common. It is our job to try to see how much one can generalize, what the similarities and differences may be and to try to learn something about radiation and about ageing from these similarities and differences.

MOLE: All I want to do is to make a plea to dissociate ageing from death. I think mental deterioration may lead to my having to retire, but it is more than likely that I will die from something else. I think that ageing, if you really think about it seriously, ought to be dissociated from what are really the accidental things that kill people.

MULLER: In fact no one dies of old age. It seems to me that we are taking the right way, mortality is an index of ageing since mortality from so many causes rises with ageing; that we can judge from morphological and physiological criteria. It seems to me it is quite legitimate to make this association. Why not?

MOLE: If you take this seriously you have got to say that if there is a peak of mortality from boys falling off their bicycles, or being involved in traffic accidents, at 13, there are some sections of the population that will already have peaked at 12 and their grandfathers who die of thrombosis will also die a year earlier, and their little brothers who swallow ferrous sulphate tablets will also die a few months earlier.

MULLER: I don't thing so because there are both environmental and genetic causes, and certainly the environmental causes may change. Actually you will find that if you subject populations of different ages to test whether they are accident prone, that rises in the same way as everything else.

MOLE: Oh, no I beg to differ; if you look at the results of the London Passenger Transport Board investigation of bus drivers, you will find that accident proneness goes down with increasing age.

BERENBLUM: Would it be fair to put it this way, that with ageing a whole number of processes, all of them lethal in the end, are likely to kill, but only one of them will kill in one animal. The earliest one is the cause of death in that animal. When you then have a form of stress like X-rays, you may then introduce one of the others. Obviously as an index it is hopeless because it is a summation of all sorts of things; they all contribute to it, but only one of them in any one animal is the real cause. Isn't that it?

MOLE: Oh yes, but I think if you take post-mortems seriously you will find it very difficult not to say that there are a lot of contributory lesions.

MULLER: I think just because mortality is influenced by so many things, all of which show an increase with age, or practically all of which, maybe I was wrong about accident proneness, that such a multiplicity makes for better criteria.

BACQ: I want to take up the point that Prof. Berenblum raised before. One should take as a test of ageing some kind of continuous, biochemical process known to appear in a practically continuous way. Something not discontinuous such as death or the appearance of an infection or sterility or an accident or things of that kind. So far as I know there are two tests and one is calcification of cartilage. Then there is another one which is not so regular, and that is the cholesterol content of the arteries, also a generally rather continuous process. I am quite certain that if we could get the turnover of these macromolecules we spoke of this morning, these mucopolysaccharides and so on, we should also find that this represents a continuous process. If we could investigate this kind of process we might get a better idea of what happens.

UPTON: I would sympathize with this approach too, but it seems to me that when we talk about senescence we really are talking about a process which is not a constant function of time. When we plot survivorship against age, we don't refer to random loss of individuals. We refer instead to a type of mortality which is relatively abrupt in its onset late in life, and I would wonder if necessarily any of these functions will be found to have mechanisms in common with mechanisms that delimit the life-span and cause the evolution of tumours and so on. This is not to argue that we shouldn't look at all age-dependent changes, because a time-dependent change which is constant in rate may eventually be found to be related to senescence. They may be strongly correlated but then again they may not be.

BACQ: But even if they are not correlated they may be more characteristic of the process of slow senescence than any accidents occurring at the end.

ALEXANDER: The mere fact that the death curve rises like this doesn't mean that it is the result of progressive change, because this change will be a threshold phenomenon, and in fact the animal doesn't die until the damage has reached a certain level. The interesting thing about radiation ageing is that in the few cases where tests have been made there has been absolutely no effect of radiation; calcium deposition has already been referred to and then of course the mucopolysaccharides to which you referred have been studied rather indirectly. The strength of the rat tail tendon increases with age. This is almost certainly due to the fact that the amount of mucopolysaccharide decreases as the amount of collagen increases; it has something to do with cross-linking too, if you cross-link you don't get this affect at all. The progressive strengthening of that rat tail tendon parallels in fact a progressive fall in turnover of mucopolysaccharides. Now the interesting thing is that progressive strengthening of rat tail tendon is not accelerated in the slightest by radiation.

BERENBLUM: I don't think I should really participate in this discussion at all because I have never worked in the field and I merely want to give my impression as an onlooker. I must say that the earlier discussions struck me as being somewhat unreal, and unreal in this sense that what we were interested in was ageing, what we were talking about was death rates, and the reason for that, I suggest, is that we haven't got a means of studying ageing and therefore we are substituting something else. That is the real problem, as I see it. Now to substitute death rate merely because we haven't got just the right experimental conditions for ageing as yet, doesn't necessarily help. This is not a criticism. It may well be that a biochemical system or something of that sort may one day be discovered and then we will, I'm quite sure, forget the problem of death in its variations and go back to the real problem which is the effect of X-rays on ageing.

SCOTT: There is one point which I hope is fairly relevant to what we have been talking about. It has been found that growing fish under adverse conditions shortened their survival time, but that, by every other test of senility, such as the rate of tail regeneration under adverse conditions the pace of ageing appeared to be reduced.

COTTIER: It may perhaps be useful to study a specific group of animals, namely the oldest surviving non-tumour-bearing animals that are killed in an apparently good condition. Compare these animals with the oldest irradiated animals under the same conditions. If you do that, you find that a number of things look very different with these animals. For example, the degree of senescent atrophy of the liver is much higher in natural senescence than in the oldest surviving irradiated animals. I have already mentioned the difference in cartilage calcification. On the other hand the one thing, I think it is almost the only thing, which was really visible in the irradiated animals that lived longest was vascular change. I think that it will be worthwhile to learn more about this change, to learn more about the generation time of epithelial cells, in order to evaluate when they would be expected to reach the next mitosis, and so on.

CASARETT: I have another example of difficulties in this field. Take the heart for example. The periodic examination of hearts, weighing them and measuring fibres and so on, will show that one gets in some animals decrease in heart weight and size and in other animals increase in heart weight and size with advancing age. It appears that if you measure the blood pressures of the animals you will find large hearts in animals that have hypertension, the small hearts are the true aged hearts uncomplicated by hypertension, and only found in animals that do not have hypertension. In most of the animals that have hypertension there is kidney disease. In serial sacrifice studies throughout life for the study of the kidney, one sees, before symptoms are present, certain vascular connective tissue changes, which until symptoms are present are not recognized by clinical pathologists or clinicians as representing a disease; nevertheless this is sub-clinical change. I would like to emphasize again that one has to know the pathogenesis and mechanisms of everything that is measured, and take into account the arbitrary lines drawn by clinicians as to what is a disease, and the gerontologist as to what is the change leading up to the disease in order to make sense out of this complicated process.

MOLE: Surely you don't make measurements after you know the pathogenesis, you make the measurement in order to find out what it is. You said that you couldn't understand what was wrong until you knew all about its pathogenesis and mechanisms, but what you want to know is what the pathogenesis and mechanism is.

CASARETT: I was interested in why some ageing rats have large hearts and some ageing rats have small hearts at the same age, and I found out. There was a disease involved. Before hypertension was recorded as a clinically recognizable disease of the kidney, there were basic changes going on which, for the gerontologist, represent ageing changes, but not all the animals will get clinically recognized nephrosclerosis and not all of the animals will get hypertension, and therefore not all of the animals will get a large heart. The distinction between what is a disease and what is a basic ageing change is what I would like to make in this remark. In other words, a plea for gerontological science.

POCHIN: There is a bit of a separation coming here isn't there? Would you agree with this first proposition that all deaths are due to some terminal change, call it a disease if you like, and that all diseases have a pathogenesis. (Yes.) Now, are we looking for something beyond these two, or are we looking for various ways in which radiation can affect the pathogenesis of various diseases, which would add up to the various causes of death? (Yes.) Because I think one can get into a muddle if one says I am not concerned with these specific diseases of radiation because I am concerned with the rest. One has got to say I am not concerned with the ways in which radiation affects the pathogenesis of the whole variety of diseases, I am concerned with something else. I don't think anybody is really concerned with that something else. I think we are thoroughly concerned with the way in which radiation affects the pathogenesis of various diseases. Is that right or is that all wrong?

MAISIN: In connection with what Dr. Pochin says, do you not think that it would be interesting to irradiate locally and thus investigate the importance of local irradiation in ageing? One could irradiate, for example, kidney or liver tissue, or even intestinal tissue.

GRAY: So far in this meeting our thoughts have been much occupied with considerations of the factors which influence the survival of proliferating cells. From the point of view of the long-term effects of radiation, possibly including carcinogenesis, it is important to consider the impairment of function which the survivors may have suffered, which may interfere with their capacity to play a normal role in their natural environment or in any other environment in which they may find themselves. In this connection I should like to refer to experiments carried out by my colleague, Dr. D. L. Dewey [X-Ray inactivation of inducible enzyme synthesis and the effect of oxygen and glycerol. *Nature, Lond.* **194**, 158–160 (1962)] on the capacity of irradiated bacteria to synthesize inducible enzymes. Briefly, Dr. Dewey's experiments show (1) that the maximum capacity of a non-proliferating population of *Pseudomonas* to synthesize the enzymes necessary for the metabolism of histidine when this substrate is introduced into the culture medium is depressed by radiation, (2) that the plot of survival of capacity for adaptive enzyme synthesis against dose is strikingly similar to that of survival of proliferative capacity, and (3) that the data implied an over-riding control of adaptive enzyme synthesis by a structure which may be inactivated by one, or at most a few, ionizing particles, and which is of rather considerable size, say around 6×10^{-16} g, which is half the size of the structure which controls proliferation. The chemical aspects of the inactivation of both structures have much in common. I would suggest that a similar situation may exist in relation to the synthesis of adaptive enzymes—among which thymidylic kinase may be of special interest—in mammalian cells. If the phenomena observed by Dr. Dewey with micro-organisms do have their counterpart in mammalian radiobiology, they might be revealed when the survivors of an irradiated population are called upon to function in a new environment.

MULLER: This is tissue culture again?

GRAY: No. The experiments were made with micro-organisms, viz. *Pseudomonas*, which were tested for the capacity to synthesize the enzymes necessary for the metabolism of histidine. In bringing these experiments to your notice I had very much in mind the effect of radiation on regenerating liver, in which the adaptive synthesis of thymidylic kinase appears to be an early step in the transition of the cells from the resting to the proliferating state.

MAISIN: Do you think that your criteria will vary with the tissue that you irradiate?

GRAY: Yes, I do. I suggest that the effect might be observed when the cells which survive an irradiation are called upon to synthesize adaptively, enzymes which are needed in order that the cell should be able to undergo a normal maturation process.

SESSION VI

Chairman: L. H. GRAY

THE MODIFICATION OF THE LATE EFFECTS OF IONIZING RADIATION

By O. C. A. SCOTT

Apparent Radiation Protection Against Local Ageing Effects in the Skin of Mice

By HERMAN B. CHASE

Dependence of Radiation-induced Life-shortening on Dose-rate and Anaesthetic

By PATRICIA J. LINDOP AND J. ROTBLAT

Effects of Post-irradiation Injection of Yeast Sodium Ribonucleate and its Nucleotides on the Differential Count of Bone-marrow

By H. MAISIN

The Effects of Total-body X-Irradiation on the Reproductive Glands of Infant Female Rats

By D. SLADIĆ-SIMIĆ, N. ŽIVKOVIĆ, D. PAVIĆ, AND P. N. MARTINOVITCH

Peripheral Blood Studies Upon Some Isogenic Chimaeras

By A. J. S. DAVIES, ANNE M. CROSS, AND P. C. KOLLER

THE MODIFICATION OF THE LATE EFFECTS OF IONIZING RADIATION

O. C. A. SCOTT

Radiobiology Department, Mount Vernon Hospital and the Radium Institute, Northwood, Middlesex, England

The literature on the delayed somatic effects of radiation has recently been reviewed by Odell et al. (1960).

The difficulties encountered in this field have already been emphasized at this meeting. There have been differences of opinion as to the very nature of the ageing process. Can we study "protection" if we do not know what we are protecting against?

There are many difficulties in the planning of these experiments. For instance, it is a fairly generally accepted principle that a dose-response curve should be established, for a given biological system, before complicating factors such as protective agents are introduced. In the case of one long-term effect, carcinogenesis, every type of dose-response curve has been described. Increasing the dose of radiation may give a steady increase in tumour yield, an increase followed by a plateau, a steady reduction, or an increase followed by a reduction (Furth et al., 1959; Hollcroft et al., 1957). It may be difficult to elucidate the nature of the interaction of a protective agent with such complex systems. Even when there is an apparently simple relationship between dose and response, carcinogenesis experiments present planning difficulties. For instance, Andervont (1943-4) showed that spontaneous tumour incidence is dependent on the number of animals in a cage, and this number may fall off in an experiment with total-body irradiation. Experiments using split doses may also be difficult to interpret, if the incidence of the observed effect varies with the age of the animal.

If the criterion of radiation effect is shortening of life-span, the dose-response relationship may be fairly simple, but interpretation is still complicated by one's ignorance of the target-organ concerned. As Dr. Upton has pointed out, we do not know whether cells, or non-cellular elements, are primarily involved. Professor Brinkman and Dr. Lamberts have shown extremely interesting immediate effects of radiation on non-cellular elements, but perhaps one should keep in mind the fact that all non-cellular elements have been produced initially by a cell. Long-term effects on collagen may, therefore, be secondary to cell death. Odell et al. (1960) have reviewed the rather scanty data on the specific problem of chemical protection against long-term effects. It may be of importance that cysteine alone in the diet, prolonged the life of mice (Harman, 1957).

Hollcroft et al. (1957) investigated the effect of cysteine combined with partial anoxia (6% oxygen) on tumour incidence in mice after 400 and 900 r. The protective treatment appeared to give a DRF of 2·5, in so far as the protected group receiving 900 r, showed the same tumour incidence as that in the 400 r sham spleen-protected mice. However, complete dose-response curves are not available and a third group, receiving 900 r + spleen protection showed a lower tumour incidence than the 400 r spleen-protected group. This effect may be due to earlier deaths at the higher radiation dose. There is a need for a correction factor to allow for the age effect, and Dr. Upton may be able to explain Dr. Kimball's correction factor, as applied to their own data.

Mewissen and Brucer (1957) investigated the effect of cysteamine and cystamine on the incidence of radiation-induced lymphosarcoma, and lymphatic leukaemia in C57 black mice. The incidence of tumours was higher, at low γ-ray doses, in the treated than in the control groups, but the meaning of this result is not clear.

Maisin et al. (1957) found that mercaptoethylamine had no effect in irradiated rats, but the dose they used may have been too small (10 mg/145–160 g rat).

Upton et al. (1959) tested MEG in irradiated mice. There was a slight, but as they themselves point out, not statistically significant, reduction in the incidence of granulocytic leukaemia. The reduction in numbers of thymic lymphomas was rather more convincing.

The general picture of chemical protection *vis-à-vis* carcinogenesis is thus far from clear. On the other hand, chemical agents give definite protection against other long-term effects; von Sallman et al. (1952) and Pirie and Lajtha (1959) have shown that cataract formation can be inhibited by cysteine given before irradiation.

The many interesting examples of chemical protection of the skin will no doubt be discussed by Professor Bacq at this meeting.

Partial anoxia has been shown to protect animals against some long-term effects in mammals. Lamson et al. (1958) consider that the effect of 1,000 r delivered to rats breathing oxygen is equivalent to 425 to 500 r delivered in air. Krebs and Brauer (1961) also investigated the effect of 5% oxygen, using mice and found no significant protection against the non-recouperable element of radiation injury.

Wright (personal communication) points out that the failure to protect a particular system by means of 5% oxygen may be due to the fact that the target-organ is not rendered anoxic by this procedure. Lindop and Rotblat (1962) have subjected mice to 25 seconds complete anoxia. They found very good protection against the acute effects (DRF of $\sim$ 2·8 in 4 to 6 week-old mice). A lower DRF in newborn mice can be explained, if the marrow in these animals is partially anoxic even when the animals are breathing air. Older animals also show a lower DRF with 25 seconds anoxia, but in this case the result may be due to a longer time requirement for the production of the anoxic state. Lindop and Rotblat found a lower DRF for long-term effects than for acute effects.

Hornsey (1959) using severe hypothermia, also observed less protection for long-term than for acute effects.

Dr. Lamberts has mentioned at this meeting that the non-cellular effects which he has investigated show no oxygen effect.

The presence or absence of the oxygen effect might, therefore, be used to distinguish between cellular and non-cellular effects.

Other protective techniques, such as feeding of alkoxyglycerol esters, hormone treatment, and marrow transplantation have been reviewed by Odell et al. (1960) and will, no doubt, be discussed further at this meeting.

REFERENCES

ANDERVONT, H. B. (1943–4). *J. nat. Cancer Inst.* **4**, 579.
FURTH, J., UPTON, A. C., and KIMBALL, A. W. (1959). *Radiation Res.* Suppl. 1, p. 243.
HARMAN, D. (1957). *J. Gerontol.* **12**, 257.
HOLLCROFT, J. W., LORENZ, E. MILLER, E., CONGDON, C. C., SCHWEISTHAL, R., and UPHOFF D. (1957). *J. nat. Cancer Inst.* **18**, 615.
HORNSEY, S. (1959). *Gerontologia* **3**, 128.
KREBS, J. S., and BRAUER, R. W. (1961). *Radiation Res.* **15**, 814.
LAMSON, B. G., BILLINGS, M. S., EWELL, L. H., and BENNETT, L. R. (1958). *A.M.A. Arch. Pathol.* **66**, 322.
LINDOP, P. J., and ROTBLAT, J. (1962). *Brit. J. Radiol.* **35**, 23.
MAISIN, J., MALDAGUE, P., DUNJIC, A., and MAISIN, H. (1957). *J. Belge Radiol.* **40**, 346.
MEWISSEN, D. J., and BRUCER, M. (1957). *Nature, Lond.* **179**, 201.
ODELL, T. T., COSGROVE, G. E., and UPTON, A. C. (1960). In "Radiation Protection and Recovery" (A. Hollaender, ed.), p. 303. Pergamon Press.
PIRIE, A., and LAJTHA, L. G. (1959). *Nature, Lond.* **184**, 1125.
UPTON, A. C., DOHERTY, D. G., and MELVILLE, G. S. (1959). *Acta Radiol.* **51**, 379.
VON SALLMAN, L., MUNOZ, C. M., and BARR, E. (1952). *A.M.A. Arch. Ophthal.* **47**, 305.

DISCUSSION

UPTON: I would certainly endorse what Dr. Scott has said about the paucity of evidence on protection against tumorigenesis by radiochemicals; the data of course, speak for themselves. We did endeavour to determine the effects of AET with mercaptoethylguanidine (MEG) in the active form in RF mice at two dose levels of 150 r and 300 r whole-body X-irradiation early in life, plotting the percentage of leukaemia against dose with myeloid leukaemia; the data with 300 r showed that there was a small difference, as I recall there was also the same kind of a difference at 150 r; this wasn't a 1 : 1 dose comparison, the differences were not large enough certainly to be statistically significant. They were in the right direction for protection and from this we infer, as Dr. Scott emphasized, that the data were at best suggestive; we certainly didn't see anything like a two-fold reduction of radiation dose under the conditions of this experiment that we would have expected if AET had protected against leukaemia induction to the same extent as it protected against acute lethality. As regards the way in which you try to correct the incidence of leukaemia or any other disease to adjust for intercurrent mortality from other causes, I have very serious doubts as to whether this kind of statistical juggling really gets us anywhere. There is a substantial latent period, or induction period, and then the disease occurs, the cumulative incidence increases as the population ages. Obviously, if you start killing animals before the disease appears not as many animals are at risk and so not as many would have developed the disease, even though the radiation were in fact acting on that tissue in a carcinogenic manner. What we have done, Dr. Alan Kimball and I, is to try trunking the probability by sub-dividing the population and adding at the different periods under consideration, animals to compensate for those that were lost from other diseases and to calculate the probability of the incidence of the disease assuming the population at risk had not been decimated. This presupposes independent probabilities apart from the interaction of various lesions within the animal. We must not assume that all tissues are independent of one another and that if we simply had larger numbers of animals we would see a larger number of tumours irrespective of effects on other organs. So I don't really know how much one could trust this sort of thing. I was interested to see that Prof. Rotblat and Dr. Lindop used a similar kind of statistical correction and perhaps he would like to comment on this point too. In regard to effects on leukaemia induction and life-shortening, we have had under observation now, a large number of F1 hybrids of this strain derivation. These preliminary observations have been presented before. I show them again to emphasize that when one compares the extent of life-shortening calculated as the percentage of survivors per roentgen, then it would seem that animals exposed to LD_{50} (30 days) when irradiated in the presence of MEG do show substantially less life-shortening per roentgen than animals given radiation alone at the LD_{50} (30 day) level without any protective treatment.

Now the complication here is that you can't really give this kind of a dose without protection and have any animals alive at the end of 30 days to follow, so unless you are willing to assume that there is in fact this kind of a life-shortening effectiveness of irradiation over the entire dose-response curve then you really don't know that this is, in fact, protection. What I am saying is that if one were to plot median survival time against dose and do an experiment and show that curve was in fact linear after 1,400 r, then this would be indicative of protection, but it could well turn out that the curve flattens out anyway so that there is less life-shortening per roentgen at a dose-level of this order of magnitude without any protection. So we have extended the experiment to include animals at 700 r with and without AET, bone-marrow, or both and 500 r with, and without the treatments, and then another group of unprotected animals at about 350 r, and before I came away from Oak Ridge, Dr. Carlsgrove told me that there was in fact an appreciable dose reduction at the 700 r level but that, for the animals at 500 r, the dose reduction was not significant. This would suggest, perhaps, that in this strain-combination under this kind of experimental condition, dose-reduction may be dose-dependent; the amount of protection one gets may vary with the radiation dose-level. I think this simply points to the need to do experiments of this sort to compare curves and not single points on a curve; unless you have a dose-response curve you really don't know whether your treatment is moving it one way or another.

ALEXANDER: It seems to me that in your data you had one indication which suggests that this way of handling the data isn't fully justified in that bone-marrow also gave you a life-span shortening reduction, whereas I think other people, giving doses under more strict conditions, find that bone-marrow provides very little protection, so that would seem to indicate that the linear dose-response curve might not be equitable.

UPTON: Yes, I'm not really sure in my own mind that bone-marrow has any effect outside the haemopoietic system. Again, this implies that you can take a mouse and carve it into pieces and look just at the haemopoietic system and forget about everything else.

ALEXANDER: I thought there were actual experimental data showing that with dose regimes which didn't produce acute deaths then bone-marrow therapy didn't help a great deal in life-span shortening.

UPTON: No, I think that this will depend a lot on the extent to which the animal is prone to leukaemia induction because it's well established now that bone-marrow does afford a substantial protection against leukaemia induction and in a strain in which leukaemia is an important lethal event following radiation then shielding bone-marrow, or bone-marrow infusion would be expected to reduce the life-shortening effectiveness of radiation.

BACQ: Someone in Liège has done a good deal of work on protection and I know that Dr. Hollaender doesn't like it very much, but he has tried to overcome this difficulty of having no unprotected controls by giving two irradiations at weekly intervals so as to increase the numbers of leukaemias in the controls. He has also used a very elaborate type of statistical treatment for his results. He has tried several doses, and he came to the conclusion that cysteamine in doses which are quite effective in reducing acute lethality do not decrease, in a statistically significant way, the incidence of leukaemia.

UPTON: I think mention has been made of protection of lens against cataract induction. We have tried very hard to protect the mouse with AET and we have been able to get no protection whatsoever. We haven't studied the uptake of labelled AET in the lens and, conceivably, we just don't get enough AET across the aqueous humour into the lens, but I suspect that one should be on guard against tissue variations and drug variations in the extent to which one can expect protection.

BACQ: The best way to do this is not to give the protector intravenously but to give it locally to the eye; in this way you can probably get a much higher amount of the protector within the lens. The second comment I would like to make on Dr. Scott's introduction about the work of Harman on the increase of life-span simply by feeding the animal constantly with cysteine, 1%, I believe in the food, is that the great objection to this type of study is that he has not measured the food intake of the animal, and it is quite likely that the addition of 1% cysteine to the food may make it so much less palatable that the animal eats much less.

ROTBLAT: You asked whether there is any evidence about hypocaloric feeding of mice. We haven't got any but Dr. Lindop and a colleague are doing a study of the acute effects. We have got no data on the long-term effects but it would appear that it doesn't protect at all, on the contrary the LD_{50} goes down. The question of anoxia; we have done various experiments with mice breathing pure nitrogen for about 25 seconds, and then given a very brief pulse of radiation. We measured the $LD_{50}(40)$, the acute effects and the long-term effects. What one has to bear in mind is that here too there is an age factor; the protection which you get varies with age, and we believe now that this is due to the fact that the older the animal the longer it will take to reduce the oxygen in its tissues to the same level, and therefore for older animals you probably need a longer time. For example, with 4-week-old animals we get a protection factor of $2 \cdot 8 \pm 0 \cdot 2$, which is about the maximum because if you go to 40 seconds you don't get any change. But now, if you are going to consider the long-term protection, this will of course again vary with age for the same reason, but we find in all cases that the protection against life-shortening is less than for the acute effects. We get a factor of 2·8 at 4 weeks for the acute effects, we only get about 1·8 to 2 for the long-term effects.

BERENBLUM: I just have one comment about the caloric restriction effect. About 20 years ago Tannenbaum in Chicago did an extensive series of experiments on the effect of caloric restriction and studied all systems of tumours both spontaneous and induced. He found, I believe without exception, that the tumour induction was reduced by at least 50% for spontaneous mammary and lung tumours, and I think also for spontaneous hepatomas and a whole series of chemically induced tumours. But since there was no exception with either spontaneous or chemically induced tumours, I would say that, arguing from this results, that there would almost certainly be a very significant reduction in carcinogenesis by a degree of caloric restriction which not only maintains healthy life but actually extends the life-span.

UPTON: Coming back to the question of organ differences, repeatedly in the course of this meeting the inter-relation between the thymus and the bone-marrow has come under discussion. It may be noteworthy that AET, at least, seems to afford very little protection to antibody-forming cells, either as measured by the production of circulating antibody or as measured by the ability of such cells to produce a transplantation reaction, so, this may be an important point in connection with leukaemogenesis.

CASARETT: In regard to experiments with restricted caloric intake it is well to remember that in Mackay's classical experiments on non-irradiated rats one of the chief killing diseases that was eliminated by caloric restriction was pneumonia. Sexton reported this in a separate paper on the pathology of these animals and from our investigations of the pathogenesis of pneumonia in that strain and in our own strain, it was clear that starting at about 18 months in the rat there is an increasing incidence of over-growth of lymphatic tissue into bronchioles causing obstruction with the production of emphysema and then a high incidence of killing penumonia. Caloric restriction prevents this over-growth of lymphatic tissue, emphysema and the high incidence of pneumonia so it's possible that in some species this restrictive caloric feeding does not have a generalized, so much as a specific, effect on certain killing diseases. The effect of radiation might therefore be based on something more specific.

UPTON: In his summary Dr. Scott mentioned the pertinent question of split-dose studies and I think this points up an important approach in methodology. Some years ago Henry Kaplan studied the incidence of lymphomas and thymic lymphoid tumours in C57BL mice, given whole-body radiation early in life, at about 33 to 35 days of age. He gave different total doses, and I have shown the results for just one dose-level, as I recall it about 570 r. If he gave the 470 r in one treatment then he got a final incidence of thymic lymphomas of about 40%. If he gave the same dose in four fractions, each one separated by 1 day, the lymphoma incidence was not significantly different. If given over 2 days, he got a reduction in yield; over 4 days he got a higher yield of lymphomas; significantly higher with 8 days and finally, with a 16-day interval between fractions, the yield dropped way down. Comparable results were observed for higher total doses, but with higher dose levels it was impossible to give the animal the total dose in one treatment. But the interval effects were observed. From this he inferred that there was a very complex relationship between total dose, number of fractions, dose per fraction and duration of interval. If, for example, one were doing an experiment on tumorigenesis and had to contend with this kind of complication, without the necessary controls one might be badly misled by the end result. For example, if you chose a 2-day interval between doses and your protective agent enabled the animals to repair the injury more rapidly so that the interval was actually more like 1 day, then a higher incidence would be expected, again if you observed a higher incidence without knowing this you would have inferred that the "protection" had actually worsened the effects of irradiation and so on. It is a pity I think that this kind of experiment hasn't been more adequately extended, since the inital work of Kaplan and Brown. Dr. Mole has referred to similar complexities under conditions of fractionated or long-term irradiation and until the kinetics of repair is elucidated by the sort of work that Dr. Lamerton mentioned, until we know more about repair rates, trying to assess the influence of protective agents without the necessary control information may be quite misleading.

COTTIER: I should like to mention that the importance of the duration of the interval may even be different for different cell lines. But it also depends on what you actually look for. If you look for thymic lymphomas you may choose different intervals than if you look for myelogenic leukaemias.

LAMERTON: I think it is probably true to say that the effect of fractionation has only been shown for thymic lymphomas. I would like to ask in fact if any fractionation work has been done with regard to other tumours, or whether there is any evidence that fractionation does affect considerably the incidence of other types of tumours and other types of leukaemia too.

UPTON: I have the impression that data were presented here on bone tumours. Was there a fractionation effect there? (Loud laughter.)

LAMERTON: Well, of course, our work on the effects of fractionation with external radiation, which is the cleanest case didn't show very much, you see. That is what I meant. An effect of fractionation has been shown very markedly with thymic lymphoma. I would like to ask Dr. Upton whether he has looked at the effect of fractionation on myeloid leukaemia, and, if so, whether there is any similar effect?

UPTON: We have not seen any similar effects in connection with myeloid leukaemia, but I regret to say that we haven't done such a nice systematic study. I think this kind of work badly needs to be done. There are really very few adequate experiments on the influence of fractionation. I don't think one could say categorically that similar phenomena won't turn up for other tumours.

APPARENT RADIATION PROTECTION AGAINST LOCAL AGEING EFFECTS IN THE SKIN OF MICE

HERMAN B. CHASE
Brown University, Rhode Island, U.S.A.

In connection with comments made on ageing at this conference, I would like to mention first a few results which seemed peculiar at the time and then give the explanation. A peculiar phenomenon was observed in that in order to protect against local ageing effects, one protection was radiation itself, a rather anomalous situation, to say the least.

The experimental set-up, obviously for different reasons at the beginning, was to use the two sides of an animal, one side irradiated, the other as the control, a standard technique. On the irradiated side the hair greying effect occurs and this was the primary interest at the time. Our study was then extended to observe the other effects of irradiation, the other effects on the grey side as compared with the control side. The first results were peculiar in that for the characteristics being studied, the skin appeared younger on the grey side than on the control side.

Six different traits or effects were examined. One was the irregular patches of keratosis which occur after irradiation and with ageing. This characteristic was less pronounced on the irradiated side than on the control side, but it had first increased then decreased. Another trait was excessive epidermal pigmentation. Mice normally do not have much pigment in the epidermis, but potential pigment cells are there and they flare up at times, especially after irradiation and also with ageing. On the X-rayed side there was first a little increase then a decrease to a level where no pigment appeared. Over the same period, however, epidermal pigmentation on the control side continued to increase slowly. Even the density of hair showed the same difference, the density on the control side decreasing with age, the density on the irradiated side remaining comparatively high. Wound healing was not as clear a trait, but the irradiated side was not at a disadvantage, except for a period immediately after irradiation. The collagen study is not as complete, but again the irradiated side appeared at least as "youthful" as the control side, possibly more so. The sixth trait is of course the greying itself. There is some increase in the number of grey hairs with time on the irradiated side, but also a considerable average increase on the control side.

What did all this mean? The radiations varied from 200 to 1,000 r, the 1,000 r giving the maximum greying effect. A major part of the explanation is attributable to a fault in the design of the experiment. The results are true but there was an extenuating circumstance. On the irradiated side frequent pluckings were made to study the change in whiteness with time. The irradiated, or white side, was plucked in some cases as many as ten times, whereas the control side was plucked only two or three times. The irradiated area was therefore maintained in a relatively rejuvenated condition by the continued induced hair regenerations. In later experiments it was found that the rejuvenation by continued plucking did indeed cause less keratosis, less epidermal pigmentation, etc. By keeping the skin working, there was no doubt that it remained "younger" by the criteria employed.

Another point to be made is that ageing is an increase in variance, an increase in the variability of a response. In studying the ageing responses already mentioned, it is observed that there is a decided increase in variability of all of them, including epidermal pigmentation, the test for collagen, etc. Most interesting is that even in some very old animals, the level for epidermal pigmentation, for instance, may remain at the low level but other individuals will have a considerable amount, consequently there is a great increase in the variability of the response. The average response and the variance increase, but some animals for a particular response will remain at the "young" level.

At a much higher level of irradiation, 2,000 r, there is more damage which is also reflected as an increase in variance. At 750 to 1,000 r, even without plucking to initiate new hair growth, radiation will result in more skin replacement, therefore rejuvenation, than on the control side.

DISCUSSION

BACQ: We have been working on local injection in mice with a variety of protectors, mainly for epilation tests in extremely young mice between 5 and 8 days old. They epilate very nicely something like 6 to 8 days after irradiation. It is an acute not a late effect but what we have seen is this, if you inject cysteamine here with a needle you have significant local protection and this is a way to investigate also the mechanism of action of various protectors. Histamine gives complete protection of the animal against total-body irradiation and this enables one to differentiate between local effects and general pharmacological effects. Recently we have been interested, because of the many contributions of Prof. Brinkman and Dr. Lamberts, in the possible effects of macromolecules on this test. Synovial fluid which contains a high proportion of hyaluronic acid, fresh human serum, or serum from the same species of mice are extremely effective in decreasing the epilation, not only when injected before the irradiation but also when injected something like 20 seconds or 30 minutes after total irradiation of the animal with a dose of 550 r at a rate of 150 r per minute hard X-rays. Now, normally this is an acute experiment and I have killed all the animals, but after this discussion and after the paper given by Prof. Chase, I am going to keep some of these animals for a long time and see what happens to the skin as compared to unprotected, but similarly irradiated, skin. As far as pigmentation is concerned there is also a remarkable protective effect. There is no whitening at all so that the protection is not only on hair growth but also on hair pigmentation.

BERENBLUM: You will recall that in one of the experiments that I have already described we gave C57BL mice a single dose of 400 r; the next day we took tissue cells from them and injected large amounts of homogenates into normal C57BL, some of which were then given urethane and so on. We found that the recipients of these injected tissues developed greying of the hair very early.

CHASE: Yes, I am quite aware that this greying phenomenon can be caused by many things about which we know very little. There has been some indication that occasionally one can give a certain type of poison and get some greying, but it doesn't happen consistently. In regard to Prof. Bacq's story, remember that the greying effect is greater if you irradiate the resting follicles; these follicles are more sensitive and there will be a much greater effect. On growing follicles the effect is less. However, since you didn't get any greying in that area with 550 r, were these hairs actually replaced at all? Was it possible that these were the hairs that were there already and simply staying as resting hairs? These mice were 5–8 days old. At 5 days they had no hair, but at 8 days they had a nice cover of hair. But at 5 days the hair is growing. Some hairs remained dark and the rest disappeared. If hairs are growing and a dose of 550 r is then given, I would expect you to get about 15% white hair.

CURTIS: I would like to ask Prof. Bacq about his synovial fluid. Does he actually think that a compound in synovial fluid penetrates to all the cells in the skin?

BACQ: We believe that some type of extra-cellular protection is operating. Apparently, but I think Dr. Brinkman knows more about this, the growth of the follicle is dependent, partly at least, on these extra-cellular macromolecules which surround the hair bulb. I do not think they do penetrate the cell, it is probably a different mechanism of protection unless you think that this is not simply protection since it can occur even after irradiation.

BRINKMAN: I think it is a general experience that when the hair growth cycle starts the first thing that you can see is an accumulation of mucopolysaccharides of a very special kind in the hair bulb and if this is depolymerized by irradiation, then the growth of the hair is changed. If you the ninject mucopolysaccharides, the hair starts to grow in the injected places, as has already been shown by Karl Meyer. The best mucopolysaccharide is heparin monosulphate. It may be the function of the mucopolysaccharide layer to convert globular into fibrous molecules.

MOLE: Prof. Brinkman earlier suggested to Prof. Lamerton that the reason why he got acquired radioresistance in the intestine was a local libertation of serotonin. May I ask Prof. Bacq if serotonin works in this system?

BACQ: Yes, it does.

DEPENDENCE OF RADIATION-INDUCED LIFE-SHORTENING ON DOSE-RATE AND ANAESTHETIC

P. J. LINDOP AND J. ROTBLAT

St. Bartholomew's Hospital Medical College, Charterhouse Square, London, England

The amount of life-shortening produced by a given dose of radiation depends on many factors, including the dose-rate. Thus, it is known that less life-shortening per unit dose is produced when an animal is exposed to chronic irradiation than when given a single dose. Some of this difference may be attributed to an age factor (Lindop and Rotblat, 1962) rather than to a true dose-rate effect, but even with single exposures life-shortening appears to depend on dose-rate.

In connection with investigations on the protective action of hypoxia (Lindop and Rotblat, 1960) it became necessary to expose mice at very high dose-rates, so that a dose of over 1,000 r could be delivered in less than a second. These mice had also to be anaesthetized before making them anoxic in order to avoid convulsions. It became, therefore, necessary to investigate the possible effects of these two factors, high dose-rate and anaesthesia, on the sensitivity of animals to radiation.

The primary criterion was the acute effect of radiation, i.e. the LD_{50} for 30 days mortality. For this purpose SAS/4 mice, male and female, all 4 weeks of age, were exposed to single whole-body doses of radiation in the range of 600–1,000 rads. The radiations used were 15 MeV electrons or X-rays from a linear accelerator. The dose-rate was varied over a very wide range, from 77 to 160,000 rads/min. From these data the LD_{50} were determined with an accuracy of about 3%.

Many of the mice, particularly in the lower dose groups, survived the first 30 days and were kept in the cages until the end of their lives. The ages at death were recorded and life tables plotted; from these the median life-spans were determined. By comparing with the life-span of the control mice, and assuming a linear relationship between life-shortening and dose, which had been previously established for these mice at this age group (Lindop and Rotblat, 1961), the amount of life-shortening per 100 rads was calculated with an accuracy of about 10%.

The results obtained for both the LD_{50} values and the life-shortening are given in Table I. It is noted that over the whole range of dose-rates studied, covering several orders of magnitude, there is very little variation of the LD_{50}. In fact, the data are compatible with the assumption that the LD_{50} remains constant. However, starting from the second line there appears to be a systematic increase in the LD_{50} with increasing dose-rate. This impression is strengthened when one compares our results with those obtained by other workers at lower dose-rates. In Fig. 1, in which the dose-rate is plotted on a

TABLE I. *Effect of dose-rate*

Dose-rate (rads/min)	LD_{50} (rads)	Life-shortening (weeks/100 rads)
77	758 ± 24	3·9 ± 0·4
480	686 ± 16	5·7 ± 0·2
1,730	715 ± 24	6·2 ± 0·5
6,000	726 ± 19	4·7 ± 0·4
31,000	748 ± 26	4·3 ± 0·4
158,000	752 ± 22	3·9 ± 0·4

log scale to accommodate the large range of dose-rates, the broken line represents the variation of LD_{50} with dose-rate obtained by other workers (Corp and Neal, 1959; Lawrey and Fowler, 1959) and normalized to the values obtained by us. It is seen that after an initial rapid decrease of the LD_{50} there is a flat minimum at about 1,000 rads/min, followed by a very gradual increase.

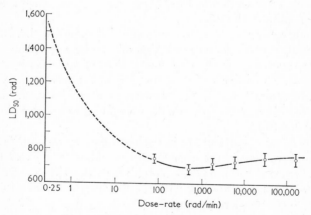

FIG. 1. Variation of LD_{50} with dose-rate.

In the case of life-shortening the variation with dose-rate follows the same trend but is much more pronounced (Table I). The results are plotted in Fig. 2, where again the broken line is based on the results of other workers. It is seen that there is a definite maximum of the life-shortening effect at about 1,000 rads/min.

Similar investigations have been carried out with the mice exposed while anaesthetized. The anaesthetic used was Nembutal, 60 mg/kg, administered intraperitoneally. Table II shows the results for both the LD_{50} and life-shortening. This time there seems to be no variation with dose-rate either for the acute or for the long-term effects of radiation.

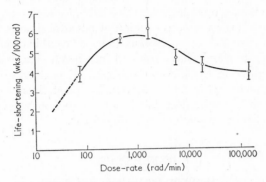

FIG. 2. Variation of life-shortening with dose-rate.

A comparison of the values in Tables I and II shows that the protective action of the anaesthetic against both acute and long-term effects of radiation appears to diminish with increasing dose-rate, becoming undetectable at very high dose-rates.

TABLE II. *Effect of anaesthesia*

Dose-rate (rads/min)	LD_{50} (rads)	Life-shortening (weeks/100 rads)
480	800 ± 29	4·2 ± 0·4
2,060	796 ± 44	3·3 ± 0·4
6,750	805 ± 49	5·1 ± 0·5
31,700	795 ± 33	4·3 ± 0·4
162,000	791 ± 32	4·5 ± 0·4

The interpretation of these results is not easy. The obvious explanation would be an oxygen effect. Thus it is possible that reduced sensitivity to radiation in the anaesthetized animal is the result of the lowering of the oxygen tension in tissues by the anaesthetic (Mack and Figge, 1952). A similar explanation might be offered for the dose-rate effect. If the dose is delivered in a very short time, the oxygen in the cells of vital tissues consumed by the first fraction of the dose is not replenished quickly enough, and the latter part of the irradiation is thus given under hypoxic conditions. Such an effect was established in bacteria by Dewey and Boag (1959). A detailed analysis shows, however, that to explain the observed variation of sensitivity with dose-rate it would be necessary to assume that the vital tissues in the animals are initially at a very low oxygen tension. This conflicts with the finding that by making the mice breathe nitrogen for 25 seconds the LD_{50} can be reduced by nearly a factor of 3 (Lindop and Rotblat, 1960), which means that the tissues must have been initially at a high oxygen tension.

Furthermore, if the protective action of anaesthesia and dose-rate both resulted from a lowered oxygen tension, then the combination of anaesthetic with high dose-rate should have resulted in a much greater increase in the LD_{50}, or a decrease in the life-shortening effect, particularly at very high dose-rates; the reverse is in fact observed. We must thus conclude that the variation of sensitivity with dose-rate is not due to an oxygen effect and that some other mechanism must be responsible for it.

REFERENCES

CORP, M. J., and NEAL, F. E. (1959). *Int. J. Rad. Biol.* **3**, 256.
DEWEY, D. L., and BOAG, J. W. (1959). *Nature, Lond.* **183**, 1450.
LAWREY, J. M., and FOWLER, J. F. (1959). *Brit. J. Radiol.* **32**, 630.
LINDOP, P. J., and ROTBLAT, J. (1960). *Nature, Lond.* **185**, 593.
LINDOP, P. J., and ROTBLAT, J. (1961). *Proc. roy. Soc.* **B154**, 332.
LINDOP, P. J., and ROTBLAT, J. (1962). *Brit. J. Radiol.* **35**, 23.
MACK, H. P., and FIGGE, F. H. J. (1952). *Science* **115**, 547.

DISCUSSION

LAMERTON: At these very high dose-rates, I'm wondering if you are approaching some of the effects of high LET radiation? What I would like to ask is whether you noticed any difference in grouping of deaths for the LD_{50}, whether at the very high dose-rates you are getting more intestinal deaths or more in the first phase than in the second phase?
ROTBLAT: I suppose that by high LET you mean that at very high dose-rates the clusters of ions from two independent particles tend to overlap. We have observed such an effect on the yield of chemical reactions but it occurs at much higher dose-rates (over 10^8 r/sec) than are used in the present work. We have not noticed any significant difference in the causes of death or in their time distribution in mice exposed at different dose-rates.
ALEXANDER: About the dosimetry, how do you compare doses at rates of 100,000 r/min and at 100 r/min? Can you use the same dose meter for these two?
ROTBLAT: The dosimetry at high dose-rate isn't very easy. We had to build special ionization chambers with very small air gaps to avoid the loss of efficiency due to recombination. Of course we always calibrate the ionization chambers by a calorimetric method.
ALEXANDER: So you use two different dose meters for the lower and higher rates?
ROTBLAT: No, we used the same chambers but to make sure that they were working properly, we always calibrated them by an absolute method.
ALEXANDER: But you use the same machine. How do you manage to slow it down, to make the same machine give you 100,000 r/min and 100 r/min?

ROTBLAT: The output from the linear accelerator can be varied over a very wide range by changing the current through the gun filament or the r.f. power.
ALEXANDER: And it won't change the quality of the radiation?
ROTBLAT: Only very slightly. To make sure we have a magnet with which we can check the energy of the electrons.
ALEXANDER: The thing seems difficult. I would not like to compare X-ray machines with that order of accuracy.
ROTBLAT: It is not really difficult, although it is not a job just for a technician.
MOLE: The results are very interesting indeed. We tried to get up to 1,000 r/min and didn't get over the hump, but I think the quantitative change between say 50 and 1,000 r a minute is extraordinarily close to what you have observed. But I would like to ask if you really can't stick to the oxygen idea. Supposing that normally there is a nearly zero oxygen tension at the vital target and oxygen is continually being used up and diffusing in, then if you produce anaesthesia and reduce the concentration difference, you will reduce the rate of diffusion, so you may get some degree of protection. Similarly, the shorter the time in which the dose is delivered the less time there is for diffusion. Could you not use the oxygen effect as a basis?
ROTBLAT: The only difficulty about this explanation is that in this case, if we give the exposure in nitrogen we shouldn't get a much larger protective effect because we already started off with a very low oxygen tension.
MOLE: But you haven't tried nitrogen at 10^5 r/min.
ROTBLAT: Yes we have tried and we get a very large protective factor, approaching 3.
BACQ: I think that mice are not very suitable for this kind of work. You should use chickens, which show remarkable dose-rate effects, very remarkable from 10 to 100 r/min. Now if you have some effect of anaesthesia with this animals which is so much more sensitive to the dose-rate in the narrow range then you might confirm your conclusion.
ROTBLAT: I quite agree, but in this work we did not set out to study the effect of anaesthesia in general, but in relation to the effects of high dose-rates and anoxia.
BRINKMAN: I want to say something about what you call oxygen tension, because that of course is very different in the various tissues of the body. If a mouse dies of anoxia it has still half its amount of oxygen and the only reason it dies is that the brain cannot get enough, although there may be very much oxygen still in other organs. That depends mainly on two factors. One factor is the presence of carbon dioxide; everything depends, not on the oxygen the animal is breathing but on the oxygen dissociation of the haemoglobin and that is much influenced by carbon dioxide. If you have asphyxia, where carbon dioxide rises and oxygen tension goes down, then you will have very much higher tension than if you had the animal in a low mixture and no asphyxia. That makes a great difference, it might make a crucial difference. For instance if one wants to hibernate animals you have to put them not in a low oxygen mixture but in a closed chamber so that the carbon dioxode goes up when the oxygen goes down and they can survive very much longer with very low oxygen contents because of this carbon dioxide effect which shifts the oxygen dissociation curve very much to the right. The other factor is capillarization of the tissues which is very much different. For instance 1 cm^3 of brain tissue has 2 km of capillary, but 1 cm^3 of cardiac tissue has 12 km, so when you use oxygen tension it is too vague. You should say a bit more about it if you can.
ROTBLAT: If I understood it correctly it would appear that if you give the animal anaesthetic then the process causing protection would be different than if you make the animal breathe nitrogen. This may explain why we get differences in the protection factors against acute and long-term effects under various conditions. For example, we found that for the animals breathing nitrogen the long-term effects are less protected than the

acute effects, while in the case of anaesthesia it is the other way round. It may well be that the anaesthetic acts on different organs. I am afraid I cannot tell you anything about the levels of oxygen tension in the animal under anaesthetic because the changes are probably too small. We did try to measure them using a polarographic method by sticking electrodes in different tissues. We did find that when animals are made to breathe nitrogen there is a very rapid drop in the oxygen tension in various tissues, but with an anaesthetic the drop is slow and small.

EFFECTS OF POST-IRRADIATION INJECTION OF YEAST SODIUM RIBONUCLEATE AND ITS NUCLEOTIDES ON THE DIFFERENTIAL COUNT OF BONE-MARROW

H. MAISIN

Cliniques Universitaires St. Raphael, Institut du Cancer, Louvain, Belgium

SUMMARY

We have counted the total number of marrow cells per mg of marrow, the number of white and red cell precursors and the number of reticular cells at different time-intervals after a LD_{50} and a LD_{100} X-ray dose in rats of our strain, control irradiated, or injected with yeast RNA or its nucleotides directly after irradiation. Nucleotides are toxic at LD_{100}. In the ribonucleate- or nucleotide-injected rats we were able to show a general faster regeneration of the marrow red and white precursors but particularly of the white; the regeneration of the red precursors is no longer possible after LD_{100}, that of the white is reduced. It was not possible to show a definite regeneration of the reticular cells but their transformation into white and red precursors must have been facilitated.

The significance of these results is discussed.

Maisin et al. (1960) were the first to establish that post-irradiation injection of yeast ribonucleic acid and its nucleotides was sufficient to decrease the mortality of rats irradiated with $LD_{85(30)}$ of X-rays. Detre and Finch (1958) had observed the same fact in mice also with yeast RNA. Formerly Panjevac et al. (1958) had shown that isologous highly polymerized nucleic acids extracted from spleen and liver improve significantly the survival of lethally irradiated rats. Yeast RNA was only active against the medullary syndrome (Maisin et al., 1960). The dose reduction factor is not very high, it is $\sim 1\cdot 12$ (Maisin et al., 1962). Experiments of Soska et al. (1958–59) with deoxyribonucleotides injected in mice after irradition had furthermore shown prevention of the decrease in red cells, induced a more rapid regeneration of white cells and stimulated the mitotic index of bone-marrow.

Personally, we were interested in confirming this medullary protection in the bone-marrow of our rats by cytological evidence and to see which precursors would be better protected. We irradiated rats with 500 r–$LD_{50(30)}$ and 600 r–$LD_{100(10)}$, the former were injected, after irradiation, with yeast nucleotides, the latter, with yeast sodium ribonucleate. Indeed from a survival point of view, yeast nucleotides were toxic at 600 r. We did not examine the marrow of our rats later than the 6th day at 600 r (because they die) or 500 r (because the regeneration of the marrow is already started).

EXPERIMENTAL CONDITIONS

We used homogeneous albino rats of our L strain. They were 4-month-old males weighing from 130 to 145 g. The rats were irradiated with a Genreal Electric Maxitron 250 X-ray at 250 kV 25 mA filter, 0·25 mm Cu + 1 mm Al, distance: 80 cm, output in air with a Victoreen Radocon was 47 r/min. The animals were irradiated four at a time on a turntable and fasted for 24 hours before, and after, exposure. The X-ray doses were 600 r–$LD_{100(10)}$ and 500 r–$LD_{50(30)}$. After irradiation, daily records were kept of the rats.

We irradiated two groups of 24 rats at 600 r and 500 r. After each X-ray dose, 12 of the rats were kept for controls. Fifteen minutes after the 600 r, the other 12 rats were injected intrapetitoneally with 2 ml of saline containing 100 mg of sodium ribonucleic acid (Merck). After 500 r, 12 rats also were injected intraperitoneally with nucleotides obtained from 100 mg of the sodium ribonucleic acid by controlled alkaline hydrolysis (Maisin et al., 1960).

After 600 r, 4 rats of each group were sacrified at 2-, 4- and 6-day intervals, after 500 r, the same number of rats were killed at 2-, 3- and 6-day intervals.

From each rat, we removed one femur for counting the number of bone-marrow nucleated cells per mg marrow following a method published by Maisin (1959), and one tibia for marrow air-dried smear. The marrow smear was stained by the May-Grünwald-Giemsa method. The results of the number of the bone-marrow nucleated cells per mg of marrow were combined with differential counts of the marrow smears and the absolute number of each type of cell per mg of marrow obtained. We just mention here the total number of white and red precursors and of reticular cells. We also examined under the same conditions marrow of 4 normal, non-irradiated, rats of the same age and same weight.

RESULTS AND DISCUSSION

After 500 r (Table I), in contrast to the irradiated controls, the total number of marrow cells started to increase from the 3rd day in the rats injected with nucleotides. If we look at the number of red and white precursors, we see that the white precursors in the nucleotide-injected rats are regenerating first, the difference from the white precursors of the controls being statistically significant. At the 6th day after irradiation, the red cell precursors are regenerating both in the controls and in the nucleotide-injected rats but the red precursors of the nucleotide-injected rats are regenerating faster, the difference also being highly significant. In other words, the yeast nucleotides influence first the white precursors. Curiously the total number of reticular cells of both controls and nucleotide-injected rats do not differ except perhaps on the 3rd day. However, following previous work, we feel

TABLE I. *Total number of nucleated marrow-cells, white and red precursors and reticular cells per mg of marrow*

Experimental conditions	Time after irradiation (days)	Total number of cells	Total number of white precursors	Total number of red precursors	G/E ratio	Total number of reticular cells
500 r + Nucleotides	2	104,000	32,500	8,300	3·85	13,400
500 r		97,000	29,000	10,200	2·9	11,800
500 r + Nucleotides	3	208,000 n.s.†	97,000 $P < 0.05$†	30,750	3·15	11,500
500 r		129,000	46,000	34,000	1·32	20,500
500 r + Nucleotides	6	836,000 $P < 0.01$†	194,000 $P < 0.02$†	247,000 $P < 0.01$†	0·78	6,900
500 r		246,000	93,500	94,000	1·01	6,300
Normal rat		1,777,000	526,000	453,000	1·15	10,000

† t-test.

TABLE II. *Total number of nucleated marrow-cells, white and red precursors and reticular cells per mg of marrow*

Experimental conditions	Time after irradiation (days)	Total number of cells	Total number of white precursors	Total number of red precursors	G/E ratio	Total number of reticular cells
600 r + NaRNA	2	136,000 $P < 0.01$†	50,500 $P < 0.01$†	17,500	2·9	8,400
600 r		100,000	26,500	17,500	1·51	14,700
600 r + NaRNA	4	104,000 $P < 0.1$†	47,000 $P < 0.05$†	10,800	4·35	8,400
600 r		82,000	33,500	12,600	2·65	9,600
600 r + NaRNA	6	107,000 $P < 0.1$†	48,000 $P < 0.05$†	9,400	5·1	7,100
600 r		72,000	33,000	8,300	4	11,300
Normal rat		1,777,000	526,000	453,000	1·15	10,000

† t-test.

that a faster regeneration of the precursors is accompanied by a reduction in the number of reticular cells, which are transformed into the precursors (Maisin, 1959).

After 600 r (Table II) which is a $LD_{100(10)}$, the marrow of the sodium ribonucleate-injected rats do not regenerate before the 6th day, neither the white precursors nor the red. Thus yeast RNA is unable to induce any regeneration in lethally irradiated rats within the 10th day. In fact, RNA is able to reduce the lesions in the white precursors, indeed the number of cells in the RNA-injected rats are always significantly higher than in the control irradiated, rats. There is no difference in the lesions in the red precursors. The granulo-erythropoietic ratio confirms this discrepancy. The number of reticular cells is the same in both the ribonucleate and control irradiated, animals.

In conclusion, we have arguments to support the view that the better survival of the ribonucleate- or nucleotide-injected rats can be explained by a general faster regeneration of the red and white marrow precursors but particularly of the white. The regeneration of the red precursors is no longer possible with a LD_{100}, it is thus limited by the size of the X-ray dose used; the regeneration of the white precursors is reduced. It was not possible to show a definite regeneration of the reticular cells, but even though it is not demonstrable, the faster regeneration of the white and red precursors cannot be explained without at least an easier transformation of the reticular cells into white and red precursors. Considering the fact that yeast RNA was protecting, or treating, only medullary death (Maisin et al., 1960) these data were very probable (Maisin et al., 1962) but now we have a clear and definite answer. These experiments support the general concept of the action of RNA and its nucleotides on marrow regeneration (Soska et al., 1958; Karpfel et al., 1959) and they specify and prove the action of yeast RNA.

REFERENCES

DETRE, K. D., and FINCH, S. C. (1958). *Science* **128**, 656.
KARPFEL, Z., SOSKA, J., and DRASIL, V. (1959). *Nature, Lond.* **183**, 1600.
MAISIN, H. (1959). Syndrome médullaire après irradiation, Arscia, Bruxelles.
MAISIN, J., DUMONT, P., and DUNJIC, A. (1960). *Nature, Lond.* **186**, 487.
MAISIN, J., DUNJIC, A., and DUMONT, P. (1962), to be published in Strahlenschutz in Forschung und Praxis, Band 2, Verlag Rombach, Freiburg im Breisgau.
PANJEVAC, B., RISTIC, G., and KANAZIR, D. (1958). Second Internat. Conf. Peaceful Uses of Atomic Energy, Geneva **23**, 64.
SOSKA, J., DRASIL, V., and KARPFEL, Z. (1958). Second Internat. Conf. Peaceful Uses of Atomic Energy, Geneva **23**, 34.

DISCUSSION

UPTON: Did the nucleotides affect recovery in the small intestine?

MAISIN: We did not look at intestine but anyway nucleotides do not protect against intestinal mortality only against bone-marrow mortality.

DRÁŠIL: We have studied the stimulation of regeneration in irradiated bone-marrow by the application of deoxynucleotides. On the basis of *in vitro* experiments in which the DNA synthesis of a single bone-marrow cell was studied we are coming to the conclusion that the precursors of DNA are capable of stimulating or initiating the DNA synthesis preferably in reticular cells. They are not capable of increasing the DNA synthesis in heavily damaged cells of the "blast" type such as myeloblasts, erythroblasts and so on. Since the reticular cells are normally either incapable of synthesis or the ability to synthesize deoxynucleotides is very low, the addition of deoxynucleotides into the medium leads to increase in the DNA synthesis. But what is the mechanism? It seems that only a very small proportion of added deoxynucleotides gets into the cells and it seems that irradiated cells have a greater permeability for deoxynucleotides than non-irradiated cells.

ALEXANDER: I would just like to ask Dr. Drášil and Dr. Maisin whether they could summarize for us their experience on the effects of giving nucleic acids and nucleotides afterwards on actual survival. There have been so many claims (which other people have not repeated) that the actual LD_{50} is affected, that I wondered whether they could summarize for us their own views on whether any of these nucleotides of DNAs or RNAs have any influence on the LD_{50}.

MAISIN: I think we know that they are really active given afterwards; the dose reduction factor is something like 1·1.

ALEXANDER: DNA or RNA?

MAISIN: Both of them. It is a little bit better maybe with RNA nucleotides than with RNA alone but for higher doses of radiation no, certainly not.

ALEXANDER: 1·1 must be fairly marginal.

MAISIN: Yes, it is not very much; with mercaptoethylamine given before irradiating we got the same dose reduction factor.

ALEXANDER: But a reduction factor of 1·1 can be obtained non-specifically by injecting almost any irritant or something that stimulates the haemopoietic system.

MAISIN: Anyway we always inject our irradiated controls with saline without protecting at all. When we combine RNA and marrow the dose reduction factor remains the same if we compare the results obtained with those rats and marrow-injected rats alone.

COTTIER: I would like to ask how this RNA was prepared; I wonder if anybody has ever used isogenic RNA?

MAISIN: We use yeast RNA. Some have used isogenic RNA. The results were no better.

DRÁŠIL: I should like to answer Dr. Alexander's question. Survival experiments after treatment of irradiated mice by DNA precursors showed that the survival is higher, but the results are very variable. There are uniform results so far as bone-marrow regeneration is concerned but very variable results for survival. It often seems that the application of DNA precursors leads to premature exhaustion of bone-marrow, that for instance when we complete the start of marrow regeneration in injected mice and in controls then the regeneration of bone-marrow starts two or three days earlier in experimental animals, but after several days it stops again.

BACQ: After irradiation you have the DNA in the nucleus and the RNA in the cytoplasm being hydrolysed by the enzymes RNAase and DNAase which are activated by irradiation. The cell does not necessarily die, but it is flooded with nucleotides. Now what may happen is that, before a cell undergoes repair, certain of these nucleotides may leak into the blood more rapidly than others so that the cell is confronted by a sample of the various nucleotides which is quite different from the normal distribution of nucleotides. Now if the sampling is too abnormal so that no regeneration is possible, maybe these nucleotides can be used by another adjacent cell. M. Errera, for example, proposed a long time ago, that the abnormal proportion of various nucleotides within the cell after irradiation may actually be the biochemical mechanism of mutation.

DRÁŠIL: I am afraid that the cells in which the DNA is hydrolysed must die and that we can help only those cells which were only damaged reversibly by radiation. This is one thing; the products of hydrolysis of DNA, the single nucleotides, may be quite different from those deoxynucleotides needed for synthesis.

UPTON: All of us are familiar with the healing of chromosomes, chromosome breakage and restitution. Is there any evidence that these phenomena involve depolymerization of nucleic acids and then re-polymerization or are they explicable in simpler chemical terms?

COTTIER: The only thing I can say about this is that after thymidine labelling the label stays in the chromosomes and I don't think that anybody has ever seen leakage of label into the cytoplasm. We also examined tissue that had been irradiated after labelling with thymidine and I could never see any evidence of leakage of label from the nucleus to the cytoplasm except in dead cells.

THE EFFECTS OF TOTAL-BODY X-IRRADIATION ON THE REPRODUCTIVE GLANDS OF INFANT FEMALE RATS

D. SLADIĆ-SIMIĆ, N. ŽIVKOVIĆ, D. PAVIĆ, AND
P. N. MARTINOVITCH

Institute of Nuclear Sciences, "Boris Kidrič", Beograd, Yugoslavia

SUMMARY

For the study of the effects of total-body X-ray irradiation on the reproductive glands of infant rats, 8- and 17-day-old females were irradiated with doses of 400 r, 200 r, 100 r and 50 r. Three-month-old adult females, serving as controls, were irradiated with doses of 400 r and 200 r. The effects of the various doses upon the reproductive glands, starting with the prepubertal period up to the first ovulation were checked by the method of counting the germ cells; by counting the mitotic divisions within the follicles; by measuring the volume of the ovaries; and by observing the fertility of the exposed rats as well as their F_1 and F_2 generations.

The reproductive glands of 8-day-old rats proved to be most sensitive to X-ray exposure. Less sensitive are those of 17-day-old rats, whereas the ovaries of 3-month-old irradiated rats are least sensitive. In 8-day-old rats irradiation with a dose of 100 r causes destruction of a large number of primary oocytes, but an accelerated development of a few of them. A dose of 50 r will induce the growth of a large number of follicles and thus reduce the existing number of primary oocytes. As a result, the litter size at first partuition of irradiated rats is larger than in controls. The F_1 and F_2 generations of the exposed females show a high percentage of dwarfism and mortality.

The mammalian ovaries are very sensitive to ionizing radiation. Total-body exposure to a dose of 400 r sterilizes adult mice in a few weeks (Russell and Russell, 1956). The X-ray sterility effects in adult organisms are due to the destruction of primary oocytes a few days after irradiation (Mandl, 1959). Their sensitivity depends on the age of the irradiated animals. According to the published data on mice, the primary oocytes are most sensitive in the second and the third week post-partum, and more resistant during the first few days after birth (Peters, 1961; Russell *et al.*, 1959) and also in adult organisms (Mandl, 1959). The sterilizing dose for infant rats has not yet been determined. Few data concerning the differences in the effects of sterilizing and sub-sterilizing doses on the ovaries have been published so far. We made an attempt to determine the sterilizing X-ray doses for infant rats of different ages and to follow the effects of irradiation on the histological picture of the ovaries. We assumed that the effect of X-irradiation on primary oocytes may range from killing them outright to causing lasting injuries which find their expression in the F_1 and F_2 generations. Having this in mind we decided to include in our studies on the ovaries the effects of sub-sterilizing doses as well.

In order to determine the effects of X-rays on the fertility of female rats irradiated on 8th, 17th, and 90th day *post-partum*, the animals were exposed to total-body irradiation of 400, 200 and 100 r under the following conditions: 200 kV, 14 mA, 0·5 Cu, with a dose-rate of 42 r/min. Three months after the exposure they were caged with male rats and their fertility was observed. The results are shown in Table I.

TABLE I

Age of animals	400 r			200 r				100 r			
	No. of rats	No. of fertile rats	Litter size	No. of rats	No. of fertile rats	% Sterile rats	Litter size	No. of rats	No. of fertile rats	% Sterile rats	Litter size
3 months	5	1	3·0	10	7	30	7·2	—	—	—	—
17 days	5	—	—	17	1	94	1·0	10	9	10	6·4
8 days	5	—	—	20	—	100	—	15	2	86	5·5

Eight-day-old rats exposed to a dose of 100 r became sterile 3 months after the exposure. Doses of 200 r and 400 r respectively are necessary to produce sterility in female rats 17 days and 3 months old.

Changes in the ovaries of rats irradiated with a smaller dose of X-rays were the subject of a special study.

Eight-day-old female rats were exposed to doses of 50 r and 100 r whole-body irradiation. These rats were sacrificed when 17, 24, 42, 46, 48 and 50 days old. The ovaries were fixed in Bouin's fluid, cut into sections 6 microns thick and stained with haematoxylineosine. In the serial sections of both ovaries the primary oocytes and the growing follicles (including the oocytes with a layer of cuboidal granulosa cells and all larger follicles) were counted, by the method of Mandl and Zuckerman (1951). A separate count of follicles exceeding 180 Å in diameter was made in every fifth section and on this basis the total number was calculated. The number of mitotic divisions in granulosa cells was estimated by counting their number in every fiftieth section. The volumes of the ovaries were also computed. The mean values were obtained from 5 rats for each estimation. The results for 17- and 24-day-old animals are given in Table II. The late histological changes were followed in rats of the age of 42, 46, 48 and 50 days. For each age mentioned we sacrificed one rat irradiated with 50 r, one irradiated with 100 r and one control (Table III)

TABLE II

Age of rats	Dose of irradiation	No. of rats	Total No. of oocytes	No. of primary oocytes	No. of growing follicles	% growing follicles	No. of follicles $\phi > 180\mu$	% follicles $\phi > 180\mu$	Granulosa cells No. of mitotic divisions	Volume of individual ovaries in mm^3
17 days	controls	5	12064 ± 1406	9634 ± 1005	2509 ± 427	20.3 ± 1.2	20 ± 9	0.1 ± 0.1	16080 ± 1572	2.02 ± 0.21
	50 r	5	8296 ± 1006	5053 ± 740†	3245 ± 216†	40.2 ± 2.8†	80 ± 8†	1.0 ± 0.3†	29700 ± 2475†	2.78 ± 0.21†
	100 r	5	1875 ± 93†	345 ± 20†	1529 ± 88†	81.5 ± 1.1†	53 ± 16	2.9 ± 1.0†	22550 ± 2040†	1.99 ± 0.26
24 days	controls	5	9732 ± 837	7693 ± 827	2038 ± 188	21.5 ± 2.3	273 ± 27	2.8 ± 0.1	55020 ± 2405	7.79 ± 0.84
	50 r	5	4749 ± 449†	2976 ± 292†	1772 ± 182	37.3 ± 1.8†	278 ± 19	6.1 ± 0.8†	61760 ± 5953	6.56 ± 0.82
	100 r	5	1138 ± 215†	251 ± 61†	888 ± 181†	77.6 ± 5.1†	192 ± 24†	19.6 ± 4.3†	47100 ± 5613	5.36 ± 0.57†

† Differences between the given values in experimental animals and the corresponding controls are statistically significant.

TABLE III

Age of rats	Dose of irradiation	Total No. of oocytes	No. of primary oocytes	No. of growing follicles	% growing follicles
42 days	control	6404	5929	575	8·9
	50 r	5325	4455	870	16·3
	100 r	365	99	266	72·8
46 days	control	7431	6273	1168	15·7
	50 r	4001	3014	987	24·6
	100 r	366	121	245	66·9
48 days	control	7166	6391	775	10·8
	50 r	3089	2435	654	21·2
	100 r	131	44	87	66·4
50 days	control	7333	6215	1118	15·2
	50 r	3552	2673	879	24·7
	100 r	176	33	143	81·2

Effects of doses of 100 r

The total number of germ cells in the ovary of 17-day-old rats, exposed when 8 days old, decreased by 85% and the number of primary oocytes by more than 96% as compared with control ovaries of the same age. On examining the histological sections, many empty nests were found, left by degenerated primary oocytes. Apparently, most of the primary oocytes were heavily damaged and killed by the dose of 100 r within the first 9 days following exposure. After 9 days, the number of germ cells kept on decreasing at the same rate as before. By the time these rats are 50 days old they may be considered sterile. In all the histological sections of the ovaries, a considerably higher percentage of growing follicles was noticed, ranging from 62 to 81%. The high percentage of growing follicles in the late period after irradiation could only suggest that the primary oocytes are stimulated to grow and develop.

Effects of doses of 50 r

In 17-day-old rats, irradiated when 8 days old, a statistically non-significant decrease in the total number of germ cells was observed, as compared with the controls. The number of the primary oocytes was also reduced, but, in comparison with the controls, the difference in estimated numbers proved to be highly significant. The decrease in the number of primary oocytes was counterbalanced by a significant increase in the number of growing follicles. The number of mitotic divisions in the granulosa cells increased as well. As a result, the number of follicles in the advanced stages of growth ($\phi > 180\mu$), and the volumes of the ovaries are increased. Besides, mean values for 480 follicles with antrum were found in the controls and for 819 in the exposed rats. Apparently, during the first nine days after irradiation the oocytes in the ovaries are stimulated to develop by the total-body X-irradiation. On the other hand, there are data in the literature, which register only the destructive effect of X-rays on the primary oocytes, but with no stimulating effect (Spalding et al., 1957). We are inclined to believe that this can be explained

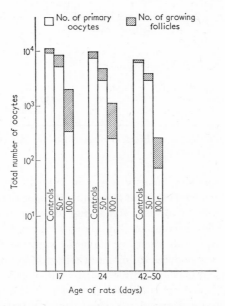

Fig. 1. This graph shows the total number of ova cells present and the relationship between the number of primary oocytes and the growing follicles as a function of time in the ovaries of rats exposed when 8-day-old to 50 r and 100 r of X-rays.

by the doses applied (750 and 3,000 r), which destroyed most of the primary oocytes, the remaining number of primary oocytes being too small to allow the phenomenon of stimulation to express itself.

In 24-day-old female rats, i.e. 16 days after exposure, the total number of germ cells decreased by 42 per cent. Nevertheless, the number of growing follicles, the number of follicles in advanced stages of growth and the number of mitotic divisions in granulosa cells in the ovaries of exposed rats did not significantly differ from the control ovaries. Throughout our observation on the rat ovaries up to 50 days of age, the percentage of growing follicles was higher in the exposed than in the control rats.

When exposed females were 50 days old we caged them with male rats and their fertility was observed. The results for the fertility of the X-ray treated rats and for their progeny are given in Table IV.

TABLE IV

Dose of X-rays	No. of rats	1st parturition Litter size	Total No. of progeny	No. of dwarfs	No. of death before maturity	Micro- and anophthalmia No. of rats	Percentage of defective progeny
controls	11	$7 \cdot 6 \pm 0 \cdot 3$ $P < 0 \cdot 02$	84	—	—	—	—
50 r	18	$9 \cdot 2 \pm 0 \cdot 5$	165	13	16	—	9·7
F_1 of rats irradiated with 50 r	37	$7 \cdot 2 \pm 0 \cdot 4$	268	18	14†	3	10·4

† Among these 14 animals, 7 were dwarf; 4 animals died immediately after delivery. Three rats which appeared normal also died.

The litter size was significantly larger in the exposed rats than in the control animals, which suggests that the stimulation of the oocytes represents a lasting effect to X-ray exposure. According to our knowledge, the process of growth of the oocyte is completed in about 12 days. If this is true for the irradiated oocytes as well, then we have to suppose that the exposure stimulates the primary oocytes to initiate their growth and further development. This stimulation may be a direct one on the ovaries or indirect *via* the pituitary gland. This problem could be solved only by local irradiation either of the ovaries or the pituitary gland. Russell and Russell (1956) irradiated adult animals with the sterilizing dose of 400 r and according to them in the exposed mice "... the mean number of fertile eggs per female is significantly increased at short irradiation-to-fertilization intervals and decreased just before sterility sets in ". The litter size in their experiments was smaller in the irradiated females than in the controls due to pre-natal death. In our experiment, after exposure to a much lower dose of X-rays, the litter size was larger in comparison with the controls, but so far we have no records on pre-natal death of embryos. On the other hand, 13 out of 165 descendants were dwarfs. The body weight of the dwarf animals was half the weight of the controls of corresponding age. One of them was sacrificed and the remaining 12 died before reaching maturity. In F_1, 9·7% of mortality was recorded, whereas there was not a single case of dwarfism or mortality before maturity in the controls.

We are not certain if the metabolic changes in the irradiated host, or true mutations in the germ cells, are responsible for the defective progeny. The data obtained for the F_1 are not sufficient to give an answer.

Following the F_2 we found that the percentage of dwarfs and the mortality rate was not reduced. Moreover, in the F_2 generation we observed in some cases anophthalmia and microphthalmia which was not recorded in F_1 (Table IV). Our breeding experiments are still in progress. Out of 405 rats belonging to F_2, 268 have reached the age when their defectiveness could be seen. They are the progeny of 13 exposed females. It seems that the damage to the germ cells is not of the same type in all irradiated rats. Only one exposed female out of 13 had progeny without a case of dwarfism or mortality in F_1 and F_2 generations. Five of them showed a decreased percentage of defectiveness in F_2, compared with F_1. Five exposed females gave defective progeny only in the F_2 generation. Two irradiated females produced some defective rats in F_2 which were different from all others. One of these two has 11 rats in F_1 and 40 animals in F_2 with no case of dwarfism, but we found 3 animals in F_2 with anophthalmia and microphthalmia. The other exposed female has, in the F_2 generation, 2 female dwarfs with heavy anaemia accompanied by the anisocytosis of red blood corpuscules. The circulatory blood of all other normal and dwarf rats was examined and no similar case was found. On the basis of the incomplete data obtained for the F_2 we are inclined to believe that serious genetic damages to the germ cells had been produced by exposure to the dose of 50 r. We hope that a follow-up of the F_3 and F_4 will help us solve the problem.

ACKNOWLEDGMENT

The authors want to express their gratitude to Prof. Z. M. Bacq for encouraging them in their work.

REFERENCES

MANDL, A. M. (1959). *Proc. roy. Soc.* **B150**, 53.
MANDL, A. M., and ZUCKERMAN, S. (1951). *J. Endocrinol.* **7**, 112.
PETERS, H. (1961). *Radiation Res.* **15**, 582.
RUSSELL, L. B., and RUSSELL, W. L. (1956). In "Progress in Radiobiology" (J. S. Mitchell, B. E. Holmes, and C. L. Smith, eds.) p. 187. Oliver and Boyd, London.
RUSSEL, W. L., RUSSEL, L. B., STEELE, M. H., and PHIPPS, E. L. (1959). *Science* **129**, 1288.
SPALDING, G. F., WELLNITZ, J. M., and SCHWEITZER, W. H. (1957). *Fert. and Steril*, 8, 80.

DISCUSSION

UPTON: Were these the first and second generation or the first and second litters? This was something I wasn't quite sure about.

MARTINOVITCH: The first and the second generation—the first litter of the first filial and the first litter of the second filial.

RUSSELL: Did you get the increased litter size in the second filial?

MARTINOVITCH: The litter size in the second filial generation was the same as in the control animals.

RUSSELL: By irradiating adult mice some time ago Mrs. Russell got an increase in their litter size with certain doses, certainly an increased ovulation—a super-ovulation effect from irradiation—in the first litter following irradiation, if the radiation was timed at the right interval before this particular ovulation. This was for higher doses, with some effect at the lower doses too. I wasn't quite sure of the data on the damage, deaths after birth and so on, but we haven't seen anything like this in mice of the filial generation. There is also an increase in the death of embryos, which does not offset the total effect. There are more eggs ovulated. Some of these are killed in embryo so that there is an excess death of embryos as compared with controls, but the net effect on litter-size is still an increase. This death is all in embryos at the time of implantation—we've not seen any significant increase in the effect after this time, even with fairly high doses.

MARTINOVITCH: A difference was found in the sensitivity of the ovaries in respect of the onset of sterility in rats of different ages, and a difference in the effect on the germ cells is also supposed. The progeny of female rats irradiated when 8-days-old was the only one followed, as no progeny from irradiated adult rats were available for comparison.

UPTON: I have the impression that the female rat was relatively radioresistant as regards the sterilizing of the ovary. I was especially interested in your Table III. I'm not sure I fully understood it but there seems to be quite extensive killing of early oocytes.

MARTINOVITCH: The two processes go hand in hand, dying off of some of the primary oocytes and stimulation of the others.

UPTON: It would seem that 50 days following the dose of 100 r you have only 176 oocytes left, as opposed to 7,300 in the control. Surely this would indicate a very high radiosensitivity of these primary oocytes.

MARTINOVITCH: This only shows the high sensitivity of primary oocytes at this particular stage of development.

MOLE: I would like to emphasize something that Russell touched on and that is that there is a very large degree of species difference in sensitivity of the ova which is unlike any other radiobiological response in mammals that I know of, because there is something like a 100-fold difference at least in radiosensitivity. Different laboratory animals (and man) are certainly at least ten times more resistant than the mouse, and possibly 100 times more resistant. This is, I think, a matter of very great interest as well as possible practical importance.

PERIPHERAL BLOOD STUDIES UPON SOME ISOGENIC CHIMAERAS

A. J. S. DAVIES, ANNE M. CROSS, AND P. C. KOLLER

Chester Beatty Research Institute, Institute of Cancer Research, London, England

The successful use of tissue therapy after total-body irradiation (Lorenz et al., 1952) and the fact that success is commonly due to the inception of a chimaeric state (Lindsley et al., 1955) are now well-known. There remain, however, many problems both from an experimental and an applied viewpoint (Koller et al., 1961). Experimentally, though we know most of the factors which determine the establishment of chimaerism, we know little of the details of the process.

In the present study the dose of radiation to the host animal and the number of cells injected therapeutically within 24 hours after irradiation have been systematically varied. Survival and speed of recovery of haematopoiesis, as assessed by peripheral blood counts, have been measured. Inbred BALB/C strain mice were used throughout. For these animals the LD_{50} 30 days is about 620 r, the LD_{99} about 700 r.

TABLE I. *The survival and peripheral blood values (as percentages of control) of BALB/C mice 10 days after various doses of total-body irradiation followed or not by intravenous injection of 10^7 isogenic bone-marrow cells*

Radiation dose	Without bone-marrow therapy			With bone-marrow therapy		
	300	500	700	300	500	700
PCV.	88	75	57	102	97	98
hb.	84	66	50	88	92	95
Reticulocytes	280	53	0	240	260	216
Platelets	29	15	7	127	76	89
Total whites	18	3	3	48	48	63
Mononuclears	15	2	4	36	36	47
Polymorphs	55	4	0	119	117	151
Survival	100	100	0	100	100	< 100

The effect of marrow therapy upon survival and peripheral blood counts is illustrated in Table I. It is noteworthy that the depression of all elements of the peripheral blood was proportional to the dose of irradiation and that in all cases injection of bone-marrow hastened recovery towards normal. It is also apparent that as early as 10 days after irradiation, if bone-marrow were given, the level of recovery was almost independent of the radiation dose. It should however, be remarked that the number of cells given was high in relation to the number required to ensure survival (see Table II) and therefore the number of cells available was unlikely to have been a factor limiting speed of recovery.

TABLE II. *The survival at 30 days of BALB/C mice given first 700 r, total-body, then within 24 hours various numbers of isogenic bone-marrow cells*

Cell dose	% Survival
0	1·0
1×10^3	0·0
5×10^3	7·4
1×10^4	65·5
$2·5 \times 10^4$	89·75
5×10^4	82·0
1×10^5	79·5
$2·5 \times 10^5$	80·0
5×10^5	86·5
1×10^6	76·75
1×10^7	100

Table II records an example of experiments in which the number of cells injected after irradiation was varied. There were at least 30 mice in each group.

TABLE III. *The peripheral blood values (as percentages of control) of BALB/C mice that had 10 days previously received 700 r, total-body, followed closely by various numbers of isogenic bone-marrow cells intraveneously*

	Cell dose			
	1×10^4	1×10^5	1×10^6	1×10^7
PCV.	64	67	85	98
hb.	61	63	79	95
Reticulocytes	13	77	167	216
Platelets	11	15	20	89
Total whites	0	4	5	63
Mononuclears	0	4	3	47
Polymorphs	0	5	18	152
Survival	65·5	79·5	76·75	100·0

It appears that survival is more or less independent of the number of cells injected between 10^4 and 10^6 but if 10^7 cells are given survival can be 100%. A possible explanation for this finding can be adduced from the data in Table III in which the effects, upon the peripheral blood values, of varying the number of cells injected after irradiation are shown. Clearly, the more cells injected the more rapidly and effectively are nearly all components of the peripheral blood restored to near normal values. Further, injection of 10^7 cells appear to have a disproportionately large effect in comparison with injection of 10^6 cells. From this it is tempting to suppose that the period at risk after radiation is reduced by rapid restoration of one or all components of the peripheral blood and that injection of 10^7 cells is most effective in reducing the period at risk.

It might also be supposed that the period at risk would generally be inversely proportional to the number of cells injected. From which it follows that, irrespective of percentage mortality, the mean time of death might be later after injection of a small number than after a large number of cells. Evidence at hand, though as yet numerically inadequate, gives no support to this corollary.

ACKNOWLEDGMENTS

This work has been supported by a grant from the International Atomic Energy Agency and also by grants to the Chester Beatty Research Institute (Institute of Cancer Research: Royal Cancer Hospital) from the Medical Research Council, the British Empire Cancer Campaign, the Anna Fuller Fund, and the National Cancer Institute of the National Institute of Health, U.S. Public Health Service.

REFERENCES

KOLLER, P. C., DAVIES, A. J. S., and DOAK, S. M. A. (1961). *Advanc. Cancer Res.* **6**, 181.
LINDSLEY, D. L., ODELL, T. T., and TAUSCHE, F. G. (1955). *Proc. Soc. exp. Biol. N.Y.* **90**, 512.
LORENZ, E., UPHOFF, D. E., REID, T. R., and SHELTON, E. (1952). *J. nat. Cancer Inst.* **12**, 197.

DISCUSSION

MOLE: If with 25,000–1,000,000 donor cells you can send up the survival from 0 to 80% and then you have to give 10 million cells to increase survival from 80 to 100%, the observation would indicate that there must be an obvious difference in the way the animals die which received a low or high number of cells. Is there a difference in the time of death?
KOLLER: We found no difference in the way the animals die after the various number of donor cells injected.
LAMERTON: Professor Koller gives his figures as percentages of controls. Now if your control values remain fairly steady this is all right but in our rats, for instance, we have found a very considerable variation. We suspect seasonally, or from year to year, we do

not know why, but this of course makes quite a difference to this sort of analysis. I wonder if I could ask Prof. Koller or Dr. Davies whether they have in fact observed considerable systematic alteration in the absolute numbers of counts in their controls.

DAVIES: We have not observed any seasonal variation. The variation which has been observed has been taken into consideration using standard statistical methods.

SESSION VII

Chairman: Z. M. BACQ
GENERAL DISCUSSION

GENERAL DISCUSSION

Fertility: Intracellular Recovery

BACQ: I thought that it would be better to concentrate on certain of the points which cover most of the facts and systems that have been discussed in this meeting. First I should like to have a brief discussion on fertility because we have had no possibility yet of discussing Dr. Russell's paper, and I would like to start the discussion in this way. In the female mouse where the effect of protection is so marked and where there is no division of the cells we have a very good instance of cell recovery; an instance of the capacity of even a very radio-sensitive cell to recover, provided that not too large a single dose is given. Could Dr. Russell tell us if this, in his opinion, is comparable to the reduction in the frequency of induced mutations with decrease in dose-rate which he has shown so beautifully?

RUSSELL: Well, I think we should be very cautious about comparing one kind of recovery with another; the recovery from damage leading to cell death may not be similar to recovery from what is probably pre-mutational damage. We believe that the dose-rate effect on mutation does represent recovery from pre-mutational damage. It seems unlikely that one can have reversal of the completed mutation. And since the dose-rate effect does indicate recovery of some sort, we conclude that it is probably recovery from pre-mutational damage. But whether there is any biochemical similarity between the recovery from a pre-mutational damage and the recovery from the type of damage that would have led to cell death, I think is just guess-work at the present time. It is, however, interesting that in new data that we hope to get out soon on mutation frequency in oocytes, it still looks as though the dose-rate effect is much greater than in spermatogonia. In other words as we go from a high to a low dose-rate there is an even greater depression in mutation rate than was found in spermatogonia. So there is at least some correlation between the apparently marked rate of cell recovery that you so rightly pointed out for the oocytes and the rather large dose-rate effect for mutations. Perhaps these cells do have some remarkably good capacity for recovery that can express itself in both systems.

MOLE: I would like to ask why the word recovery is used. Why don't we just say there's less damage? This seems to me to be of quite fundamental importance, because when we use the word "recovery" we think of some biological process, and when we say less damage we mean less physical damage in the first instance.

BACQ: Yes, but one might think that in a cell there are two processes in competition, one, the normal capacity for recovery of every cell and the other, the rate of injury. If the rate of injury is not too great then you've a long time for the recovery capacity to express itself.

UPTON: This point that Dr. Mole raises is one which bothered me for a long time, too. It is this: one could distinguish, I think, between injury as such and recovery from that injury, and at a low dose-rate one might have less injury, not merely more recovery. But I wonder whether this isn't a semantic question now, in that if one has the same number of ionization events from a given total dose of radiation, then one is looking at the effect of that number of ionization events. One must repair injury at the ionization level, even though this is never expressed in terms of any biologically detectable injury.

ALEXANDER: This is really the same point. If we consider that the primary injury is a chemical reaction, and we have every reason for believing this to be the case, then we know of almost no chemical reactions that are highly dose-rate dependent except for polymerization and other chain reactions. I think we can conclude safely that polymerization isn't likely to be a very important lesion in damaging the cell. There might be oxidation changes involved where dose-rate does come in, but this seems improbable. I would say that the most probable thing is that the initial injury, defined as a chemical event, is dose-rate independent. When you have 10^8 or so r/sec then there is a slight dose rate dependence but for many radiation-induced chemical reactions in the doserate region about which we are talking the initial injury is almost certainly the same, whatever the rate.

RUSSELL: I would say essentially the same thing, although I would distinguish between pre-mutational damage and the actual completion of the mutation. We don't claim that there is recovery from mutation. Therefore, in one terminology this is less damage. But if one is referring to the pre-mutational effect, there must be recovery, as I see it, in the sense expressed by Dr. Upton and Dr. Alexander. The primary damage presumably is the same, from the point of view of the number of ionizations, but since the total effect of this on mutation is less at low dose-rates, there must have been recovery in the intermediate processes.

BERENBLUM: I think it is important to consider both injury and recovery as separate entities because one can surely visualize two kinds of injury—one which is reparable and one which is not. If you give a dose of radiation such that the injury is irreversible and the cell dies you will obviously get no mutations. Are you talking of recovery of the cell itself, or are you talking of recovery of the tissue?

BACQ: The cell itself.

BERENBLUM: Yes, but is it always possible to distinguish this? Can't you conceive a situation where, with a particular dose of radiation, all the cells that are injured are killed. The cells that are not injured, presumably they are less sensitive, will also show no mutations. One can visualize here a large dose producing less mutations because it has killed off all the cells, whereas a smaller dose will produce a higher incidence of mutations because not all the cells have been killed. Is that possible? I'm just putting it forward as a thesis.

RUSSELL: Yes, we have postulated that it is exactly this effect which accounts for a reduction in mutation rate when the dose increases from 600 r to 1 000 r.

ROTBLAT: This work was done at a high dose-rate. Do you think you would get a drop at a lower dose-rate?

RUSSELL: No. At a lower dose-rate there is no drop.

ROTBLAT: This fits in with what has just been said about the difference in killing effect and mutation.

Damage to Vessels

BACQ: May we take up now this question of injury to blood vessels? Just to start the discussion I will tell you of one human case which has been presented by A. Massart in January of this year in Essen, at the meeting of the Vereingung deutscher Strahlenschutzartze † where I happened to be present. A man received accidentally during manipulation of radioisotope a rather big dose on a hand. He got the usual oedema which subsided, but one year and a half (or two years) afterwards, there began to show the usual atrophy of the skin, beginning of gangrene even, and the surgeons were ready to amputate at least two fingers. The physician in charge thought that before amputation one should try one long-lasting vasodilator substance kallicrein which is present in an extract of the pancreas called Padutin. He injected the drug three times a week regularly.

† See "Strahlenschutz in Forschung und Praxis", Band 2. Verlag Rombach, Freiburg-im-Breisgau, 1963.

Very slowly this gangrenous condition disappeared. Healing was obtained after several months. At the time of the meeting, the man was present and showed his hand. It was very difficult to distinguish between the normal fingers and the fingers which had had such a severe dose. It seems that the therapeutic effects of Padutin could only be due to some action on the blood vessels and that this hand has slowly received more blood, more oxygen, and could recover. This is just to show that in chronic late stages, blood vessels have certainly every importance.

MOLE: I was just going to question the causal relation between the treatment and the natural history because in radiation nephritis the condition comes to a kind of climax in 6 months to a year after exposure and if the patient doesn't die he improves spontaneously (Luxton, R. W. (1961). *Lancet ii*, 1220). It is probable that underlying radiation nephritis in man is a vascular lesion. It is in a sense self-limiting.

BACQ: A similar condition can be induced experimentally by cobalt salts, and you get the same effects with Padutin on very carefully controlled cases. You cannot naturally make controlled experiments like the one reported. It may be an extrapolation from other casuses of atrophy in gangrene localized to the extremities. This unique observation, although not entirely convincing, is nevertheless useful because, so far, no effective treatment has been found for these late vascular effects.

BRINKMAN: I also have seen and read that in many cases the capacity of blood vessels, especially the smaller ones, for regeneration is large, but you can read in books on gerontology that that is not the case for the blood vessels in the bone-marrow, and I should very much like to know if anybody knows more about this, because this is a very suggestive statement in books on gerontology.

BACQ: Two Russian authors, L. Zhinkin and A. Zavarzin, have shown that cysteamine is especially very highly concentrated in the wall of the blood vessels.

MAISIN: Is this lack of regeneration of the vessels of the bone-marrow so marked? In irradiated rats, we know that directly after irradiation the capillaries are dilated. Later on I believe that the capillaries in the marrow must regenerate because otherwise how can you explain the return to normal of erythropoiesis in the marrow?

BRINKMAN: It depends of course on how much they have been damaged.

DEVIK: May I ask if there were any visible signs of loss of hair or keratinization of the skin? If not, the dose might not have been too excessive.

BACQ: Oh yes, it was a large dose and the man had a very marked oedema in the days following accidental irradiation. The dose calculated was something like 2,000 to 3,000 rem if my memory is correct.

CASARETT: I would like to say that there is plenty of evidence from observations of wound healing at all ages, that small vessels can maintain their regenerative capacity, but if the tissue is not disturbed and no stimulus to regeneration is provided, then the gerontological studies, including some of our own with microangio-radiography, show that there is generally a net reduction in fine vasculature with age. We have done some work with radiation of grafted and normal skin and studied the influence of the vascular changes on the disappearance of radioisotopes injected into the skin and on occasion we have seen that there is supra-vascularization in the regenerative phase in irradiated skin. However, the disappearance curves for the isotopes are delayed just as in the case where there is decreased vascularization. The reason for this is the great increase in the histohaematic barrier, i.e. the connective tissue barrier between vessels and the dependent tissue. So we think that the stimulus for the supra-vascularization resides in the histohaematic barrier which is impeding the vascular functions, rather than something primarily connected with endothelial cells themselves. With regard to Dr. Mole's statement about the nephritic cases, patients who died or recovered in 6 months, I think that he is referring here only to acute cases. We see cases that present themselves clinically at varying times during life after having received therapeutic radiation involving the kidney regions. They simulate other clinical syndromes as far as the kidneys are concerned—benign or malignant hypertension, acute or chronic nephritis and so on. There are changes in some cases that can be seen early, but the patients may or may not die acutely. We are working on localized radiation of the kidneys trying to simulate these clinical conditions and we find the same thing. We do renographic, angiographic, and histopathological studies in animals sacrificed periodically after irradiation to observe the development of nephrosclerosis, and whether it presents clinically as an acute or sub-acute case or chronic case depends on the rate of the vascular change. In this regard our observations are in agreement with Luxton's excellent description of human patients with radiation nephritis.

COTTIER: I should like to comment shortly on the possible pathogenesis of small vessel damage in various stages of irradiation. As Dr. Casarett mentioned, one finds a reduction in the number of small blood vessels, which we observed also. In addition to that, this is a well-known arteriolocapillary hyalinosis, which is a discontinuous process—you find it at specific sites along the vessel between apparently normal parts, and then you find an increased incidence of capillary telangiectasis. Now, we produced by heavy irradiation of the pituitary within 24 hours microthrombi of the sinusoids of the pituitary, and it is interesting to see that in these late changes of arterioles and capillaries the obstructive lesion—the intimal hyalinosis—is also focal in nature, not diffuse, and therefore one has to consider microthrombi forming on endothelial lesions or on other forms of vessel well lesions to be one, if not the most important, effect.

BRINKMAN: In rabbits which have received about 1,000 r the agglutination rate of the corpuscles about two hours after irradiation is very much increased. This is caused by the release into the circulation of some large macromolecules, and we do not know what they are. But the rate of agglutination is increased so much that this could well explain the microthrombi.

Changes in Connective Tissue

BACQ: I think we will now pass on to another point. As far as the circulation within the bone-marrow is concerned we have very little physiological data indeed. Technically it is very difficult if not impossible to separate in the bone the circulation to the bone-marrow from that of the bone tissue itself. But one may also believe that the injury to the vessels is not so much within the endothelium—or some factor in the blood which accelerates the possibility of thrombi formation—but within the elastic tissue of the vessels. This point: the very important changes which may be induced immediately after irradiation or a very long time after irradiation in the physiology or biochemistry of the whole connective tissue has been already discussed yesterday. One must remember a few things—first that the biochemistry of this tissue—the rate of renewal of the molecules—is very tightly controlled by the neuro-endocrine system, mainly the pituitary and the adrenal cortex. On the other hand, there is one tissue which is very peculiar. It's the brain, where one hasn't got any connective tissue; the connective tissue is replaced by glial cells which apparently are quite sensitive to irradiation according to some work done by Gerebtzoff in Liège. It seems that one doesn't know enough about what happens a long time after irradiation, about the delayed biochemical effects on these connective tissues, on the macromolecules and especially the mucopolysaccharides which play such an important role. I thought that just to start the discussion, it might be a useful research project in this field to have *in the same laboratory*, working on the same strain of animals with the same techniques of irradiation, parallel studies of histological conditions and the biochemical condition of the connective tissue. If one does only anatomical work it is dangerous because—even with the electron microscope—everybody works with a certain idea and very likely, consciously, or unconsciously, chooses the picture which suits best the idea that one has. So that it would be interesting to have some biochemical test just similar to the one we have at the present time for haemoglobin synthesis. One doesn't do simply a histological examination of the bone-marrow with ^{59}Fe it is easy to estimate the total capacity of the bone-marrow to synthesize haemoglobin. Has anybody done experiments on the rate of renewal of the mucopolysaccharides immediately, or a long time, after irradiation?

BRINKMAN: I would remark that it would not be sufficient to study mucopolysaccharides biochemically because you cannot thus study the degree of depolymerization and I think that this is very important. You must study it in a functional way by measuring mechanical resistance and so forth. If you want to know something about regeneration, you can learn it very simply by the injection pressure technique I described and, in this way, you generally find rather a rapid regeneration in say two to three hours. To the collection of hormones which you already mentioned which have an influence on this process, I would like to add tri-iodothyronine which is, in my opinion, the most potent of them all. We know that tri-iodothyronine has very interesting properties in accelerating the rate of mucopolysaccharide production, and this might be of some importance for extra-cellular protection.

CURTIS: We have tried tri-iodothyronine as far as the long-term effect of radiation is concerned, and Dr. Nixon of the Sloan-Kettering Institute, with whom we worked, has tried it in clinical experiments on recovery from radiation burns. In both cases—in our experiments with animals and his with patients, there was no effect at all with tri-iodothyronine.

UPTON: What late effects were you examining, Dr Curtis?

CURTIS: Simply longevity. We were trying to see whether there was perhaps a difference in tumour production, but we found nothing there. Dr. Nixon, of course, was simply looking for recovery in the burned area of skin.

BACQ: It may be that the normal concentration of thyroxin and tri-iodothyronine circulating in the blood is already optimal so that if you give more you cannot have any improvement. Your control must not be a normal animal but a thyroidectomized animal. We have done so, with success, in another type of experiment involving no radiation.

UPTON: I understand you've called for comments on the effects of radiation on connective tissue. I find myself wondering whether the apparent lack of effects of less then optimal amounts of radiation on the age-dependent changes in collagen may not be correlated with the fact that collagen is a relatively inert material and not subject to renewal throughout life. I believe that the effects on the mucopolysaccharides of connective tissue cited by Dr. Brinkman have also been observed by Sobell. This, on the other hand, is a moiety of connective tissue which is renewed periodically in the life of the individual. Do you think that there could possibly be some correlation between the lack of effect on non-renewable materials and the effect on the renewable materials of the connective tissue system.

BRINKMAN: It has been shown by Neuberger with the aid of labelled glycine that the collagen in the cells is renewed all the time. It is only the fibrous form which is not active in this way and which disappears; but this is replaced by new fibres from the fibroblasts.

UPTON: Do I understand then Dr. Brinkman that extra-cellular collagen fibres are in fact renewed periodically, that there is a turnover?

BRINKMAN: Yes. In older animals it would be very slow, but in young animals it is quite rapid. This holds also for elastic fibres. They are renewed rapidly from the ground substance. Perhaps I might be allowed to say something here about the relation of extracellular to intracellular changes. This morning it was suggested that possibly some so-called extracellular changes might really be caused intracellularly. This does not hold anyway for the immediate effects seen during irradiation, they must be of extracellular origin, and it does not hold either for the many things you can observe in dead tissues.

ALEXANDER: I wonder whether Prof. Brinkman would like to make a guess as to whether the immediate effects on the extracellular tissue of which he has just spoken have any lasting influence—he's mentioned the fact that after two or three hours the damage seems to have been repaired. Is it his guess that this recovery is incomplete and that some of the permanent damage in connective tissue results from this, or are the permanent changes in connective tissue due to the changes in the biochemistry and turnover of the ground substances?

BRINKMAN: That depends very much on what type of connective tissue you are considering. If it is the skin, I think that repair is complete, and that there is no residual damage at all. But if you consider the mucopolysaccharide and connective tissue fraction in the aortic wall where they have an ultrafiltration function, breakdown of polymerization for only 1 hour can be enough to let in macromolecules from the blood, and then you have secondary consequences.

BACQ: Cholesterol is not synthesized in the tissues of the blood vessels but mainly in the liver. It is brought to the vessels by the blood from the liver, or from the food by intestinal absorption. If one find an increased amount of cholesterol in a tissue other than the liver (and the adrenals) it means that for some time the capacity of this tissue for storing cholesterol has been increased. I do not know if the various tissues of the blood vessels metabolize cholesterol.

BRINKMAN: I would say that the ultrafilter layer of the wall of the large vessels is breaking down all the time. It has already started with young children. You can see small plaques especially round the places where other vessels join. It is a real struggle to keep the macromolecules out. This battle is lost if you add sufficient radiation.

COTTIER: I just wanted to emphasize that there is quite a long latent period after irradiation before these small vessel changes appear. This is somewhat hard to reconcile with the assumption that the initial short-lived increase of permeability should be the main factor responsible. I think that since the ground substance and fibrillar material is continuously renewed by cells, one should again emphasize that we need to know more about the life-span of the endothelial cells and the function of these cells.

UPTON: I would simply like to emphasize that we may be confronted by a multiplicity of effects. The initial effects on the ground substance lining the aorta may indeed be initiating events for the occurrence of atheromatous changes locally, which may or may not be related to the changes in fine vasculature.

Chromosome Breaks and Cell Death

BACQ: I would like now to call on Dr. Muller, who wishes to make some comments as an addition to his paper.

MULLER: I thought I would give you an indication of how some of these matters that I discussed with you yesterday might be checked quantitatively. I left the matter very vague as to what proportion of cells were killed in a given case, but there are formulae which Dr. Ostertag and I have worked out which should apply to this matter and which we have put to a little testing.

Suppose we have as a standard a chromosome of such length and other characteristics that when it is subjected to a unit dose of radiation, it has a certain effective breakability, represented by S. In other words, among a large number of these chromosomes subjected to unit dose the proportion S will undergo a breakage which is not restituted and which therefore leads to the loss of the chromosome (resulting either in hypoploid descendant cells or lethal bridge-formation between them). Then the proportion of these chromosomes not affected in this way, i.e. the potential chromosome-survivors among them, is $(1 - S)$. If now the dose is d instead of unity the proportion of these chromosomes in which there is no effective breakage is $(1 - S)^d$. Moreover, if we take, instead of chromosome S, one whose effective length or breakability is LS, the proportion unaffected by unit dose is $(1 - S)^L$ and the proportion unaffected by dose d is $(1 - S)^{Ld}$. Following the customary procedure, we may represent $(1 - S)$ by e^{-a}, where $-a$ is the natural logarithm of $(1 - S)$. Then the proportion of chromosome-survivors, for a chromosome of effective length LS at dose d, is simply e^{-aLd}, and the proportion of these chromosomes lost is $1 - e^{-aLd}$. When the chromosome dealt with is the single X-chromosome of man or *Drosophila* this same expression also represents the proportion of cells lost in the critical tissue and stage under consideration.

When *Drosophila* chromosomes that occur in pairs are dealt with, cell death occurs only if both members of the same pair are lost from the same cell. The probability or proportion of these killed cells may be taken as approximately $(SLd)^2$, for any chromosome type of effective length SL, at dose d, so long as the product SLd, that is, the total chance of a given chromosome being lost, is not more than about 0·1. Thus the proportion of cells surviving from this cause of death will in that case be about $[1 - (SLd)^2]$. For greater accuracy one may reckon as follows. Since the frequency of loss of any given chromosome is $(1 - e^{-aLd})$, the frequency of lethal loss involving both members of that pair is $(1 - e^{-aLd})^2$. Hence the frequency of avoidance of this cause of cell death is $1 - (1 - e^{-aLd})^2$. This can be reduced to $2e^{-aLd} - e^{-2aLd}$.

To get the total frequency of cell survival from all chromosome losses one multiplies together the survival frequencies found for each chromosome-type, using the formula e^{-aLd} for the single X of males and one of the formulae just given for all chromosome pairs which have two separately viable homologues present. If, however, one or both members of a pair contains a deficiency or other recessive cell-lethal, the same expression is used as for the X of males (with suitable change of the value of L), provided that only one member of the pair is thus affected. If both members are affected the frequency of survival from damage to the chromosomes of this pair is the square of the frequency obtaining when only one member is affected, i.e. it is e^{-2aLd}.

Where two or more chromosome types are sensibly alike in their effective chromosome length—as is the case for the second and third chromosomes of *Drosophila* and probably in some cases for different chromosomes belonging to a given size-group in man—the calculation is of course simplified by grouping the like ones together. One then raises the figure representing the frequency of survival based on one pair to a power representing the number of these pairs.

By these means one arrives at an algebraic expression representing the frequency of surviving *cells* for any given dose. In this expression, only the value of the constant a (or S) is unknown, when an organism such as *Drosophila* is dealt with—one in which the relations between the effective breakage frequencies, or "lengths", of the different chromosome types are known. Of course it has been assumed in all these calculations that the total dose and dose-rate have been low enough to allow the effective breakages in any two chromosomes to have been independent of one another in somatic cells of the given kinds.

It is possible to get empirical evidence regarding the validity of these expressions and, if they prove not invalid, to solve for a and for the frequencies of cells killed by (or surviving) given doses, if one has observations on the frequencies of *individuals* surviving given doses. For it can be assumed that (other things being reasonably equal) two groups of individuals that show an equal frequency of survival from irradiation have had the same proportions of cells killed within the individuals. Thus when the induced mortality of two chromosomally different groups, such as males and females, is studied in relation to radiation dosage, it will become evident what dose when applied to, say, the females, causes the same amount of damage as a given dose, d, applied to males. Suppose, for instance, that this dose for females is $2d$. These actual doses can then be substituted for the term d in the two respective expressions for the frequency of *cell* survival in females and males. Then when these two expressions are equated to one another they can be solved for a. In other words, the frequency of cell death is thus discovered, if the formulae are correct.

Now a check on the correctness of the formulae may be obtained by testing whether with the value for a thus deduced, there is the expected correspondence at other points along the two empirically determined dosage-mortality curves. That is, when some other dose, such as $1 \cdot 5d$, is applied to males, does one now find that the dose which, applied to females, gives the same mortality as $1 \cdot 5d$ does for males, bears the same relations to $1 \cdot 5d$, as would be found by these formulae (using the a already determined)?

Dr. Ostertag, who is writing up these methods, calculations, and results in detail, has made some tests of this kind, using his *Drosophila* results, and has found that the formulae check as closely as could be expected, that is, they give results consistent with the data derived from different doses applied to the two sexes or to groups with a deficient chromosome. As for the actual values indicated, the example may be given that the dose of 3,500 r applied to third-instar male larvae, which results in some 50% of them failing to reach "viable maturity" (in which life has persisted till eclosion and for at least 5 additional days), can be reckoned to have caused about 84% of the cells to have effective breaks in their X-chromosome, and that about a third this dose, which keeps 12% of the males from maturity, causes effective breaks in about half their X-chromosomes. This very high proportion of cell deaths must of course be understood to be concentrated in certain critical tissues and cell stages.

Once results like these have been obtained, estimates can be made of the amount of mortality of individuals to be expected for some other structural constitution of the genotype, such as one having a deficiency in a given major autosome. For the latter case may be treated as if the individual were haploid for that autosome, and one would reckon with the autosome as one does for the single X-chromosome of a male, except that the effective length, L, would be correspondingly greater. Another method of checking that Ostertag is applying, is the cytological study of the chromosomes as observed at the first mitoses that follow irradiation. In these ways the general theory can be tested and can be tried out in diverse organisms.

ROTBLAT: Is the constant a the same for both sexes?

MULLER: Yes, it is based on the effective breakability for unit length, S, with $\ln S = -a$, but the choice as to what constitutes a unit of length is arbitrary. The results do not indicate a difference in a for the two sexes. L of course represents the number of the units present in a chromosome. If one wished to do so one could dispense with the constant a by incorporating it along with L into a variable that one might term B (for effective breakability), but one would still have to deal with the ratios between different B's, preferably in terms of a standard B, so that we would return to a term equivalent to a after all.

ALEXANDER: If chromosome damage were a major or predominant cause of cell death in mammalian cells, would you not expect the bone-marrow of a male mouse to be more sensitive than that of a female mouse? We believe that the LD_{50} for an acute dose is largely a result of bone-marrow damage—should not one then expect a very pronounced sex difference?

MULLER: Not so pronounced a sex difference. When we did some figuring with this we found that the male and female—and the tests showed it too—were somewhat more alike than we had offhand judged that they should be. In the case of a mammal, the X-chromosome material is a much smaller part of the chromosome mass than it is in *Drosophila*. Moreover, you must remember this—that just where your curve will lie in any given case depends on other factors too, because with a given genotype and given environmental conditions an individual might be killed by a smaller number of cell-deaths than under other conditions, and even the sex difference might in itself influence things.

ALEXANDER: But if one could culture female and male cells *in vitro* then one would expect a real difference. May I take this one step further. If one did have two bottles full of cells which have been cultured from male and female animals of the same strain, and one then measured their direct dose: response curve to cell killing, clone formation or something like that—would you then agree that if there was no great difference between the two, that this would indicate that chromosome breakage was not a major factor contributing to cell death?

MULLER: As you may remember, I did conclude that chromosome loss by breakage, giving hypoploidy, was probably not the main cause of death when *Drosophila* embryos were irradiated. However, I don't yet feel certain that chromosome breakage was not the cause of death since in some cells, death may be caused by chromosome bridges resulting from the breaks, and this influence works the other way than ordinary chromosome loss, being positively correlated with degree of ploidy. If that is the case then at a certain stage the two mechanisms of mortality by chromosome breakage might balance one another. There is in fact evidence by Clark and others that in the very early embryo of *Habrobracon* the mortality is positively correlated with degree of ploidy. One has to look out for all sorts of complications like that.

UPTON: I know very little about the P6 chromosome, and I wondered whether experiments might be done with mice having such translocations in an effort to investigate the influence of ploidy—the total chromosomal length so to speak?

MULLER: Investigations along these lines certainly need to be carried out on vertebrates. As long ago as 1941, I applied for a grant to study the influence of ploidy in salamanders on the amount of radiation damage, as measured by effects on growth and regeneration, carcinogenesis, etc. But the organization to which I applied, led by Peyton Rous, had a quite different viewpoint on such matters. Such studies can still be done with salamanders and was there not some report from Sweden of triploid rabbits or has this been discredited? At any rate, the material for work of this type is now becoming richer.

Systems for Carcinogenesis Studies

BACQ: Thank you, Dr. Muller. I think we might now proceed to a further question. I would take for a start a remark by Prof. Lamerton who said that bone was not a satisfactory system to work with in carcinogenesis. We have seen that skin also has many obvious difficulties. Now I put the question, what kind of system might be better? Leukaemia in mice is a peculiar system because viruses are likely to be a major problem in this type of carcinogenesis. Can it be compared with others? Prof. Lamerton, would you make some comments on this?

LAMERTON: One of the difficulties of using bone in carcinogenesis studies is, of course, that its radiation response is so very complex because of the many different types of cell present quite apart from the change with age, and other factors, of proliferation patterns.

A more homogeneous tissue would have many advantages. Bone has been used a great deal in radiation carcinogenesis work because with bone-seeking isotopes, one can give a very large radiation dose compared with the rest of the body and obtain a high yield of tumours. However, this in itself, is not necessarily a good basis for carcinogenesis studies.

There is another point about which I feel more and more strongly. If radiation dose is the only variable in carcinogenesis studies, it is going to be very difficult to interpret results because dose alters so many things at once. It is a great help if one can work with variables other than the total dose given. The progress in the understanding of radiation-induced thymic lymphoma has come about largely as a result of finding a significant effect of dose fractionation. So far our own fractionation studies on bone with external radiation have not demonstrated any large effect. Another factor which we have investigated is that of mechanical injury but we have shown that such injury to the metaphysis has not effect on the tumour yield when radioactive phosphorus is given subsequently.

One factor which can be varied is the quality of the radiation and it might well be that useful conclusions would follow from a careful study of the change of RBE for tumour production, of radiations of different quality over a range of dose levels.

Of course, the direct attack on radiation carcinogenesis is not necessarily the best and we ought perhaps, to be adopting a much more fundamental approach.

BACQ: Thank you very much. The ideal tissue is still to be discovered. Dr. Berenblum?

BERENBLUM: Leukaemia is not only an ideal, but actually two ideal systems, involving two different leukaemias with independent sources of origin: (1) the myeloid system with the bone-marrow as the source, and (2) lymphoid leukaemogenesis, with the thymus as the common source in mice. I believe Metcalf and others are doing some very precise work in investigating the changes in the maturation and development of lymphocytes in relation to the thymus in two strains of animals, one where the leukaemia arises spontaneously (AKR) and the other in which it does not.

MAISIN: There are many organs in which we are interested in studying carcinogenesis; but in order to do that one has to give local irradiation, otherwise you do not obtain any local tumours except leukaemia in certain strains. In our strain, for example, we get lymphosarcoma pretty easily but not leukaemia. But to get lymphosarcoma you must just irradiate the ilio-caecal lymph nodes.

BACQ: Do you think that locally-irradiated lung would be a better target than bone?

MAISIN: That would be a very good organ to irradiate because, if you are just irradiating one lung and giving enough radiation you get a tremendous incidence of cancer. It is just the same for the kidney and it is quite enough to irradiate one part of the kidney, you don't need to irradiate all of it.

ALEXANDER: What animal did you use? (The rat.) Do you get more kidney tumours if you irradiate the kidney locally than if you irradiate the whole animal? (Yes.) So the ascopal factors here are therapeutic and not carcinogenic.

UPTON: Could not the greater tumorigenic effect of this local renal irradiation be ascribed to a decrease in survival of the animals given whole-body irradiation, which thus interferes with the expression of damage?

MAISIN: I guess so, yes.

UPTON: Without trying to be pessimistic and cynical, I wonder whether there is any ideal tumour system. Cancer is, I think a generic term. We are likely, in trying to elucidate the mechanism of carcinogenesis, to end up with many mechanisms, it seems to me. It would be wonderful if the results for one tumour could be generalized to everything that is neoplastic, but if this is not the case, then do we not need to study patiently, tumours of many different kinds, however difficult they may be.

BACQ: In this connection I might come back to the suggestions brought forward by Dr. Martinovitch. One way of doing a maybe even cleaner experiment would be to irradiate an autologous graft *in vitro* at certain doses and then put it back in an organism. You then know that not a single cell of that organism has been irradiated. Or you could graft it back in the organism irradiated in a certain way, so the two factors, the origin of the tumour and the influence of the organism can be worked out independently.

POCHIN: I want to come back to this single organ irradiation in looking for an organ in which you can get clean irradiation without involving others; lung obviously might be a possibility with inhaled β-emitters, although one might run into trouble from generalized lung fibrosis. Liver is obviously a possibility. Has anybody tried gut with absorbed short-range soft β-emitters?

MAYNEORD: I think the answer again is that the dosimetry is extremely complex.

UPTON: The ovary may offer an interesting system. It has been emphasized already in the course of the meeting that tumorigenesis in the ovary depends on pituitary activity, but it is quite possible experimentally by exteriorizing the ovary to irradiate it without irradiating the rest of the animal. One has to irradiate both ovaries admittedly, but very small amounts of radiation will induce tumour formation. I don't think anyone has carefully worked out the dose-response relationship in detail for the kinds of mechanism involved here.

MOLE: Some of the answers to these questions depend on whether we feel we want to get results fairly quickly or not. If we do, let us work on tumours that are hormone-dependent. Perhaps we ought to stop using the laboratory mouse. We have learned a very great deal about it and if you accept that there is the possibility that, underlying tumours, there are chromosomal changes then why not do comparisons between different kinds of species with different kinds of chromosomes to see if there is any kind of correlation. In this connection, perhaps I should mention that, in the last few years, we have been using Chinese hamsters, partly with this idea in mind, and we have found it impossible (with but relatively small numbers) to produce any appreciable amount of leukaemia by giving single doses or multiple fractionated doses of the kind that are so effective in the laboratory mouse.

BACQ: What about Syrian hamsters?

MOLE: We have not tried these. All I am suggesting is that concentrating on the laboratory mouse will only tell us perhaps something more about what we know already. If we try using different species we might learn something quite different.

CURTIS: In the United States there are rather extensive studies along these lines with dogs and with primates. These results are very very slow in coming—it will be years before one gets anything there, but I think you are quite right, that this is something that has to be done.

LAMERTON: I am not sure whether Dr. Curtis is referring to work with bone-seeking isotopes. If he is, of course, you have this very great problem of what you mean by the dose and where you are to measure it. I would doubt very much in fact what one is going to learn from comparative studies on different species with the bone-seeking isotopes, particularly with the α-emitters. On the other hand, whole-body uniform X-irradiation— if you can get it—will probably tell you much more.

BACQ: In order to get on with the discussion I want to ask Professor Berenblum if it is possible for him to go a bit further than he so far has? He was very cautious in his paper. Is it not possible to try at least to translate in other terms what he calls initiating and promoting action? I know it is a way to understand, to express the results, but it is a concept and a concept has to be translated into some kind of biochemical or histological terms.

BERENBLUM: Mr Chairman, you say that I have been very cautious. I am afraid, on the contrary, that I have gone much further than I should have done with results which were from experiments not completed. If these results are confirmed and extended; if it is established that by radiation you can produce within 24 hours something which is transmissible; and if the transmissible entity proves to be not a living cell not a *complete* virus, yet something which becomes leukaemogenic when urethane is given subsequently, then we will certainly have a new system in leukaemogenesis. We have the feeling that the earlier, exciting work of Kaplan and his colleagues seems to have come to an end. We hope that the two-stage system might serve as a new technique, to permit us to explore the problem further.

BACQ: Can you tell us if there is some peculiar property of urethane which can be linked with that promoting effect?

BERENBLUM: We have also been following up the problem from the urethane angle and whereas from the biological angle it seems to be straightforward, from the metabolic angle is the very opposite. Whatever you find out about how urethane acts, the results are always negative. We have, for instance, failed to confirm Rogers' claim of an intermediate metabolite of urethane, and were able to account for his results as being due to residual unchanged urethane. His further claim that urethane might act by interfering with the biosynthesis of pyrimidine also led to negative results in our hands. My colleague, Dr. Kaye, went to the trouble of testing urethane on each of the individual metabolic steps in the biosynthesis of pyrimidine and found that not a single one was hindered by urethane.

BACQ: Is it true that urethane increases the tolerance to a homograft if applied to the host?

COTTIER: I should like to mention just one point, very little has been said about immunological processes in connection with carcinogenesis and leukaemogenesis. It would be worthwhile to check the action of urethane on immune response.

MOLE: This is a biological footnote to what Berenblum has said. There have been some recent experiments which suggest that there is another sort of breakthrough in the Kaplan system. In Kaplan's experiments the thymectomized irradiated animal had the leukaemogenic potentiality restored by a subcutaneous graft of thymus. It has been recently reported that putting the thymic graft into the spleen or into the capsule of the kidney does not restore the leukaemogenic capacity of the animal. (O'Gara, R. W., and Ards, J. (1961). *J. nat. Cancer Inst.* **27**, 299.)

LAMERTON: Perhaps, Mr Chairman, the action of urethane may be non-specific, in which case to look for specific biochemical pathways will get nowhere and in fact what you have to do is look for its damaging action on the thymus or on the marrow and recognize that the effect may be via some essential cell damage or cell destruction.

BERENBLUM: It means really widening our scope of search for substances that have the same properties. It is interesting to note that other anaesthetics, or other analogues of urethane, do not act in the same way. So, from the chemical-structure point of view, there appears to be no correlation.

BACQ: I know that our colleague, Professor Mayneord, has a beautiful quotation about life-span with which to conclude this Symposium and, if no-one has any further contribution to make, I will ask him to give it.

MAYNEORD: One of the most distinguished workers in this subject was no less a person than the great Lord Chancellor, Francis Bacon, who wrote a book "*Historia Vitae et Mortis*" published in 1623. There is one sentence I would like to quote: "Touching the length or shortness of life in beasts, the Knowledge which may be had is slender, the Observation negligent, the Tradition fabulous."

AUTHOR INDEX

Numbers in *italic* indicate the page on which the reference is listed

A

Abbatt, T. D., 10, *14*
Abrahamson, S., 241, *242*
Adams, K., 214, *218*
Alex, M., 182, *186*
Alexander, P., 116, *117*, 171, *174*, 196, *203*, 236, *242*, 259, 260, 262, *263*, 278, 279, 280, *282*
Amos, D. G., 75, *78*
Andervont, H. B., 301, *303*
Andre, R., 75, *79*
Armitage, P., 13, *14*, 161, *165*
Arnason, T. J., 255, *256*
Assenmacher, I., 221, *227*
Astaurov, B. L., 239, 241, *242*, *243*
Atwood, K. C., 241, *242*
Axelrad, A. A., 72, *78*

B

Bacq, Z. M., 116, *117*, 221, 226, *226*, *227*
Baikie, A. G., 74, 75, *79*, 101, *102*, 104, *105*
Bardawill, C. J., 268, *271*
Bargmann, W., 183, *185*
Barnes, D. W. H., 14, *14*, 59, 63, *64*
Barr, E., 302, *303*
Baserga, R., 28, 31, 32, *32*
Bateman, J. A., 27, *32*
Bauer, W., 114, *117*
Baxter, R. C., 139, *142*
Bayreuther, K., 74, *78*
Beard, J. W., 69, *78*
Beaudreau, G. E., 69, *78*
Becker, C., 69, *78*
Bender, M. A., 68, 69, 72, 76, *78*, *79*
Benedict, W. H., 13, *15*, 107, *110*, 172, *174*, 278, 280, *283*, 285, 286, 292, *294*
Bennett, L. R., 185, *186*, 302, *303*
Benoit, J., 221, *227*

Bensted, J. P. M., 150, *156*, 213, *218*
Berenblum, I., 41, 42, 43, 44, 46, *51*, *52*, 72, 73, *78*
Bergstrand, B. S., 140, *142*
Bergstrand, P. J., 164, 165, *165*
Bernhard, W., 115, 116, *117*
Bernyer, G., 104, *105*
Betz, H., 17, 21, *25*
Bielka, H., 71, *78*
Billings, M. S., 185, *186*, 302, *303*
Biskis, B. O., 140, *142*, 164, 165, *165*
Blackett, N. M., 150, 151, *156*, 214, *218*
Blair, H. A., 191, 192, 195, 196, 200, *203*, *204*, 278, *282*
Blieri, A., 77, *78*
Bloom, M. A., 213, *218*
Bloom, W., 116, *117*, *118*
Boag, J. W., 316, *316*
Boche, R. D., 236, *242*
Bond, V. P., 6, *14*, 27, 28, *32*, 77, *78*
Bonorris, G., 183, *186*
Borgese, N. G., 69, *79*
Borstel, R. C. von, 241, *242*
Bostian, C. H., 241, *243*
Boveri, T., 73, *78*, 83, *90*
Braams, R., 182, *185*
Brambell, F. W. R., 229, 230, *232*
Brauer, R. W., 302, *303*
Brecher, G., 25, *25*, 27, *32*
Brill, A. B., 6, 7, 8, *14*
Brinkman, R., 208, *210*
Brown, M. B., 14, *14*, 31, *32*, 35, 38, *38*, 44, *52*, 59, *64*, 72, *79*, 107, *110*, 122, *129*
Brucer, M., 302, *303*
Brues, A. M., 5, 6, 7, 12, *14*, 75, 76, *78*, 163, 164, *165*, 191, *203*
Brunet, P., 181, *185*
Buckton, K. E., 74, 75, *79*, 104, *105*
Bunting, C. H., 183, *185*
Bunting, H., 183, *185*
Burch, P. R. J., 7, 13, *14*
Burrows, H., *133*

C

Caldecott, R. S., 251, *256*
Carnes, W. H., 44, *52*, 59, *64*
Carsten, A. L., 192, 200, *204*
Carter, C. O., 75, *78*
Carter, T. C., 83, *90*
Casarett, G. W., 139, *142*, 183, *185*, 192, 196, 198, 200, 201, 202, 203, *204*
Casarini, A., 174, *174*
Catchpole, H. R., 183, 184, *185*, 208, *210*
Cattanach, B. M., 174, *174*
Chandley, A. C., 27, *32*
Chaikoff, I. L., 208, *210*
Christenberry, K. W., 13, *15*, 107, *110*, 172, *174*, 278, 280, *283*, 285, 286, 292, *294*
Cicak, A., 240, *243*
Clark, A. M., 237, 239, 240, 241, *242*
Clark, E., 229, 231, *232*
Cleveland, L. R., 116, *117*
Clunet, J., 121, *129*
Congdon, C. C., 25, *25*, 31, *32*, 301, 302, *303*
Conklin, J. W., 285, *294*
Connell, D. I., 171, *174*, 236, *242*, 259, 260, 262, *263*, 278, 279, 280, *282*
Comfort, A., 190, 191, *204*
Corp, M. J., 314, *316*
Cornfield, J., 27, *32*
Cosgrove, G. E., 301, 303, *303*
Court Brown, W. M., 4, 6, 7, 9, 13, *14*, 74, 75, 76, *78*, *79*, 104, *105*
Cronkite, E. P., 6, *14*, 25, *25*, 27, 28, *32*, 77, *78*
Crowley, Cathryn, 270, *270*
Curtis, H. J., 171, *174*, 178, *185*, 251, 252, 255, *256*, 267, 270, *270*, *271*

D

Dacquisto, M. P., 17, 21, *25*
Dakin, R. L., 207, *210*
David, M. M., 268, *271*
Davies, A. J. S., 335, *337*
Day, T. D., 182, *185*
de Harven, E., 116, *117*
Delihas, N., 252, *256*
Deringer, M. K., 121, *129*
Detre, K. D., 319, *323*
Devik, F., 135, *137*
Dewey, D. L., 316, *316*
Doak, S. M. A., 335, *337*
Doan, C. A., 121, *129*
Dobzhansky, T., 181, *185*
Doherty, D. G., 302, *303*
Doll, R., 4, 7, 13, *14*, 76, *78*, 161, *165*
Dorfmann, A., 183, *185*
Doty, P., 181, *185*
Dougherty, T. F., 161, *165*
Drasil, V., 319, 323, *323*
Dreyfuss, B., 75, *79*
Dulbecco, R., 73, *78*
Dumont, P., 319, 323, *323*
Dunjic, A., 302, *303*, 319, 323, *323*
Dunn, T. B., 281, *282*
Duplan, J. F., 94, *102*
Duran-Reynals, M. L., 73, *78*
Dustin, P., Jr., 116, *117*

E

Eddy, B., 69, *79*
Engel, M. B., 183, *185*
Ephrussi, B., 3, *15*, 241, *243*
Evans, K., 75, *78*
Evensen, A., 135, 136, *137*
Ewell, L. H., 302, *303*

F

Faber, M., 10, *14*
Fahmy, O. G., 260, 262, *263*
Fahmy, M. J., 260, 262, *263*
Failla, G., 181, *185*, 237, *242*
Farquhar, M. G., 184, *185*
Ferrary, V. A., 260, 263
Fey, F., 71, *78*
Figge, F. H. J., 315, *316*
Finch, S. C., 319, *323*
Finkel, M. P., 140, *142*, 149, 156, 161, 164, 165, *165*
Fischer, P., 221, *226*
Fisher, J. C., 13, *14*
Fitzgerald, P. H., 104, *105*
Fitzpatrick, C. T., 181, *185*

Fliender, T. M., 27, 28, *32*, 77, *78*
Ford, C. E., 14, *14*, 59, 63, *64*, 74, *78*, 83, 84, 89, *90*, 93, 101, *102*
Forkner, C. E., 121, *129*
Fortune, D. W., 103, *105*
Fowley, J. F., 314, *316*
Frieben, 121, *129*
Friend, C., 68, *78*
Fröier, K., 241, *242*
Furth, J., 13, *15*, 31, *32*, 59, *64*, 72, 73, 77, *78*, *80*, 89, *90*, 107, *110*, 172, *174*, 191, 196, 201, *204*, 278, 280, *282*, *283*, 285, 286, 292, *294*, 301, *303*
Furth, O. B., 59, *64*

G

Gabay, S., 183, *186*
Gafosto, F., 77, *78*
Gambino, J. J., 185, *186*
Gardner, W. U., 121, *129*
Gebhard, K. L., 171, *174*, 267, *270*
Gelin, O., 241, *242*
Gersh, I., 183, 184, *185*, 208, *210*
Ghose, T., 7, 8, *14*
Gilbert, C. W., 17, 21, *25*
Gimmy, J., 71, *78*
Glass, B., 181, *185*
Glauser, O., 113, *117*
Glicksman, A. S., 182, *185*
Gochenour, A. M., 69, *79*
Gold, H., 207, *210*
Goldfarb, A. R., *133*
Gooch, P. C., 76, *78*
Good, R. A., 75, *79*
Gorer, P. A., 75, *78*
Gornall, A. G., 268, *271*
Goscienski, P. J., 73, *79*
Goutier-Pirotte, M., 226, *227*
Grace, J. T., 68, *79*
Grahn, D., 195, *204*
Graffi, A., 68, 71, 72, *78*
Greer, S., 181, *186*
Grisham, J. W., 27, *32*
Gross, L., 44, 45, *52*, 68, 70, 71, 72, 73, *78*, 93, 100, 101, *102*
Grubbs, G. E., 69, *79*
Gustafsson, A., 241, *242*
Guttman, P. H., 184, *185*

H

Haddow, A., 259, *263*
Hamerton, J. L., 63, *64*, 74, *78*, 83, 84, *90*, 93, 101, *102*
Hampton, I. C., 113, *117*
Hansen, G. F., 73, *79*
Harada, T., 162, *165*
Haran, N., 42, *51*, *52*
Harman, D., 302, *303*
Harnden, D. G., 74, 75, *79*, 104, *105*
Harris, G. W., 221, *227*
Hauschka, T. S., 4, *14*, 74, 77, *78*, *79*
Haynes, R. H., 116, *118*
Healey, Ruth, 267, *270*
Heidenthal, G., 239, *242*
Hempelmann, L. H., 6, 9, *15*, 18, 22, 25, *25*
Henkel, J. F., 252, *256*
Henshaw, P., *133*, 181, *185*, 196, *204*, 236, *243*
Heslot, H., 260, *263*
Hewitt, D., 103, *105*
Heyne, E. G., 241, *243*
Heyssel, R., 6, 7, 8, *14*
Hirsch, B. B., 14, *14*, 44, *52*, 59, *64*, 73, *79*
Hofman, J., 27, *32*
Hollcroft, J. W., 25, *25*, 31, *32*, 301, 302, *303*
Hornsey, S., 303, *303*
Hoshino, T., 7, 8, *14*
Howarth, S., 260, *263*
Howland, J. W., 192, 200, *204*
Hughes, W. L., 28, *32*
Hungerford, D. A., 74, *78*, *79*, 101, *102*, 103, 104, *105*
Hursh, J. B., 192, 200, *204*
Hutchinson, F., 182, *185*
Huxley, J., 11, *14*

I

Ida, N., 42, *52*
Ilbery, P. L. T., 14, *14*, 59, *64*, 84, 89, *90*, 146, *146*
Illig, L., 183, *185*
Ingram, D. L., 229, 230, 231, *232*
Ishida, M., 162, *165*

Ishihara, T., 74, *79*
Iversen, O. H., 135, 136, *137*

J

Jacobs, P. A., 74, 75, *79*, 101, *102*, 104, *105*
Jagie, N. von, 67, *80*
Jee, W. S. S., 153, *156*
Jellinek, S., 182, *185*
Jellinke, N., 18, *25*
Jenkins, V. K., 68, 69, 72, *79*
Jennings, H. S., 236, *242*
Jew, J., 207, *210*
Johnson, H. A., 27, *32*
Johnson, R. R., 68, 69, 72, *79*
Jones, H. B., 196, 197, *204*
Jones, K. W., 14, *14*, 59, *64*
Joseph, N. R., 183, *185*

K

Kanazir, D., 319, *323*
Kaplan, H. S., 14, *14*, 31, *32*, 35, 38, *38*, 44, 46, 50, *52*, 59, *64*, 68, 72, 73, *79*, 88, 89, *90*, 93, *102*, 107, 108, *110*, 121, 122, *129*
Kaplan, W. D., 104, *105*
Karpfel, Z., 319, 323, *323*
Kawamoto, S., 42, *52*
Kelly, E. M., 241, *242*
Kember, N. F., 151, 152, *157*, 214, *218*
Kigner, G., 181, *185*
Killmann, S. A., 27, 28, *32*, 77, *78*
Kimball, A. W., 13, *15*, 31, *32*, 72, 73, *80*, 107, *110*, 172, *174*, 191, 196, 201, *204*, 278, 280, *282*, *283*, 285, 286, 292, 293, *293*, *294*, 301, *303*
King, M. J., 74, 75, *79*, 104, *105*
Kinlough, M. A., 74, *79*
Kinosita, R., 104, *105*
Kirk, J. E., 208, *210*
Kirschbaum, A., 42, *52*, 73, *79*, 121, *129*
Kisieleski, W. A., 28, 31, 32, *32*
Kitagawa, T., 182, *185*
Knowlton, N. P., 18, 22, 25, *25*
Koch, R., 207, *210*
Kohn, H., 184, *185*, 207, *210*

Koller, P. C., 101, *102*, 146, *146*, 174, *174*, 335, *337*
Korenchevsky, V., 180, 182, *185*
Krebs, C., 121, *129*, 302, *303*
Krischke, W., 72, *78*
Krivit, W., 75, *79*
Kukita, A., 181, 182, *185*
Kuzma, J. F., 149, *157*

L

Lacassagne, A., *133*
Lajtha, L. G., 28, *32*, 302, *303*
Lamberts, H. B., 208, *210*
Lamerton, L. F., 150, *156*, 161, *165*, 213, 215, *218*
Lamson, B. G., 185, *186*, 302, *303*
Lamy, R., 239, *243*
Lang, D. A., 185, *186*
Langham, W. H., 185, *186*
Lantz, H., Jr., 268, *271*
Lansing, A. I., 182, *186*, 236, *243*
Latarjet, R., 73, *79*, 94, *102*, 241, *243*
Law, L. W., 14, *14*, 70, *79*
Lawrey, J. M., 314, *316*
Lea, A. T., 10, *14*
Lejeune, J., 104, *105*, 247, *250*
Lewis, E. B., 6, *14*
Lewis, F. J. W., 103, *105*
Lick, L., 121, *129*
Liebelt, A. G., 73, *79*
Lieberman, M., 44, 46, *52*, 68, 72, *79*, 93, *102*
Linderström-Lang, K., 268, *271*
Lindop, P. J., 13, *15*, 178, *186*, 240, 242, *243*, 278, *283*, 302, *303*, 313, 316, *316*
Lindsay, S., 207, 208, *210*
Lindsley, D. L., 335, *337*
Lisco, H., 28, 31, 32, *32*, *133*, *157*
Lord, B. I., 213, *218*
Lorenz, E., 25, *25*, 31, *32*, 121, *129*, 236, *243*, 301, 302, *303*, 335, *337*
Loutit, J. F., 14, *14*, 59, 63, *64*, 89, *90*, 146, *146*
Loveless, A., 260, *263*
Lowe, M., 153, *156*
Luippold, H. E., 241, *243*
Lynch, R. S., 236, *242*
Lyon, M. F., 83, *90*

M

McBride, J. A., 74, 75, *79*, 101, *102*, 104, *105*
McClement, P., 237, *242*
McDonald, T. P., 285, *294*
Mack, H. P., 315, *316*
McKay, C. M., 178, *186*
MacMahon, B., 9, *15*
Magno, G., 183, *186*
Magnusson, S., 73, *79*
Maisin, H., 17, 18, 20, 21, 22, 25, *25*, 302, *303*, 320, 323, *323*
Maisin, J., 302, *303*, 319, 323, *323*
Maldague, P., 302, *303*
Maloney, W., 6, *14*
Mandl, A. M., 327, 328, *333*
Marani, F., 77, *78*
Marie, P., 121, *129*
Marmorston, J. J., 183, *186*
Marmur, J., 181, *185*
Martin, C. M., 73, *79*
Martinovitch, P., 221, *226*, *227*
Martland, H. S., 121, *129*
Mathé, G., 10, *15*
Matthews, J., 17, 21, *25*
Mayer, M., 247, *250*
Mayneord, W. V., *133*
Mazia, D., 116, *117*
Melville, G. S., 302, *303*
Mewissen, D. J., 302, *303*
Mical, R., 153, *156*
Michaelson, S. M., 192, 200, *204*
Miller, E., 25, *25*, 31, *32*, 63, *64*, 301, 302, *303*
Miller, J. F. A. P., 35, *38*, 73, *79*, 100, *102*
Mirand, E. A., 68, *79*
Mitchell, C. J., 239, 240, *242*
Mittwoch, U., 75, *78*
Miwa, T., 74, *79*
Mixer, H., 121, *129*
Mole, R. H., 6, 12, 13, *15*, 59, *64*, 74, *78*, 83, *90*, 93, 101, *102*, 107, 108, *110*, 161, 164, 165, *165*, 191, 193, *204*, 274, 275, 285, *293*
Moloney, J. B., 68, 70, 71, *79*
Mortimer, R. K., 241, *243*
Müntzing, A., 241, *243*
Muller, C. J., 180, *186*
Muller, H. J., 236, 238, 239, 240, *243*

Munoz, C. M., 302, *303*
Munson, R. J., 191, 193, *204*
Murray, J. M., 229, 230, *232*
Murray, R. W., 6, 9, *15*

N

Nagareda, C. S., 72, *79*, 89, *90*
Nagel, A., 184, *186*
Neal, F. E., 314, *316*
Neary, G. J., 191, 193, *204*, 285, *294*
Nelson, E. S., 68, 69, 72, *79*
Neuberger, A., 182, *186*
Nickson, J. J., 182, *185*
Nishimura, E. T., 7, 8, *14*
Noonan, T. R., 192, 200, *204*
Nowell, P. C., 74, *79*, 101, *102*, 103, 104, *105*
Nüssel, M., 184, 185, *186*

O

Oakberg, E. F., 229, 230, 231, *232*
Odell, T. T., 70, 72, *80*, 301, 303, *303*, 335, *337*
Ohno, S., 104, *105*
Oleson, F. B., 252, *256*
Oliver, R., 28, *32*
Oster, I. I., 236, 240, *243*
Ostertag, W., 240, *243*
Ostriakova-Varshaver, V. P., 239, *242*
Ottosen, P., 153, *156*

P

Palade, G. E., 183, 184, *185*, *186*
Panjevac, B., 319, *323*
Pape, R., 18, 25, *25*
Parkes, A. S., 229, 230, *232*
Parsons, R. F., 68, 69, 72, *79*
Paterson, E., 17, 21, *25*
Paull, J., 35, 38, *38*, 44, *52*, 59, *64*, 72, *79*
Pavić, D., 221, *227*
Pavlovitch, M., 221, *226*
Pearce, M. L., 25, *25*
Perez-Tamayo, R., 182, *186*, 208, *210*
Perrone, J. C., 182, *186*
Person, S. R., 252, *256*
Peters, H., 327, *333*

Phillips, R. J. S., 83, *90*
Phipps, E. L., 231, *232*, 327, *333*
Pirie, A., 302, *303*
Pontecorvo, G., 238, 239, *243*
Pontifex, A. H., 214, *218*
Popp, R. A., 72, *79*
Post, J., 27, *32*
Potter, M., 14, *14*
Poulding, R. H., 103, *105*

Q

Quastler, H., 113, *117*, 213, 214, *218*

R

Radivojevitch, D., 221, *226*
Ramasarma, G. B., 182, *186*
Ramsey, D. S., 68, *79*
Rask-Nielsen, H. C., 121, *129*
Raulot-Lapionte, G., 121, *129*
Rauscher, F. J., 68, *79*
Ray, D., 182, *185*
Reid, T. R., 335, *337*
Renson, J., 226, *227*
Rethore, M.-O., 247, *250*
Rewald, F. E., 44, *52*, 72, *78*
Rhoades, R. P., 183, *186*
Riley, E. F., *133*, 181, *185*
Ristic, G., 319, *323*
Ritchie, A. C., *133*
Roberts, E., 182, *186*
Roberts, J., *133*, 262, *263*
Robertson, J. S., 28, *32*
Robson, H. N., 74, *79*
Roe, F. J. C., 41, *52*
Rosenthal, T. B., 182, *186*
Rotblat, J., 13, *15*, 178, *186*, 240, 242, *243*, 278, *283*, 302, *303*, 313, 316, *316*
Rous, P., 12, *15*
Rubin, M. A., 237, *242*
Rubini, J. R., 27, 28, *32*, 77, *78*
Rudali, G., 94, *102*
Ruffie, J., 104, *105*
Russ, S., 196, *204*
Russell, L. B., 230, 231, *232*, 255, *256*, 327, 332, *333*
Russell, W. L., 230, 231, *232*, 237, *243*, 255, *256*, 327, 332, *333*

S

Sabin, F. R., 121, *129*
Sacher, G. A., 191, *203*, *204*
Salaman, M. H., 41, *52*
Sallman, L. von, 302, *303*
Salmon, C., 75, *79*
Sandberg, A. A., 74, *79*
Sauer, M. E., 27, *32*
Schenk, G. O., 207, *210*
Schiller, S., 183, *185*
Schneider, L., 113, *117*
Schoolman, H. M., 68, *79*
Schrader, F., 116, *117*
Schugt, P., 230, *232*
Schunk, J., 184, 185, *186*
Schwartz, S. O., 68, *79*
Schwarz, F., 67, *80*
Schweisthal, R., 25, *25*, 31, *32*, 301, 302, *303*
Schweitzer, W. H., 331, *333*
Scott, G. M., 196, *204*
Shelton, E., 335, *337*
Shildkraut, C., 181, *185*
Shubik, P., 42, *52*, *133*
Siebenrock, L. von, 67, *80*
Simpson, S. M., 213, *218*
Sinex, T. Marrott, 180, *186*
Slack, H. G., 182, *186*
Sladić-Simić, D., 221, *226*, *227*
Smelik, D. S., 226, *227*
Smith, W. W., 27, *32*
Snider, R. S., *133*
Sniffen, E. P., 70, 72, *80*
Sobel, H., 183, *186*, 203, *204*
Sonneborn, T. M., 236, *243*
Sorieul, S., 3, *15*
Soska, J., 319, 323, *323*
Soule, H. D., 68, *79*
Sowby, F. D., 152, *157*
Spalding, G. F., 185, *186*, 331, *333*
Spurrier, W., 68, *79*
Stanley, B., 73, *78*
Stansley, P. G., 68, *79*
Stapleton, G. E., *133*, 181, *185*
Steel, G. G., 213, 215, *218*
Steele, M. H., 231, *232*, 327, *333*
Steffenson, D., 255, *256*
Stelzner, K., 230, *232*
Stevenson, K. G., 171, *174*, 251, 252, *256*, 270, *271*

Stewart, A., 75, 79, 103, 105
Stewart, S. E., 69, 79
Stich, H. F., 74, 79, 80, 101, 102
Stim, T., 69, 78
Stodtmeister, R., 114, 117
Stohlman, F., Jr., 25, 25, 27, 32
Strehler, B. L., 181, 186
Szanto, P. B., 68, 79
Szilard, L., 181, 186, 237, 243

T

Tausche, F. G., 335, 337
Taylor, D. M., 152, 157
Taylor, G., 42, 52, 73, 79
Thomas, A. M., 274, 275
Tobias, C. A., 241, 243
Tomonaga, M., 7, 8, 15
Tough, I. M., 74, 75, 79, 101, 102, 104, 105
Tovey, G. H., 75, 79
Trainin, N., 43, 44, 46, 51, 52, 72, 73, 78
Trentin, J. J., 73, 79
Trujillo, J. M., 104, 105, 185, 186
Tsunewaki, K., 241, 243
Tultseva, N. M., 241, 243
Turpin, R., 104, 105, 247, 250
Tuttle, L. W., 139, 142
Tyree, E. B., 182, 185

U

Uphoff, D. E., 121, 129, 301, 302, 303, 335, 337
Upton, A. C., 13, 15, 31, 32, 67, 68, 69, 70, 71, 72, 73, 79, 80, 107, 110, 172, 174, 191, 196, 201, 204, 278, 280, 282, 283, 285, 286, 292, 294, 301, 302, 303, 303
Uretz, R. B., 116, 117, 118

V

Valencia, J. I., 239, 243
Valencia, R. M., 239, 243
Valentine, W. N., 25, 25

van der Gaag, H. C., 72, 78
Verzar, F. 278, 283
Vincent, R., 133
von Borstel, R. C., 241, 242
von Jagie, N., 67, 80
von Sallman, L., 302, 303
Von Siebenrock, L., 67, 80

W

Wadel, J., 208, 210
Wagner, A., 121, 129
Wakonig, R., 74, 80, 101, 102
Walker, B. E., 27, 32
Wallbank, A. D., 69, 78
Warren, S., 196, 204
Warwick, G. P., 262, 263
Weisberger, A. S., 180, 186
Weiss, P., 208, 210
Wellnitz, J. M., 331, 333
Wendt, E., 116, 117
Westergaard, M., 260, 263
Whiting, A. R., 239, 241, 242, 243
Wimber, D. R., 213, 214, 215, 218
Wise, M. E., 7, 15
Wissig, S. L., 184, 185
Wolff, F. F., 72, 73, 80, 107, 110
Wolff, S., 241, 243
Woodbury, L. A., 7, 8, 14

Y

Yabe, Y., 73, 79
Yamasaki, M., 7, 8, 14
Yokoro, K., 89, 90

Z

Zamenhof, S., 181, 186
Zander, G. E., 149, 157
Zirkle, R. E., 116, 117, 118
Zuckerman, S., 328, 333
Zuideveld, J., 208, 210